Public Communication of Research Universities

This book analyses communication of university research institutes, with a focus on science communication. Advancing the 'decentralisation hypothesis', it asserts that communication structures are increasingly built also at 'subordinate unit' levels of research universities.

The book presents a cross-country systematic comparison of institutes' communication activities showing ongoing transformations in their communication capabilities and practices. It considers a potential 'arms race' in activities, professionalisation, motivations, and evaluation. Based on empirical evidence from an international study carried out in various countries across Europe, the Americas, and Asia, the book examines the possibilities for civic science communication in this new context.

It will be of interest to scholars and students of Communication Studies, STS, and Science Communication as well as to those taking or leading courses in the fields of Sociology, Public Relations, Marketing, Environmental and Risk Communication, Innovation Studies, and Social Psychology. It is an essential resource for funders, practitioners, teachers, and students dealing with science communication and the position of science in society.

Marta Entradas is Assistant Professor at the Department of Sociology at Iscte-Lisbon University Institute and Visiting Fellow at the London School of Economics and Political Science (LSE). She is a former Marie Curie Fellow at LSE (2016–18) and Fulbright Scholar at Cornell University (2015–16). In a current FCT-funded project (grant PTDC/COM-OUT/30022/2017), she is leading a cross-national study examining public communication at central communication offices at research universities. She received her Ph.D. in Science and Technology Studies (STS) from UCL in 2011. She is the 'European Young Researcher (EYRA) Award' 2016 winner (Euroscience).

Martin W. Bauer read Psychology and Economic History (Bern, Zurich and London) and joined London School of Economics and Political Science [LSE] in the mid-1990s, after a post-doctoral fellowship at the Science Museum London. Professor of Social Psychology and a former Head of the LSE Methodology Department (2008–2010), he currently directs the Msc Social & Public Communication. He is a former Editor-in-Chief of the international journal Public Understanding of Science (2009–2016) and a regular academic visitor in Brazil (Porto Alegre, Campinas and Rio) and recently also to China, where he co-directs the Centre for Study of Science Cultures, an LSE-NAIS-Tsinghua University venture in Beijing.

Routledge Studies in Science, Technology and Society

42 **Understanding Digital Events**
Bergson, Whitehead, and the Experience of the Digital
Edited by David Kreps

43 **Big Data—A New Medium?**
Edited by Natasha Lushetich

44 **The Policies and Politics of Interdisciplinary Research**
Nanomedicine in France and in the United States
Séverine Louvel

45 **Apocalyptic Narratives**
Science, Risk and Prophecy
Hauke Riesch

46 **Distributed Perception**
Resonances and Axiologies
Edited by Natasha Lushetich and Iain Campbell

47 **Questing Excellence in Academia**
A Tale of Two Universities
Knut H. Sørensen and Sharon Traweek

48 **Public Communication of Research Universities**
'Arms Race' for Visibility or Science Substance?
Edited by Marta Entradas and Martin W. Bauer

For the full list of books in the series: https://www.routledge.com/Routledge-Studies-in-Science-Technology-and-Society/book-series/SE0054

Public Communication of Research Universities

'Arms Race' for Visibility or Science Substance?

Edited by

Marta Entradas and
Martin W. Bauer

Routledge
Taylor & Francis Group
LONDON AND NEW YORK

First published 2022
by Routledge
4 Park Square, Milton Park, Abingdon, Oxon OX14 4RN

and by Routledge
605 Third Avenue, New York, NY 10158

Routledge is an imprint of the Taylor & Francis Group, an informa business

British Library Cataloguing-in-Publication Data
A catalogue record for this book is available from the British Library

Library of Congress Cataloging-in-Publication Data
Names: Entradas, Marta, editor. | Bauer, Martin W., editor.
Title: Public communication of research universities : 'arms
race' for visibility or science substance? / Edited by Marta
Entradas and Martin W. Bauer.
Description: Abingdon, Oxon ; New York, NY : Routledge, 2022. |
Series: Routledge studies in science, technology and society |
Includes bibliographical references and index.
Identifiers: LCCN 2021056474 (print) | LCCN 2021056475 (ebook) |
ISBN 9780367461355 (hardback) | ISBN 9780367494643 (paperback) |
ISBN 9781003027133 (ebook)
Subjects: LCSH: Communication in science. | Education,
Higher—Research.
Classification: LCC Q223 .P826 2022 (print) | LCC Q223
(ebook) | DDC 501/.4—dc23/eng20220207
LC record available at https://lccn.loc.gov/2021056474
LC ebook record available at https://lccn.loc.gov/2021056475

ISBN: 978-0-367-46135-5 (hbk)
ISBN: 978-0-367-49464-3 (pbk)
ISBN: 978-1-003-02713-3 (ebk)

DOI: 10.4324/9781003027133

Typeset in Times New Roman
by codeMantra

Public Communication of Research Universities

'Arms Race' for Visibility or Science Substance?

Edited by

Marta Entradas and
Martin W. Bauer

Routledge
Taylor & Francis Group

LONDON AND NEW YORK

First published 2022
by Routledge
4 Park Square, Milton Park, Abingdon, Oxon OX14 4RN

and by Routledge
605 Third Avenue, New York, NY 10158

Routledge is an imprint of the Taylor & Francis Group, an informa business

British Library Cataloguing-in-Publication Data
A catalogue record for this book is available from the British Library

Library of Congress Cataloging-in-Publication Data
Names: Entradas, Marta, editor. | Bauer, Martin W., editor.
Title: Public communication of research universities : 'arms
race' for visibility or science substance? / Edited by Marta
Entradas and Martin W. Bauer.
Description: Abingdon, Oxon ; New York, NY : Routledge, 2022. |
Series: Routledge studies in science, technology and society |
Includes bibliographical references and index.
Identifiers: LCCN 2021056474 (print) | LCCN 2021056475 (ebook) |
ISBN 9780367461355 (hardback) | ISBN 9780367494643 (paperback) |
ISBN 9781003027133 (ebook)
Subjects: LCSH: Communication in science. | Education,
Higher—Research.
Classification: LCC Q223 .P826 2022 (print) | LCC Q223
(ebook) | DDC 501/.4—dc23/eng20220207
LC record available at https://lccn.loc.gov/2021056474
LC ebook record available at https://lccn.loc.gov/2021056475

ISBN: 978-0-367-46135-5 (hbk)
ISBN: 978-0-367-49464-3 (pbk)
ISBN: 978-1-003-02713-3 (ebk)

DOI: 10.4324/9781003027133

Typeset in Times New Roman
by codeMantra

Contents

List of figures   ix
List of tables   xiii
List of contributors   xvii
Preface   xxi
Acknowledgements   xxiii

PART I
Introduction and Overview   I

1 Public Communication Activities of Research
   Institutes: Setting the Stage with the Decentralisation
   Hypothesis   3
   MARTA ENTRADAS AND MARTIN W. BAUER

2 Why and How to Sample Research Institutes:
   Methodological Challenges   23
   COLM O'MUIRCHEARTAIGH

PART II
Cross-National Comparisons   35

3 Professionalizing the Communication of Research
   Institutes   37
   MARTA ENTRADAS AND JOÃO M. SANTOS

4 Public Duty or Self-Interest? Public Communication
   of University-Based Research Institutes after an Era of
   Governance Reforms in Europe   57
   FRANK MARCINKOWSKI

5 Perceived Successfulness of Public Engagement at
   Research Institutes   79
   JOHN C. BESLEY AND ANTHONY DUDO

6 An emerging "Arms Race": Resourcing the Public
   Communication Effort   97
   MARTIN W. BAUER AND MARTA ENTRADAS

7 Public Engagement Profiles and Types of Research
   Institutes   116
   GIUSEPPE PELLEGRINI AND BARBARA SARACINO

PART III
National Situation and Profiles   131

8 The Communication of Research in Italy: The Efforts
   of Academia and Research Institutes   133
   GIUSEPPE PELLEGRINI AND BARBARA SARACINO

9 Public Engagement at Research Institutes in
   the Netherlands: Fertile Territory or Terra Nullius?   153
   PEDRO RUSSO, ROBERT BERGSVIK, AND JULIA CRAMER

10 US American Scholars Are Finding Paths to
    Engagement through their Research Institutes
    and Centers   168
    JOHN C. BESLEY AND ANTHONY DUDO

11 Public Communication in Japanese Research Institutes:
    Still Dark or Sunrise?   186
    ASAKO OKAMURA

12 Communicative Dispositions of British Research Institutes   205
    MARTIN W. BAUER

13 Public Engagement Activities of German Research
    Institutes: A Tale of Two Worlds   228
    TIM BELKE AND FRANK MARCINKOWSKI

14 'Research Excellence' and Public Communication in
   Portugal   247
   FERNANDO CHACÓN AND MARTA ENTRADAS

PART IV
Methodological Considerations   263

15 Studying Public Communication of Research Institutes:
   Sample Design and Data Collection   265
   MARTA ENTRADAS, MARTIN W. BAUER AND COLM
   O'MUIRCHEARTAIGH

16 Framework and Indicators of Public Communication of
   Research Institutes   272
   MARTA ENTRADAS

   Index   285

Figures

1.1 Distribution of scores of the decentralisation index
     (n=1539). Scores range from 0 to 10 in which high
     decentralised activity scores between 6 and 10 (light
     grey; 38%); partial scores between 3 and 5 (47%) and low
     decentralised activity scores between 0 and 2 (dark grey; 15%)   11
1.2 Degree of decentralisation of public communication
     compared across countries, ordered by 'high
     centralisation' (light grey) to low decentralisation (dark
     grey) [N=1539]   12
2.1 Diagram of sampling frame/target population   25
2.2 Conceptual scheme of the sampling frame   26
2.3 Enhanced inference schema, considering nonresponses   31
3.1 Frequency of the various public communication roles (N=763)   46
3.2 Comparison of specialist and nonspecialist scores across
     participant institutions   46
3.3 Specialist versus nonspecialist scores by country   49
3.4 Specialist versus nonspecialist scores by research area   49
6.1 Percent of research units that (a) report their own full- or
     part-time communication staff, (b) maintain a media
     contacts list locally, and (c) employ decentral specialist
     staff or have access to specialists at the central level and
     for different countries and disciplines   104
6.2 Estimates of annual resources allocated to PC activities in
     % of annual unit budget. Declared funding is the subjective
     estimates of current expenditure; actual funding is a
     calculation of PC expenditure including staff salaries in
     comparable currency; and expected funding is a subjective
     estimate of future communication expenditure   105

6.3 The graphics show the correlation of our three resource
    indicators (declared, future, and actual funding) with
    two competition indicators for different countries: (a)
    on the left, FT researchers in HE stands for the national
    population density chasing grants and (b) on the right, QS
    ranking stands for international competition. The dots
    represent the different countries   106
6.4 The top two graphics show the line-up of resources
    across countries for the disciplines at the extremes of the
    scale: the correlation between declared expenditure for
    'engineering' and 'agriculture'. The bottom graph provides
    a summary overview of these alignments. We considered
    three indicators of resourcing (actual, declared, and
    future). Their alignment is the strongest with the density
    of the research population across the eight countries. The
    alignment is less strong when competition is indicated
    by university rankings. For the actual expenditure, this
    alignment is the strongest across all competition indicators   107
6.5 Awareness of competition in research units expressed by
    the rationale for communication PR rationales (funding,
    support, attention, recruitment) in relation to PE rationales
    (policy compliance, dissemination, involvement). The PR/
    PE index indicates the balance between these two: R > 1
    means 'PR dominates'; R < 1 means 'PE dominates' the
    communication activities of units. Note, percentages add
    to more than 100% because of multiple mentions   109
7.1 Diffusionist institutes main features   125
7.2 Institutional group institutes main features   126
7.3 Market-oriented institutes main features   127
9.1 Public events by policy. The bars represent the sum
    of "Never" and "Annually" responses, expressed in
    percentages. Lower percentages indicate higher frequency
    of events (e.g. 32% of institutes with no policy never
    conduct or conduct annually, public events compared to
    9% of institutes with policy that do not conduct public events) 161
9.2 Traditional media channels by policy. The bars show the
    percentage of "Never" and "Annually" responses together.
    Lower percentages indicate higher frequency   162
9.3 New media channels by policy. The bars represent the
    percentage of "Never" and "Quarterly" responses together.
    Lower scores indicate higher frequency   163
9.4 Communications staff by policy   164

9.5   Public perceptions identified in Dutch institutes   165
9.6   Barriers to researchers' engagement in public communication   165
11.1   Public event activities by research area   195
11.2   Media activities by research area   196
11.3   Interaction with audiences by research area   197
11.4   Reasons to undertake communication with non-specialist
       audiences   199
12.1   The expansion of British higher education since the 18th
       century from age participation rates of 1–2%, to 48% by
       2015 (source: THE, UK historical statistics)   211
12.2   The stratified random sample of non-Russell research
       institutes (n=73) and Russell-based research institutes
       (n = 113) of our study; data collected between February
       2017 and May 2018 with probability weighted response rate
       of 31% from a sample >1000 (see Entradas et al., 2020)   215
12.3   % of units performing public events, traditional media
       channels and new media channels; the stacked bars report
       the % of units that report activities over the 'past three years'   217
12.4   Different espoused rationales for communication activities
       for Russell and non-Russell units (n=187)   218
12.5   Main communication tasks and their intensity reported in
       % of units (n=187)   219
12.6   Target audiences and the general intensity of contacts in
       percent of units (n=157)   220
12.7   Access to communications staff across research areas, and
       the mobilisation of researchers for communications, in
       percent of research units (n=187)   222
13.1   Frequency of different events by subsample (uni/non-uni)   235
13.2   Frequency of communication through traditional media
       channels by subsample (uni/non-uni)   236
13.3   Audience types by subsample (uni/non-uni)   238
13.4   Public perceptions by subsample (uni/non-uni)   238
14.1   Public communication activities reported by Portuguese
       research institutes by level of excellence. The charts
       represent the estimated number of activities   254
14.2   Frequency of use of traditional media channels compared
       between excellent and less-than-excellent institutes ($N = 208$)   255
14.3   New media frequency compared between excellent and less
       excellent institutes ($N = 208$)   256
14.4   Frequency of engagement with various non-academic
       audiences by level of excellence   256
16.1   Indicators for measuring public communication of science   273

Tables

1.1 Indicators coding of 'decentralised communication' at the
     institute level   10
1.2 Indicators of decentralised activity at the institute level by
     country (N=1,593)   14
2.1 Organizational system of research institutions   24
2.2 Response rate calculation for the Natural Sciences field
     in the UK   31
3.1 Factor loadings for a two-dimension CFA model   42
3.2 Descriptive statistics for the professionalization indices   42
3.3 Descriptive statistics of the control variables   43
3.4 Characterization of communication staff. Percentages do
     not sum up to 100% because multiple options (one person
     could be counted in more than one category for each
     dimension) (N=773)   44
3.5 Regression of roles on specialist/nonspecialist scores (N=600)   47
3.6 Hierarchical regression models   51
4.1 Classification of reported rationale for communication
     with nonscientific audiences into three categories (question
     wording: 'What would you say is the most important,
     the second most important, and least important reason
     for your research unit to undertake communication with
     nonspecialist audiences?')   67
4.2 Combination of mentions as the most important
     and second most important rationale for public
     communication of research institutes (N = 1,005)   68
4.3 Frequency of contact with different publics: mean values
     (1 = never to 3 = frequently; N between 1,009 and 1,099)   69
4.4 One-way analyses of variance for types of activities by
     institutional motivation for communication (results of post
     hoc multiple comparisons according to Tamhane-T2)   73
4.5 Measured values and rankings of seven self-promotion
     indicators for five countries   73

5.1   Descriptive statistics and correlations with perceived
      engagement successfulness efforts, by country   85
5.2   Univariate General Linear Model (Type III) regression
      coefficients (unstandardized), standard errors, and partial-
      eta^2 for perceived successfulness of public engagement
      efforts, by country   90
5.3   Univariate General Linear Model (Type III) regression
      coefficients (unstandardized), standard errors, and partial-
      eta^2 for perceived successfulness of public engagement
      efforts, all countries   92
7.1   Index construction: reliability statistics and scale statistics 120
7.2   Research institutes and three different communication
      strategies   122
8.1   Sample characteristics by research area (OECD)   137
8.2   Public events (CATPCA; n=347)   138
8.3   Public events by OECD research areas (mean of the indices
      between 0 and 1; n=347)   139
8.4   Traditional media channels (CATPCA; n=347)   140
8.5   Traditional media channels by OECD research areas
      (mean of the indices between 0 and 1; n=347)   141
8.6   New media channels (CATPCA; n=347)   142
8.7   New media channels by OECD research areas (mean of the
      indices between 0 and 1; n=347)   142
8.8   Audiences (CATPCA; n=347)   143
8.9   Audiences by OECD research areas (mean of the indices
      between 0 and 1; n=347)   144
8.10 Opinions regarding the public (CATPCA; n=347)   149
8.11 Opinions regarding the public by OECD research areas
      (mean of the indices between 0 and 1; n=347)   150
9.1   Science communication timeline in the Netherlands
      (adapted from Dijkstra et al., 2020)   156
9.2   Overview of the research institutes sampled for this study   160
10.1 Public events and traditional media channels engagement
      activities by research institutes/centers   174
10.2 New media channels engagement activities by research
      institutes/centers   177
10.3 New media channels engagement activities by research
      institutes/centers   179
10.4 Reasons for taking part in public engagement activities   180
10.5 Perceived reasons for lack of engagement   181
10.6 Perceived successfulness of public engagement efforts and
      need for public engagement resources, both by research area   183
11.1 Sampling frame and response rates (for universities)   192

11.2   Variables by type of activity   194
11.3   Public communication policies/action plans   198
11.4   Regression results   200
12.1   Participation rates by period of expansion   210
13.1   Sampling frame   233
13.2   ANOVA for the effects of an RI's communication funding
       on its communication activities   242
14.1   Number of research institutes that received FCT funds
       over the years. RI split in percentages to show the
       evaluation and qualification distribution, based on a 5
       scale from excellent to weak   250
14.2   Sample characteristics by research area   252
14.3   Percentage of agreement with statements on perceived
       images of the public by level of excellence and
       significance levels   257
14.4   Percentage of agreement with statements on
       communication policy, and percentage of allocated
       funding per level of excellence and significance levels   259
15.1   Sampling frames and procedures employed in each country   266
15.2   Number of contacted (N), number of that responded
       (N) by country and areas of research, unweighted
       (RR), and weighted response rates (WRR). For every
       country, we present the unweighted RR; for countries
       where we undertook a nonresponse mitigation approach
       by subsampling nonrespondents and approaching the
       subsample again (the United Kingdom, Germany, and the
       United States), we present the WRR (4)   269
16.1   Dimensions and indicators' definitions   274
16.2   Indicators of communication activities   276
16.3   Indicators of audiences and media relations   277
16.4   Indicators of resource allocation   278
16.5   Indicators of rationales, perceived images of media and
       the public   281
16.6   Indicators of general information   282

Contributors

Tim Belke is a Bachelor in Communication Sciences and Literature by the University of Erfurt, Master in Communication Sciences by the University of Munster. Research assistant of the Institute of Communication Science at University of Münster. He now works as a journalist at a public service broadcaster in Germany.

Robert Bergsvik has a background in political science and international relations, with a specific focus on critical political economy and global governance. Originally from Norway, he has a master's degree in political science from Stellenbosch University in South Africa. Before joining Wageningen University, he was a research fellow in the department for Science Communication & Society at Leiden University where he focused on the culture of public engagement at Dutch research institutions.

John C. Besley is the Ellis N. Brandt Chair in Public Relations at Michigan State University. He studies how views about decision processes affect perceptions of science and technology with potential health or environmental impacts. This work emphasizes the need to look at both citizens' perceptions of decision-makers and decision-makers' perceptions of the public.

Fernando Chácon has a degree in Biology and a Masters in Mestrado em Spatial Analysis and Geoinformatics. Fernando was a research assistant at CIBIO-UE, Universidade de Evora, in conservation planning (2017), and research assistant for the MORE-PE project at Iscte-IUL (2019). Currently, Fernando works for an NGO (Lisboa e-Nova), identifying hubs for cultural heritage, creative industry, and innovation for European cities.

Julia Cramer is a physicist and science communication researcher, interested in the boundary between fundamental science and society. She is currently working at Leiden University.

Anthony Dudo is an associate professor in the Stan Richards School of Advertising & Public Relations at the University of Texas at Austin. His research focuses on scientists' public engagement activities, media representations of science and environmental issues, and the contributions of informational and entertainment media to public perceptions of science.

Frank Marcinkowski is a professor, Dr. habil. He studied Political Studies, Sociology, and Economics at the Gerhard Mercator University Duisburg, where he worked as research assistant and assistant professor from 1988 to 1999. In 1992, he finished his doctorate (PhD) and habilitated in Political Studies in 1999. From 1999 to 2000, he was visiting professor at FernUniversität Hagen, followed by a research professorship for Political and Communication Studies at the Liechtenstein Institute in the Principality of Liechtenstein. From 2003 to 2006, he was a professor for Communications at the University of Zurich. From 2007 to 2017, he held the chair for Communication Studies at the Institute of Communication of University of Münster. Since October 2017, he is a full professor for Political Communication at the Department of Social Sciences at Heinrich Heine University, Düsseldorf.

Colm O'Muircheartaigh is Professor at the University of Chicago Harris School of Public Policy and served as dean of Harris from 2009 to 2014. His research encompasses survey sample design, measurement errors in surveys, cognitive aspects of question wording, and latent variable models for nonresponse. He is a senior fellow at NORC, where he is responsible for the development of methodological innovations in sample design, and a Visiting Professor in Statistics at the London School of Economics and Political Science. He is a member of the US Federal Economic Statistics Advisory Committee. O'Muircheartaigh is principal investigator on the US National Institute on Aging's [NIA] SAMLAP project, devising sampling methods and measurement strategies for a study of LGBT ageing, and co-principal investigator on a number of other NIA and National Science Foundation projects.

Asako Okamura is Senior Research Fellow at the National Institute of Science and Technology Policy (NISTEP), the Ministry of Education, Culture, Sports, Science and Technology (MEXT), Japan, since 2020. Her professional experiences include the National Graduate Institute for Policy Studies (GRIPS) as science, technology, and innovation (STI) policy specialist from 2015 to 2020, where she conducted the MORE-PE survey for Japan in 2018, and the Organisation for Economic Co-operation and Development (OECD) as analyst on indicators and impact of science and research from 2013 to 2015. She has been working closely with the government, especially on STI policy topics. Her current research topics include measuring science and society and foresight.

Giuseppe Pellegrini (Ph.D., Sociology 2004) teaches Innovation, Technology, and Society at the University of Trento. In this area of investigation, a specific attention he devoted to public engagement and public communication. His methodological skills are both qualitative and quantitative. He leads the Italian research team of the European project CONCISE and TRESCA studying the public perception of science and technology. He is the president of Observa Science in Society and members of the Public Communication on Science and Technology network. His main research interests are related to the study of science, technology, and society issues.

Pedro Russo is Assistant Professor in Astronomy & Society at Leiden University, the Netherlands, where he leads the Astronomy & Society Group. He was the global coordinator for the United Nation's International Year of Astronomy 2009. Pedro obtained his University degree in applied mathematics, physics, and astronomy from the University of Porto, Portugal. Pedro is involved with several international organisations, like the International Astronomical Union, European Astronomical Society, Europlanet (European Planetology Network), the International Astronautical Federation.

João M. Santos is a postdoctoral researcher at Instituto Universitário de Lisboa (ISCTE-IUL) in the project OPEN 'Organisational Public Engagement with Science and Technology' (2019–2023), specializing in data analysis and science studies. His main research goal is finding out more about how scientists define their research agendas so that we can better understand how the way to scientific discovery is paved. Beyond his work on research agendas, João also does scientific work in the fields of higher education, public policy, communication, data analysis, and psychology.

Barbara Saracino (PhD in Methodology and Social Research) is Senior Assistant Professor of Sociology at the Department of Political and Social Sciences of the University of Bologna. She teaches Methodology of Political and Social Research and Science, Society, and Public Engagement. Her main research interests deal with Methodology and Sociology of Science. She is expert in investigation techniques and analysis of both quantitative and qualitative data. She is a member of the steering committee of the research centre Observa Science in Society and also the coordinator of Science in Society Monitor.

Preface

Dimensions of institutional change in science – between 'truth value' and 'news value'

When glancing through the chapters in this book, it may not be immediately apparent to the majority of readers that they address what may end up as a profound and consequential change of one of modern society's central institutions: science. Institutional change is rarely rapid but rather gradual, often escaping attention for a while. Its range and dimensions only appear after some time with hindsight. When the German science ministry announced in a recent report that it would have to become part of a scientist's reputation to communicate to the public, it was probably not aware either of the implications of such an announcement with respect to the operation of science nor of its own limitations in effecting such a change. At best, it initiates a process the outcome of which is beyond its control.

The undifferentiated and unspecified call on science to communicate to the general public is just another step in a process that began at least some four decades ago. One of its intended effects was the intensified competition between scientists and among scientific institutions, i.e. universities. In political rhetoric, 'opening science to society' is now termed a 'moral' and 'civic obligation', supposed to add pressure to that process. Its significance must be understood by dissecting its dimensions in detail.

With respect to the function of communication in science, the shift is to communication to the public. In science, communication serves to scrutinize knowledge claims and thereby to certify knowledge production. Communicating science to the public serves to augment visibility and, thus, public prominence in the case of individual scientists and public reputation in the case of universities and research organizations. More generally said, the shift is from creating 'truth' to attracting attention.

With respect to the agents of communication, the shift is from scientists who are specialized experts for different research areas to professional communicators whose expertise is in communication to a general public, i.e. in public relations, marketing, and branding because the public is considered to need translation of the scientific subject of concern. However, the communicators of science are also in need of translation of the specialized knowledge they are supposed to communicate. Thus, another interface is added. In the case of organizational communication, the media environment is judged to be unpredictable and the public prone to get a wrong impression of the respective university, therefore, requiring professional address.

With respect to the contents of communication, the shift is from new research findings and their critical assessment to embellished and euphemistic reports of new achievements, to the cultivation of personal or organizational images. Of course, this is also dependent on the publics that are being addressed (cf. below). Here, the problem is in the detail. Announcements by a team of scientists or by a university of new advances in a particular research field may by framed so as to inform the public, perhaps even add to an ongoing discourse and thereby contributing to the rationality of public debate. That is the manifest function of science. Given the 'new' competitive and attention seeking context of science communication, the same announcements may be given a slant to persuade the public. This important distinction has been eroded in the traditional media and dramatically so in the social media.

With respect to the addressees of communication, the shift is from the peers within the relatively narrow confines of disciplines or research fields to the, in principle, unlimited general public. While the communication within disciplines is between actors who have a high level of specialization and competence about the contents of their communication and advance knowledge in the process, the general public does not, by definition, have that competence. Thus, either communication with the expectation of understanding and even receiving productive feedback has to address specified groups of (knowledgeable) stakeholders or 'citizen scientists' or the expectation can at best be the general interest and attentiveness of an unspecified public.

All the shifts that are implied in the general drive for communicating science can be subsumed under a shift in the underlying value orientation of science as an institution from truth value to news value. This shift is first visible in the change of actors communicating, i.e. the expansion and renaming of communication units, then in the resources spent on them and their products and activities, compared to what remains for the university's central functions, i.e. research and teaching. The shift is further visible in the contents of public discourse, the trust bestowed on scientific experts and organizations, the degree of public understanding of the nature and function of scientific methods, of the reliability of scientific knowledge. Finally, the shift becomes visible in the profile of research topics receiving funding and being pursued by scientists. The (potential) development from truth value to news value has been termed medialization of science.

If this admittedly dystopian scenario will play itself out as described is still open. As long as there is an awareness of the desirability of available options and potentially wrong turnoffs to be avoided, as is witnessed by the contributions to this volume and the line of research they represent, institutional aberration may be corrected. But we may be at a crossroads.

Peter Weingart – Bielefeld, October 2021

Acknowledgements

This work would have not been possible without the efforts of all those involved in the various phases of the MORE-PE project. We thank our collaborators in the project who co-coordinated the national studies, and who agreed to discuss them in this book. Without their contributions, this book would have not come into existence. The authors would like to thank the research assistants in the national teams, who were key in the compilation of the national databases of contacts and in the data collection. We acknowledge Manuel Valença, Joana Marcelino, António Revez, Luis Junqueira, João Santos, Fernando Chacón, Tim Belke, Maria Luisa D'Alba, Giulia Burato, Maria Gafforio, Robert Bergsvik, Julia Cramer, Anne Kerkhoven, and Selina van den Oever, Luiz Amorim, Carla Silva, Ji Zhao, Yan Li, Weiyu Yang, Jianquan Ma, Yanqiu Sun, Jia Liu. We also extend our thanks to Colm O'Muircheartaigh in the Chicago Harris School, for advising on the methodological aspects through the project. Last but not least, we thank to the Fundação para a Ciência e Tecnologia (FCT) for funding this project (grant number PTDC/IVC-COM/0290/2014), and the European Commission Marie Curie Scheme for the individual fellowship (MCIF; EU project 708434- OPEST – DLV-708434) attributed to the first Editor.

We would also like to thank the national partner institutions involved in the project: ISCTE-the Lisbon University Institute (Portugal, coordinator), London School of Economics and Political Science (LSE) (co-coordinator), Michigan State University (USA), University of Texas, (USA), Universitá degli studi di Trento (Italy), Universitá degli studi di Padova (Italy), Science and Society (Italy), Leiden University (Netherlands), and Dusseldorf University (Germany). And our affiliate partner institutions: Museu da Vida, Osvaldo Foundation (Brazil), the National Graduate Institute of Policy (Japan), National Sun Yat-sen University (Taiwan), and the National Academy for Innovation Science) (NAIS) (China).

Part I

Introduction and Overview

Chapter 1

Public Communication Activities of Research Institutes

Setting the Stage with the Decentralisation Hypothesis

Marta Entradas and Martin W. Bauer

Public communication function at research universities

For the past 40 years, universities around the world have articulated a set of democratic values that bind together research and society. The so-called Third Mission of services to society (Laredo, 2007) has been adopted by most universities around the world. Yet, this traditional 'outreach' function takes on an added urgency and many new formats in a world of global competition between universities for their reputation amongst students, staff and funding streams.

Communication beyond the lecture hall and the peers has in various forms been part of the university since medieval time (Koch, 2008); it, however, accentuated with the 'university extension movement' arising in England in the 1870s. This history is reconstructed as community engagement, the civic university, outreach, extension, extra-mural activities and 'Volkshochschule', or, in the latest turn, as 'research impact' (see Cunningham, Oosthuizen & Taylor, 2009). Today universities and research institutions engage in a global market for the 'knowledge products' of facts, discoveries and techniques created by their efforts, and they look for an attentive public on their 'pathways to impact'. As part of these transformations, the university organisation has moved from an internally fragmented "multiversity" (Kerr, 1964; Krucken, 2020) into a goal-oriented, competitive organisational actor here management is key (Maasen and Stensaker, 2019; Engwall, 2020). Modern research at all levels is increasingly 'doing' as well as 'showing and telling', i.e. doing research and drawing attention to their doing. 'Playing the trumpet' has become a part of the research activity. Delivering such messages about research requires thinking about the competence to be effective, how to build these competencies and how the effort can be resourced.

The communication function of universities is examined in the literature in organisational and corporate communication (e.g. Clark, 1998; Hallahan et al., 2007). It has hitherto received little attention in the public understanding of science (PUS) and, only very recently, in the science communication

DOI: 10.4324/9781003027133-2

literature (Fähnrich et al., 2019; Schäfer & Fähnrich, 2020). This is somewhat surprising. Research universities are key actors in science–society relationships: this community provides the resources for public communication, defines and reflects the priorities and sets agendas on how relationships with stakeholders, the mass media and wider publics are established. Our international study MORE-PE '*Mobilisation of Resources for Public Engagement*' (2016–2021), on which we draw in this book, aimed to understand this new dynamic of science communication within research universities in Brazil, Germany, Italy, the Netherlands, Portugal, the United Kingdom, the United States of America, Japan, China, and Taiwan.[1]

In this chapter, we outline an analytical framework for the study of institutional science communication and report practices of science communication at the meso level of research institutes (RIs). We start the chapter with a snapshot of the New Public Management (NPM) ideology and our basic assumption that the institutional communication analysed is a response to academic competition in this new era; we review existing literature on institutional public communication of science. After this, we present data regarding ongoing transformations at the meso level in the light of the 'decentralisation hypothesis' – i.e. the trend towards distributed communication from central to the meso-level of RIs and the forms this division of labour might take within any research-based university. These ideas on decentralised communication build on earlier work presented by Entradas and Bauer (2019a) and Entradas (2021).

Box I: On key terminology 'public communication of science'

There seems to be a certain reluctance amongst researchers and also practitioners to identifying or subsuming 'science communication' under the 'NPM' logic of branding, public affairs and knowledge marketing. Maybe distinctions and different words matter here: it may be that science communication is more than just an action to seek a specific outcome, but rather that it contributes to public debates with an open end and, thus, to the larger goal of creating a common understanding about the world in common sense (Habermas, 2001). We will refer to this as the **science communication function of universities**, i.e. the communication efforts of a university to communicate and engage publics in research for a civic purpose. This communication may involve different formats in order to communicate about research, dialogically and open-ended, and we refer to these as public communication of science activities.

In investigating communication of RIs, we face the threefold problem of a diverse vocabulary: (a) as researchers we need to work with conceptual distinctions for purposes of question framing (e.g. institution/civic focus); (b) our field of research, i.e. the people working in professional communication, might use these terms but not in line with the intended, conceptual meaning of the research and (c) furthermore, this terminology-in-use varies across different countries and languages.

There are other terms frequently used in this context such as 'institutional communication', 'organisational communication' and 'internal or external communication' and also the term 'public engagement'. In the future, it might be possible to order these terms on a continuum of communication with a focus on institutional communication (i.e. classic strategic, self-interested public relations) on the one side and a focus on public engagement open-ended (i.e. inviting and enabling public participation in science and policy development) on the other side. For now, this requires a temporary solution by an ad hoc convention.

Hence and for the present purposes, we prefer the term 'public communication of science' to that of 'public engagement' because we know that RIs are most likely to carry one-way, mono-logical formats of communication, while at the same time, espousing terminology referring to 'dialogical' communication (Entradas & Bauer, 2017; Linell, 2009). Thus, using the term 'public engagement' would not entirely reflect the actual range of institutional public communication, which is varied and embraces many activities, dissemination and engagement.

'Public communication of science' activities are then events or media activity that institutes undertake to communicate about research to any non-peer public, and this ranges from public lectures, to media interviews, open days, citizen science, public debates, policymaking, social media, amongst others. Any distinctions within this range then become an empirical matter of, indeed, differentiating or not between more strategic or more civic functions. Whenever other terms are used in chapters of this book, their meanings are explained by the authors.

NPM of research universities

Starting in the early 1980s, we saw a gradual marketisation of universities and a move from elite-peer communication to marketing communication (e.g. Hemsley-Brown & Oplatka, 2006; Krücken, 2014). The organisational literature documents this well: universities have adopted a promotional attitude of strategic communication (Brown & Carasso, 2018; Kleimann, 2019) with the goals of establishing an identity (e.g. Gioia et al., 2000) by building a brand reputation and corporate image (Wæraas & Sataøen, 2019). Several university league tables, elaborating national and international rankings, have become the currency of this reputation game (Bekhradnia, 2016).

Key to this move was a Neo-liberal shift in State Policy to reform the public sector (e.g. Slaughter & Rhoades, 2004) to more business-like public service delivery through the adoption of management models (Hood, 1995). 'Less state/more market' are according to De Boer et al. (2007) 'superficial neo-liberal slogans'; this transformation can also be seen to be part of broader changes including the rise of the 'knowledge society' and demands for 'global literacy' (Drucker, 1993), cutbacks in state funding for research, the progressive privatisation of R&D (Brown, 2011) and the off-loading of education costs as 'personal investments' to tuition fees (in many countries in Europe and in the United States). With this comes calls for more public accountability, which leads to an increasing demand for value-for-money auditing of resource allocation, including assessing academic performance (e.g. Power, 1997; REF, 2014). Quality-assurance measures of which 'impact on society' has become a part, and the requirements from funders to purpose research to solve society's problems, have put significant pressure on universities to involve citizens in research and communicate broadly (European Commission, 2013, 2016). Universities face the necessity to facilitate 'pathways to impact' for their research to compete for excellence, public visibility and media attention (Entradas et al., 2020; Peters et al., 2008; Weingart & Maasen, 2007).

Professionalisation of PR and marketing functions

The implication of NPM for university communication is reflected in the professionalisation of the communication function, following the 'holy trinity' of lobbying for policymaking (e.g. maintaining direct relations with policy circles to influence science policy and direct funding for research) (Kollman, 1998), public relations (e.g. branding to maintain a relationship of trust with the public) (e.g. Borchelt & Nielsen, 2014; Grunig & Grunig, 2008) and marketing (placing the university in the national and international student market) (Brown, 2011; Paradeise et al., 2009). With this comes the new 'university management professional', 'PR and information officer' responsible for managing relationships between the university and society in the interest of the institution to flourish (e.g. Krücken & Meier, 2006).

The emergence of this 'new' management logic has also influenced the way universities communicate science – *the science communication function*, at three levels of analysis: the central level (in the university central offices), the meso level (in RIs) and the micro level (amongst individual researchers and their teams):

- *The central level of university communication*, that is the Public Relations (PR)/Marketing/Communications offices and officers, has become common at universities and large research organisations. The empirical research with a focus on this level has mostly looked at (media) communication and products (e.g. Koso, 2021; Neresini & Bucchi, 2011; Rödder, 2020; Rowe & Brass, 2011), tasks and views of professional communicators, scientists and university leaders (e.g. Elken et al., 2018; Horst, 2013; Kohring et al., 2013) pointing to an internal conflict of science communication at this level. We shall refer to this literature further in the chapter.

- *The meso level of RIs* comprises the research units, centres and/or institutes within universities and other research institutions (hereafter RIs). This level remains under-researched with our benchmark study of the Portuguese research units being the first to map communication activities at this level (Entradas & Bauer, 2017) and providing the basis for the MORE-PE study. As our research suggests, institutes are expanding capacity for public communication of science, which is best understood as work in progress. This level is the unit of investigation in this book.

- At the *micro level*, individual researchers and scientists are communicating their work outside their peer-academic networks. Their mobilisation and motivations have received attention from a growing body of research, mostly pointing towards an increasing activity amongst scientists, increased sense of duty, but also barriers to greater involvement (e.g. Besley et al., 2018; Crettaz Von Roten, 2021). The mobilisation of individual scientists remains uneven; the activity levels are stratified by hierarchy, field of research and sometimes even geographical region (Bauer & Jensen, 2011; Entradas & Bauer, 2019b).

The PUS literature sheds some light on how the science communication function is entering and shaping the university landscape. With an empirical focus primarily on the activities of central PR offices and officers (Bühler et al., 2008; Elken et al., 2018; Peters et al., 2008), this research provides insights on how these structures operate, the task variety and how professional communicators understand their roles on science communication and public engagement.

This literature shows an intensification of media contacts and diversification of PR products in research universities (Autzen, 2018; Boumans, 2018; Vogler & Schäfer, 2020) and increased use of websites and social media channels to disseminate research results (Fähnrich et al., 2020; Metag & Schäfer, 2019). The literature points also to increasing professionalisation of science communication at universities. Some now employ 'science communicators' and 'public engagement officers' to engage publics in research and support researchers (e.g. Koivumäki & Wilkinson, 2020; Watermeyer & Lewis, 2018; Wellcome Trust, 2015). Yet, most rely on non-specialist staff who in the main carry out PR functions including branding, advertisement and public and media relations through contact with journalists and stakeholders, blogging platforms and social networks, the production of press releases, open days and institutional public event making (e.g. Entradas, 2015; Peters et al., 2008). These central communication structures medialise science, facilitating institutional self-promotion and visibility of 'star' researchers (Marcinkowski & Kohring, 2014), but also face challenges to communicate about science due to lack of resources, training and skills (Watermeyer & Lewis, 2018) (see Chapter 3 in this volume for a detailed analysis of the community of science communicators).

Scholars argue that strengthening science communication creates opportunities to defend the autonomy of the research university, and securing its 'licence to operate', because it increases the visibility of science and scientists in the public sphere (Rowe & Brass, 2011), facilitates the willingness of 'scientists to go public' (Rödder, 2011) and directs political attention (Kohring et al., 2013; Scheu et al., 2014). Yet it can also pose a potential threat to scientific autonomy (Marcinkowski & Kohring, 2014) as the legitimacy of science as 'news making' might compete with research performance (see Part 2 in this book for the 'arms race for public communication', ARPC hypothesis). The institutional focus on medialised communication and the 'push logic' adopted for promotion (e.g. Marcinkowski & Kohring, 2018; Weingart & Pansegrau, 1999) has put the marginal role of science communication at universities again into the spotlight (e.g. Claessens, 2014; Shipman, 2013).

From this literature, we intimate that science communication might be caught in internal tensions with central offices' communication tasks and priorities. This leads us to ask: could the role of science communication be at levels other than central structures? Are NPM priorities limited to central communications, or are they also affecting other levels of operations? And, are there variations in the way universities in different countries and traditions roll out science communication activities (see Mejlgaard & Stares, 2013; Mejlgaard et al., 2018)? Based on these preliminary observations, we asked:

> RQ1: Do research institutes within universities have their own communications infrastructures or do they rely on the central function to bring their research into the public sphere?
> RQ2: And if there is meso-level communication activity, what are they doing, why, for whom, and how are they resourced?
> RQ3: Are there similarities and differences across institutes in different countries?

Decentralisation of the public communication function

Our study pointed to science communication expanding at the meso level, with event making, traditional media and online contacts led and resourced locally (Entradas et al., 2020). Most institutes engage in various activities most commonly public lectures, open days or newspaper interviews; address various audiences, more often the general public and schools than industry and policy stakeholders and allocate resources to this: Four in ten report staff dedicated to communications tasks, and 50% rely on central communications offices to disseminate their news; the average institute reports around 3% of the annual research budget for public communications, and about half have policies for science–society in place. The evidence is of rising activity: in the past five years, public communication of science increased in most institutes (61%) and has just started in a further third; also, expectations are for continuous growth with about half expecting to dedicate more resources to the activity in the coming years, and 90% expecting their researchers to engage in it.

Our data show that these (ongoing) changes in the university landscape are not unique to any single country. We find similar patterns of communication across countries with institutes overall engaging in the same type of activities, showing a certain homogeneity of practice, but we find differences in the intensity of activities and in the resource allocation. Despite all this, the professionalisation of this communication effort is only beginning in most contexts.

Our main thesis for this growing activity will be the **'decentralisation hypothesis'**: the transformation that we observe is consistent with an increasingly distributed activity to the level of institutes. We contend that research institutes are taking communication responsibilities on their own and are becoming important places for civic science communication and public engagement within the university environment (Entradas, 2021). We use the term 'decentralisation' to refer to distributed public communication across different levels of activity in the university (central, meso, micro), and this includes decision-making (Mahoney, 1997) when research institutes exercise financial autonomy and local discretion, for example, over hiring professional communication staff.

Degrees of de/centralisation of communication

We developed some basic indicators of 'decentralised communication'. We can explore different strengths of 'decentralisation' and examine patterns of similarities and differences in decentralisation across countries. For these analyses, we use three outcome indicators: (1) the level of public event making (*event making*), (2) the intensity of interaction with traditional media channels (*traditional media*) and (3) the intensity of new media activities (*new media*). We also consider five input indicators of support for these activities such as (1) communications staff (*none, central level, within institute fully/ partly dedicated*), (2) the level of funding allocated to public communications (*funding*), (3) the existence of an explicit policy for public communication (*policy*), (4) the mobilisation of researchers into these communication activities (*active researchers*) and (5) the starting time and age of these communications structures (*start time*) (see Chapter 16 for description of indicators and intensity indexes).

All variables were binary recoded for activity (0/1 outcome indicators; yes/no for input indicators); staffing is an ordinal variable. Table 1.1 shows the coded variables. All variables are equally weighted; we sum the values and order institutes according to the 'degree of decentralisation' on a scale that ranges from 0 to 10, more or less decentralised (see Figure 1.1; M=4.9, SD=2.2, Median=5); 'low' values mean less decentralised activities, i.e. more centralised communication activities; 'high' means a tendency towards decentralised communication activities in an institute.

These scores allow for a division of institutes and also a division of countries (Mejlgaard et al., 2012) into three levels of decentralisation: *full decentralised*, *partial decentralised* and *centralised*; in the latter category, institutes are handing public communication over into central responsibilities, or central offices want to control. Figure 1.1 shows the distribution of institutes on this continuum 'centralisation–decentralisation'.

Fully decentralised (6–10 points; report high activities in at least six of the ten indicators) represent 38% of the sample. They are likely to have high outcomes and high level of resources, policy and professionalised staff in particular staff fully dedicated to communication tasks and might (might not) have been doing public communication for a longer time; this is the case of Japan that although is amongst the countries that reported older traditions, it is more likely to show centralised approaches (see Table 1.2). Decentralised activity may be seen in universities that choose to give decision-making power to other levels of the organisation such as the research institutes and departments; meso structures in these environments might be more likely to have full autonomy and control over their communications strategy and

Table 1.1 Indicators coding of 'decentralised communication' at the institute level

Indicator/variables	Variable
(1) Start time	Older (1) / Newer (0)
(2) Intensity of public event making	Hi (1) / Low (0)
(3) Intensity of traditional media contacts	Hi (1) / Low (0)
(4) Intensity of new media use	Hi (1) / Low (0)
(5) Communications staff	No staff (0)
	Central office (1)
	Within institute partly dedicated (2)
	Within institute/fully dedicated (3)
(7) Having policy/guidelines for public communication	Yes (1) / No (0)
(8) Having funding for public communication activities (>5%)	Hi (1) / Low (0)
(9) Percentage of researchers mobilised (>40% of active researchers)	Hi (1) / Low (0)

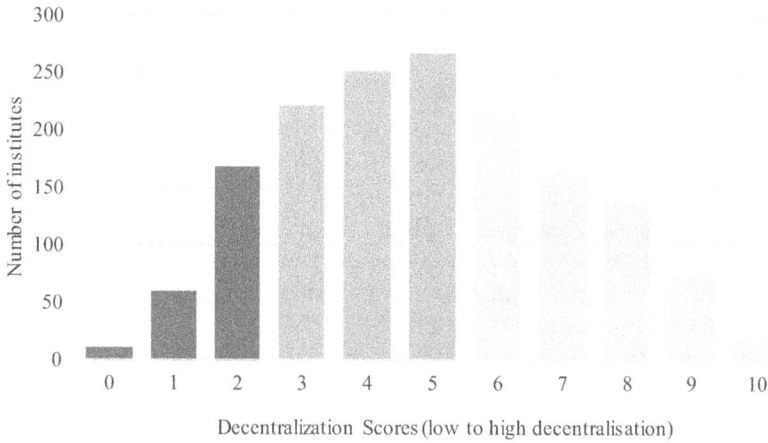

Figure 1.1 Distribution of scores of the decentralisation index (n=1539). Scores range from 0 to 10 in which high decentralised activity scores between 6 and 10 (light grey; 38%); partial scores between 3 and 5 (47%) and low decentralised activity scores between 0 and 2 (dark grey; 15%).

activities, and plan independently from the university central communications. In organisations with this approach, communication of science might become mostly a function of smaller units.

Partly decentralised (3–5 points; report intensity of activities close to the average and more mixed combinations) represent around 47% of all institutes. They are more likely to report medium to high levels of activity in the various channels of communication and might (or might not) have a policy in place and staff that is, however, not fully dedicated to communications tasks. This could be seen in organisations where communication might happen at various levels, but the central or top management of the institution might hold control over the institutional communication strategy. That is, all structures may work in a common framework which is determined by the top management of the organisation. This may be reflected in divergent trends for science communication sensitive for the organisation: a trend towards centralisation of noticeable findings, prizes, awards and decentralisation of outreach type of communication.

Institutes with low activity (0-2 points; report little activity, which is more likely centralised at the central offices). This corresponds to around 15% of the institutes. In this third category, there is little communication activity happening at the local level. Public and media activity is low, and resources are scarce. These institutes are likely to be hosted in universities, where the central communications office assume the full communication function, including science communication; decisions are taken centrally and the communications strategy is decided by the central management, whose views drive the entire institutional communication effort.

Patterns of decentralising communication across countries

Figure 1.2 shows the three categories of 'decentralisation' across countries and points to differences between countries. For example, in Japan, 24% of the institutes report high decentralised communication, while 26% rely entirely on central responsibilities. In Portugal, the situation is reversed: 51% of institutes report decentralised communication operations, while 10% rely on central provisions instead.

A common trend is that in all the surveyed countries most institutes show some degree of decentralised activity (full + partial; 85% on average), suggesting that most universities hosting these institutes adopt distributed approaches to public communication of science. Yet, there are differences in this level of decentralisation between countries. For example, in Japan, around 74% of the institutes show decentralised activity to a certain extent, while in Brazil, this reaches 91% of the institutes.

Another trend is the predominance of partly decentralised approaches, in most countries. For example, in the Netherlands and in the United Kingdom, around 51–52% of the institutes fall into this category compared to a quarter that fall into the full decentralised or centralised approaches. In Germany and Portugal, however, the predominance is for decentralisation,

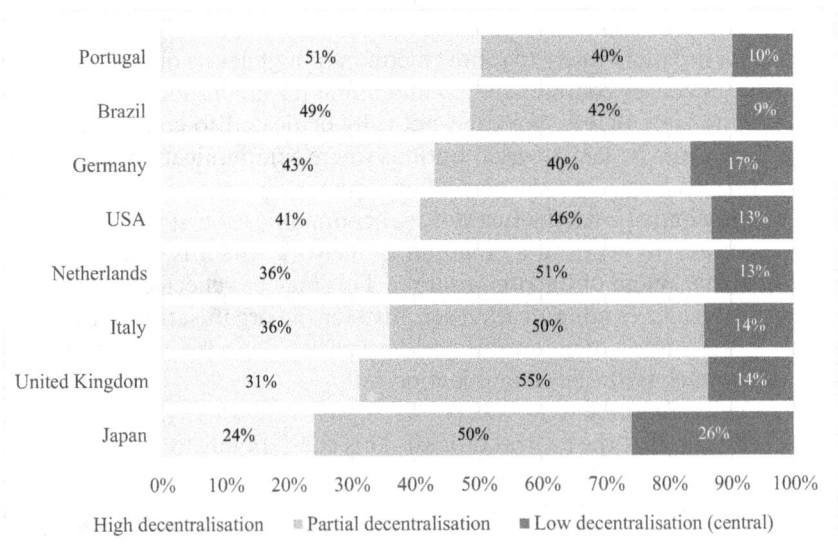

Figure 1.2 Degree of decentralisation of public communication compared across countries, ordered by 'high centralisation' (light grey) to low decentralisation (dark grey) [N=1539].

with the majority of institutes falling into what might be full decentralised approaches to communication (51% in Portugal and 43% in Germany). On the opposite end, Japan shows a slightly different pattern with most institutes pointing to signs of centralised communication with only around 24% pointing to fully decentralised environments (Figure 1.2).

Table 1.2 shows further patterning across countries. Event making is more decentralised in Brazil, Italy and the Netherlands; less so in Portugal, the United States or Germany and certainly much less in Japan. Traditional media are more decentrally served in the Netherlands, Brazil and Italy; less so in the United Kingdom, Germany, the United States and Portugal and much less so in Japan. New media is very much in the institutes' operations in the United Kingdom, Brazil, the United States and the Netherlands; less so in Portugal and Italy and much less so in Germany and Japan. Having decentralised communication in response to policy is very much in evidence in Portugal and Italy and much less so elsewhere. Being able to get support from central communication offices is very much in evidence in Italy, the Netherlands and the United Kingdom; to a lesser extent in the United States, Brazil and Japan and clearly less so in Germany and Portugal.

Discussion and specifying the decentralisation hypothesis

So far, we have reviewed the literature on the communication function of universities in order to define the issues of this book and contextualise our investigation. Public communication is clearly not a novel activity at universities, what is new is a game defined by NPM which makes universities work even more like Sports Clubs, requiring to operate with lobbying, branding and marketing (and merchandising) to compete in various 'global super-leagues' and potentially in a 'winner takes all' economy and thereby potentially loosening its local roots. This strategic re-orientation is hitting old and new practices of civic engagement. Opaque distinctions between 'public communication' and 'science communication' or between 'science communication' and 'public engagement' mark different orientations and significant differences in motive and function. But, the trend point towards a professionalisation of the university science communication function at all levels, central operations, research units and individual researchers that seems to leave little space for practical idealism.

This chapter outlined these transformations at the institute level. The data show signs of science communication expanding at the meso level, but this is best understood as a 'work in progress'. We observe the commitment of research institutes towards public communication in their portfolio of activities, in the resources, plans and (in some cases) policies for science communication and in future expectations. We also observe the lack of resources that inhibits the development of solid communication structures for

Table 1.2 Indicators of decentralised activity at the institute level by country (N=1,593)

	Brazil (N=121)	Portugal (N=184)	Germany (N=278)	USA (N=205)	Italy (N=308)	Netherlands (N=109)	UK (N=160)	Japan (N=228)	Total (N=1,593)
Events (high)	65%	46%	48%	47%	59%	56%	44%	27%	48%
Traditional media channels high	63%	41%	48%	45%	59%	66%	50%	37%	50%
New media channels high	66%	48%	39%	57%	50%	56%	67%	26%	48%
No staff	15%	26%	7%	10%	6%	4%	3%	21%	12%
Staff central level	47%	23%	39%	54%	66%	61%	71%	49%	51%
Staff within unit (partly dedicated)	37%	51%	55%	36%	28%	35%	26%	30%	38%
Staff within unit (fully dedicated)	24%	23%	32%	15%	15%	13%	12%	10%	18%
Funding (high)	10%	26%	6%	32%	10%	3%	23%	17%	16%
Policy (yes)	36%	53%	39%	27%	56%	39%	31%	32%	40%
Active researchers (high)	42%	55%	33%	49%	34%	33%	41%	38%	40%
Start time (older)	76%	77%	77%	77%	59%	75%	65%	83%	73%
Decentralisation Index Mean (SD)	5.5 (2.0)	5.5 (2.3)	5.0 (2.3)	4.9 (2.2)	4.8 (2.1)	4.8 (2.0)	4.6 (2.0)	4.0 (2.0)	4.9 (2.2)

many of our informants. The clear tendency of many institutes is to increase their level of communication activity, take charge of it and recruit their own professional staff, which we capture in the *professionalisation thesis* (see Part 3 of this volume) and in first and foremost the *decentralisation hypothesis*.

Box 2: Restating the 'decentralisation hypothesis'

Let us specify the 'Decentralisation Hypothesis' of public communication of science into five different elements including capacity building, professionalisation of staff, division of labour, rising tensions and a potential niche for civic science communication.

H1: [Capacity building] There is a tendency towards decentralised communication activity at universities and towards capacity building also at the meso level of research institutes;

H2: [Professionalisation] Public communication is becoming professionally competent at the central and meso levels, with centralisation of PR and marketing functions and decentralisation of the science communication function;

H3: [Division of labour] There will be division of labour in the university communication functions, distributing activities between central and meso levels;

H4: [Growing tensions] Tensions will be rising between central and meso-level responsibilities over resourcing, forms and functions of communication;

H5: [Potential niche] Decentralised communication is more likely to foster science communication with a civic focus within the organisational culture.

The data point also to varying degrees of decentralisation amongst institutes. In most cases, institutes are not (yet) autonomous in their public communications and are more likely working in a framework provided by the university. Yet, the fact that centralised approaches are observable, but not yet paramount, is consistent with our decentralisation thesis, i.e. institutes enjoying some autonomy over their communications. However, whether this state of affairs represents a later stage of development or an early one, whether the trend is moving away from centralisation of activities and coordination, cannot be finally concluded from our data; this will require longitudinal evidence. For the time being, and for cross-sectional comparisons, we might, therefore, ask:

- What is the level of autonomy of research institutes for their public communication activities?
- Are meso structures fully or partially independent from the central communication or integrated into any central communication strategy?

- How are relationships between various communication structures developing, negotiated and maintained?

We also sought to examine whether decentralised structures could lead to increased prominence of science communication within the university and whether the university provides space for the science communication function to develop its niche at the institute level (Entradas, 2021). Research institutes might build their own capacity for communication, filling the gap that exists within central offices in which science communication may be marginal because it is only one amongst many tasks. From this intimation, further questions emerge from our international study:

- What is the role of science communication (at all levels) and where is it most likely to be found within the organisation?
- How much of institutional communication serves the open-ended civic purposes of science communication?

Finally, it is likely that the ongoing changes bring tensions to the institutional environment. An obvious one might be tension between the meso and central levels of communication about who does what and with what resources and what values and goals motivate the communication activities. How will communication functions be distributed amongst the professionals entering the field at the central and meso level of the university? While interdependency may facilitate the activity of research institutes, it can also limit their autonomy about what can and should (or not) be communicated. Decentralised communication, on the other hand, while giving freedom at the institute level, also runs the risk of haphazard institutional communication and difficulties in harmonising the institutional marketing, branding and image making. Tension might also arise from a conflict of values. If the communication strategy is driven by ideals of visibility, impact and media 'news values', the operational requirements of feeding these values may trickle down to the institute level, as seen already dominant at the central level of universities' communication. This could counteract the 'good intentions' of communication efforts that mainly seek to endorse the 'evidential values' of responsible science and research, thus becoming yet another tool for institutional visibility. These and other issues are investigated and documented in this book.

The data of this book pertain mainly to the meso level, the research institute's communication activity. This is hitherto a focus with very little research and even less comparative evidence. However, even our data will be incomplete to fully understand the structures of universities and science communication – for this purpose, we will have to consult the central structures at universities and the micro level of individual researchers' outreach activities; the fact that an institute employs communications staff, has a policy and funds public communication are good indicators of the emerging autonomy and local decision-making consistent with our decentralisation hypothesis.

As always, more research will be needed, also into the communication function of universities and the role of science communication within the university environment. The present book sets the scene for research into institutional science communication, provides a framework to think about public communication at the institute level and serves as a benchmark for future comparisons. One thing seems clear: there is work in progress to build capacity for communication at the level of research institutes, which potentially fosters capacity for science communication beyond the university marketing function.

A short overview of what follows in the book

We end these introductory remarks with a short overview of the four parts of this book. Part 1 introduces the topic at hand and provides an overview of the literature on institutional science communication, while also defining our level of analysis – the meso level of research institutes. Chapter 1, by the editors Entradas (Portugal) & Bauer (UK), reviews the implications of a NPM regime of universities and frames the ongoing transformations in the communication function in the light of the 'decentralisation hypothesis'. In Chapter 2, O'Muircheartaigh (USA) discusses methodological challenges in conducting large-scale comparative institutional analyses, with a focus on the present MORE-PE project, while also offering practical advice for scholars wanting to design similar studies.

Part 2 of the book addresses the institutional transformations. Five chapters discuss the issues from a comparative perspective. In Chapter 3, Entradas et al. (Portugal) investigate the professionalisation of science communication at the meso level by examining their community of science communication professionals. The authors distinguish specialist and non-specialist staff, their roles and characteristics. In Chapter 4, Marcinkowski (Germany) compares the goals that motivate institutes to go public, examining 'self-interest or public duty' impulses and discusses the evidence, not in line with the imperative of self-interested communication, arguing that it might take some time for strategic communication 'to trickle down to the grassroots of research institutions'. Chapter 5, by Besley (USA) and Dudo (USA), tests predictors of perceived success in public communication using 'Excellence Theory'. The authors identify correlates of perceived success and discuss the importance of organisational structures in fostering engagement quality. In Chapter 6, Bauer (UK) & Entradas (Portugal) observe the roll out of public communication in the light of the medialisation hypothesis and specify the Arms Race model (ARPC, arms race for public communication). By examining relationships between allocation of resources to public communication and level of competition between universities, the authors point to clear evidence for an arms race, in some fields and countries more than others, which, they argue, affords increasing re-allocation of funding for communication with repercussions for research and a potential 'new baroque' in science communication. In Chapter 7, Pellegrini (Italy) and Saracino (Italy) present a typology of institutes based on their propensity for public communication. Their analysis suggests three types of research institutes characterised as 'diffusionist', 'corporate' and 'market oriented'.

Part 3 investigates what, for whom and why, institutes in the various countries engage in public communication. Chapters 8–14 describe in detail the national practice of public communication, each with a particular focus. **Italian** research institutes are compared across areas of research; **Dutch** research institutes with and without a communication policy; **US** institutes and their support for scientists' outreach activity; in **Japan,** the low level of communication activity at research units; in the **United Kingdom**, how do research institutes' communication activities compare in Russell and non-Russell universities; **German** institutes activity is compared between within and outside universities and **Portuguese** institutes are compared across levels of research excellence. Part 4 of the book brings us back to key methodological issues arising from examining communication practices. Chapter 15 offers a detailed description of the sampling design and data collection of this international project, and Chapter 16 describes the framework of analysis and the indicators used in the MORE-PE study.

All these examinations contribute to our understanding of institutional responses of growing competition in a knowledge market, and the growing imperatives to engage various publics for strategic purposes of increasing visibility and reputation management but also for civic purposes of fostering open-ended debate and creating a common understanding of an uncertain future.

Note

1   Data for Taiwan and China are not presented in any of the analyses in this book. Taiwan went through a restructuring in the research system which coincided with the period of data collection invalidating many of the contacts in the sample. In China, the data collection went in different directions compromising comparability.

References

Autzen, C. (2018). Press releases — the new trend in science communication. *Journal of Science Communication*, *13*(03), C02. https://doi.org/10.22323/2.13030302

Bauer, M. W., & Jensen, P. (2011). The mobilization of scientists for public engagement. *Public Understanding of Science,20*(1), 3–11. https://doi.org/10.1177/0963662510394457

Bekhradnia, B. (2016). *International university rankings: For good or ill?*. Oxford: Higher Education Policy Institute.

Borchelt, R. E., & Nielsen, K. H. (2014). Public relations in science: Managing the trust portfolio. In *Routledge handbook of public communication of science and technology: Second edition* (pp. 58–69). London: Routledge. https://doi.org/10.4324/9780203483794-12

Boumans, J. (2018). Subsidizing the news?: Organizational press releases' influence on news media's agenda and content. *Journalism Studies, 19*(15), 2264–2282. https://doi.org/10.1080/1461670X.2017.1338154

Brown, R. (2011). *Higher education and the market.* London: Routledge.

Brown, R., & Carasso, H. (2018). Everything for sale? The marketisation of UK higher education 1980–2012. *Higher Education Forum, 9,* 1–15.

Bühler, H., Naderer, G., Koch, R., & Schuster, C. (2008). *Hochschul-PR in Deutschland.* Ed. H. Bühler, G. Naderer, R. Koch, & C. Schuster. Wiesbaden: Springer-Verlag.

Claessens, M. (2014). Research institutions: Neither doing science communication nor promoting "public" relations. *JCOM, 13*(3), 1–5.

Clark, B. (1998). *Creating entrepreneurial universities: Organizational pathways of transformation. Issues in higher education.* https://eric.ed.gov/?id=ED421938

Crettaz Von Roten, F. (2021). How do scientists doing animal experimentation view the co-evolution between science and society? The Swiss Case. In *Communicating science and technology in society* (pp. 59–77). Cham: Springer. https://doi.org/10.1007/978-3-030-52885-0_4

Cunningham, P., Oosthuizen, S., & Taylor, R. (2009). *Beyond the lecture hall. Universities and community engagement from the middle ages to the present day.* Bar Hill, Cambridge: Victoire Press.

De Boer, H., Enders, J., & Schimank, U. (2007). On the way towards new public management? The governance of university systems in England, the Netherlands, Austria, and Germany. In *New forms of governance in research organizations: Disciplinary approaches, interfaces and integration* (pp. 137–152). Berlin: Springer. https://doi.org/10.1007/978-1-4020-5831-8_5

Drucker, P. F. (1993). The rise of the knowledge society. *The Wilson Quarterly, 17*(2), 52–72.

Elken, M., Stensaker, B., & Dedze, I. (2018). The painters behind the profile: The rise and functioning of communication departments in universities. *Higher Education, 76*(6), 1109–1122. https://doi.org/10.1007/s10734-018-0258-x

Engwall, L. (2020). The governance and missions of universities. In *Missions of universities* (pp. 1–-19). Cham: Springer.

Entradas, M. (2015). Envolvimento societal pelos centros de I&D em Portugal [Societal engagement by R&D centres]. In M. de Lurdes Rodrigues & M. Heitor (Eds.), *40 anos de Políticas de Ciência e de Ensino Superior* [40 years of science and higher education policies] (pp. 503–518). Porto: Almedina.

Entradas, M. (2021). Public communication at research universities: Moving towards (de)centralised communication of science? *Public Understanding of Science.* https://doi.org/10.1177/09636625211058309

Entradas, M., & Bauer, M. W. (2017). Mobilisation for public engagement: Benchmarking the practices of research institutes. *Public Understanding of Science, 26*(7), 771–788. https://doi.org/10.1177/0963662516633834

Entradas, M., & Bauer, M. W. (2019a). Kommunikationsfunktionen im Mehrebenensystem Hochschule. In *Forschungsfeld Hochschulkommunikation* (pp. 9–122). Wiesbaden: Springer VS.

Entradas, M., & Bauer, M. W. (2019b). Bustling public communication by astronomers around the world driven by personal and contextual factors. *Nature Astronomy, 3*(2), 183–187. https://doi.org/10.1038/s41550-018-0633-7

Entradas, M., Bauer, M. W., O'Muircheartaigh, C., Marcinkowski, F., Okamura, A., Pellegrini, G., Besley, J., Massarani, L., Russo, P., Dudo, A., Saracino, B., Silva, C., Kano, K., Amorim, L., Bucchi, M., Suerdem, A., Oyama, T., & Li, Y.-Y. (2020). Public communication by research institutes compared across countries and sciences: Building capacity for engagement or competing for visibility? *PLOS ONE*, *15*(7), e0235191. https://doi.org/10.1371/journal.pone.0235191

European Commission. (2013). *Options for strengthening responsible research and innovation*. Brussels: European Union. https://doi.org/10.2777/46253

European Commission. (2016). *Open innovation, open science, open to the world – a vision for Europe | Shaping Europe's digital future*. Luxembourg: Publications Office of the European Union.

Fähnrich, B., Metag, J., Post, S., & Schäfer, M. S. (2019). Hochschulkommunikation aus kommunikationswissenschaftlicher Perspektive. In *Forschungsfeld Hochschulkommunikation* (pp. 1–21). Wiesbaden: Springer Fachmedien. https://doi.org/10.1007/978-3-658-22409-7_1

Fähnrich, B., Vogelgesang, J., & Scharkow, M. (2020). Evaluating universities' strategic online communication: how do Shanghai Ranking's top 50 universities grow stakeholder engagement with Facebook posts? *Journal of Communication Management*, *24*(3), 265–283. https://doi.org/10.1108/JCOM-06-2019-0090

Gioia, D. A., Schultz, M., & Corley, K. G. (2000). Organizational identity, image, and adaptive instability. *Academy of Management Review*, *25*(1), 63–81. https://doi.org/10.5465/AMR.2000.2791603

Grunig, J. E., & Grunig, L. A. (2008). Excellence theory in public relations: Past, present, and future. In *Public relations research* (pp. 327–347). VS Verlag für Sozialwissenschaften. https://doi.org/10.1007/978-3-531-90918-9_22

Habermas, J. (2001). *Kommunikatives Handeln und detranszendentalisierte Vernunft*. Stuttgart: Reclam.

Hallahan, K., Holtzhausen, D., van Ruler, B., Verčič, D., & Sriramesh, K. (2007). Defining strategic communication. *International Journal of Strategic Communication*, *1*(1), 3–35. https://doi.org/10.1080/15531180701285244

Hemsley-Brown, J., & Oplatka, I. (2006). Universities in a competitive global marketplace: A systematic review of the literature on higher education marketing. *International Journal of Public Sector Management*, *19*(4), 316–338. https://doi.org/10.1108/09513550610669176

Hood, C. (1995). The "new public management" in the 1980s: Variations on a theme. *Accounting, Organizations and Society*, 20(2–3), 93–109. https://doi.org/10.1016/0361-3682(93)E0001-W

Horst, M. (2013). A field of expertise, the organization, or science itself? Scientists' perception of representing research in public communication. *Science Communication*, *35*(6), 758–779. https://doi.org/10.1177/1075547013487513

Kerr, C. (1963). The idea of a multiversity. In C. Kerr (Ed.), *The uses of the university* (pp. 1–34). Cambridge, MA/London: Harvard University Press.

Kleimann, B. (2019). (German) Universities as multiple hybrid organizations. *Higher Education*, *77*(6), 1085–1102. https://doi.org/10.1007/s10734-018-0321-7

Koch, H. A. (2008). *Die Universität – Geschichte einer Europäischen Institution*. Darmstadt: WBA.

Kohring, M., Marcinkowski, F., Lindner, C., & Karis, S. (2013). Media orientation of German university decision makers and the executive influence of public relations. *Public Relations Review*, *39*(3), 171–177. https://doi.org/10.1016/j.pubrev.2013.01.002

Koivumäki, K., & Wilkinson, C. (2020), "Exploring the intersections: researchers and communication professionals' perspectives on the organizational role of science communication". *Journal of Communication Management*, 24(3), 207–226. https://doi.org/10.1108/JCOM-05-2019-0072

Kollman, K. (1998). *Outside lobbying*. Princeton: Princeton University Press.

Koso, A. (2021). The press club as indicator of science medialization: How Japanese research organizations adapt to domestic media conventions. *Public Understanding of Science*, 30(2), 139–152. https://doi.org/10.1177/0963662520972269

Krücken, G. (2014). Higher education reforms and unintended consequences: A research agenda. *Studies in Higher Education*, 39(8), 1439–1450. https://doi.org/10.1080/03075079.2014.949539

Krücken, G. (2020). "The European University as a Multiversity." In *Missions of Universities* (pp. 163–178). Cham: Springer.

Krücken, G., & Meier, F. (2006). Turning the university into an organizational actor. In G. S. Drori, J. W. Meyer, & H. Hwang (Eds.), Globalization and organization (pp. 241–257). Oxford: Oxford University Press.

Laredo, P. (2007). Revisiting the third mission of universities: Toward a renewed categorization of university activities? *Higher Education Policy*, 20(4), 441–456. https://doi.org/10.1057/palgrave.hep.8300169

Linell, P. (2009). *Rethinking language, mind, and world dialogically*. Charlotte: IAP.

Maassen, P., & Stensaker, B. (2019). From organized anarchy to de-coupled bureaucracy: The transformation of university organization. *Higher Education Quarterly*, 73, 456–468.

Mahoney, J. T. (1997). The mechanisms of governance. *Academy of Management Review*, 22(3), 799–802. https://doi.org/10.5465/amr.1997.9708210726

Marcinkowski, F., & Kohring, M. (2014). The changing rationale of science communication: a challenge to scientific autonomy. *Journal of Science Communication*, 13(3), C04.

Marcinkowski, F., & Kohring, M. (2018). The changing rationale of science communication: a challenge to scientific autonomy. *Journal of Science Communication*, 13(03), C04. https://doi.org/10.22323/2.13030304

Mejlgaard, N., Bloch, C., Degn, L., Nielsen, M. W., & Ravn, T. (2012). Locating science in society across Europe: Clusters and consequences. *Science and Public Policy*, 39(6), 741–750. https://doi.org/10.1093/scipol/scs092

Mejlgaard, N., Bloch, C., & Madsen, E. B. (2018). Responsible research and innovation in Europe: A cross-country comparative analysis. *Science and Public Policy*, 46(2), 198–209. https://doi.org/10.1093/scipol/scy048

Mejlgaard, N., & Stares, S. (2013). Performed and preferred participation in science and technology across Europe: Exploring an alternative idea of "democratic deficit." *Public Understanding of Science*, 22(6), 660–673. https://doi.org/10.1177/0963662512446560

Metag, J., & Schäfer, M. S. (2019). Hochschulkommunikation in online-medien und social media. In *Forschungsfeld Hochschulkommunikation* (pp. 363–391). Wiesbaden: Springer VS. https://doi.org/10.1007/978-3-658-22409-7_17

Neresini, F., & Bucchi, M. (2011). Which indicators for the new public engagement activities? An exploratory study of European research institutions. *Public Understanding of Science*, 20(1), 64–79. https://doi.org/10.1177/0963662510388363

Paradeise, C., Reale, E., & Goastellec, G. (2009). A comparative approach to higher education reforms in western European Countries. *Higher Education Dynamics*, 25, 197–225. https://doi.org/10.1007/978-1-4020-9515-3_9

Peters, H. P., Heinrichs, H., Jung, A., Kallfass, M., & Petersen, I. (2008). Medialization of science as a prerequisite of its legitimization and political relevance. In *Communicating science in social contexts: New models, new practices* (pp. 71–92). Dordrecht: Springer. https://doi.org/10.1007/978-1-4020-8598-7_5

Power, M. (1997). *The audit society: Rituals of verification.* Oxford: Oxford University Press.

REF. (2014). *Research excellence framework* (Worldwide). Higher Education Funding Council for England. https://www.ref.ac.uk/

Rödder, S. (2011). Science and the mass media –'medialization' as a new perspective on an intricate relationship. *Sociology Compass, 5*(9), 834–845. https://doi.org/10.1111/j.1751-9020.2011.00410.x

Rödder, S. (2020). Organisation matters: Towards an organisational sociology of science communication. *Journal of Communication Management, 24*(3), 169–188. https://doi.org/10.1108/JCOM-06-2019-0093

Rowe, D., & Brass, K. (2011). "We take academic freedom quite seriously": How university media offices manage academic public communication. *International Journal of Media & Cultural Politics, 7*(1), 3–20. https://doi.org/10.1386/mcp.7.1.3_1

Schäfer, M. S., & Fähnrich, B. (2020). Communicating science in organizational contexts: toward an "organizational turn" in science communication research. *Journal of Communication Management, 24*(3), 137–154. https://doi.org/10.1108/JCOM-04-2020-0034

Scheu, A. M., Volpers, A.-M., Summ, A., & Blöbaum, B. (2014). Medialization of research policy: Anticipation of and adaptation to journalistic logic. *Science Communication, 36*(6), 706–734. https://doi.org/10.1177/1075547014552727

Shipman, M. (2013). Public relations as science communication. *Journal of Science Communication, 13*(03), C05.

Slaughter, S., & Rhoades, G. (2004). *Academic capitalism and the new economy: Markets, state and higher education.* Baltimore: The Johns Hopkins University Press.

Vogler, D., & Schäfer, M. S. (2020). Growing influence of university PR on science news coverage? A longitudinal automated content analysis of university media releases and newspaper coverage in Switzerland, 2003–2017. *International Journal of Communication, 14.* http://ijoc.org.

Wæraas, A., & Sataøen, H. L. (2019). What we stand for: Reputation platforms in Scandinavian Higher Education. In *Universities as agencies* (pp. 155–181). Springer International Publishing. https://doi.org/10.1007/978-3-319-92713-8_6

Watermeyer, R., & Lewis, J. (2018). Institutionalizing public engagement through research in UK universities: Perceptions, predictions and paradoxes concerning the state of the art. *Studies in Higher Education, 43*(9), 1612–1624. https://doi.org/10.1080/03075079.2016.1272566

Weingart, P., & Maasen, S. (2007). *Elite through rankings – The emergence of the enterprising university* (pp. 75–99). Dordrecht: Springer. https://doi.org/10.1007/978-1-4020-6746-4_4

Weingart, P., & Pansegrau, P. (1999). Reputation in science and prominence in the media: The Goldhagen debate. In *Public Understanding of Science, 8*(1), 1–16. https://doi.org/10.1088/0963-6625/8/1/001

Wellcome Trust. (2015). *Factors affecting public engagement by researchers – Technical report.* London: Wellcome Trust.

Chapter 2

Why and How to Sample Research Institutes
Methodological Challenges

Colm O'Muircheartaigh

Introduction

There are a number of particular challenges in describing the performance of institutions even within a single country; these challenges are magnified when the results are to be compared across countries. First, there is the definition of the universe of institutions in each country – the units and the criteria for inclusion; the wide variation in organizational structures and concentration of activity across countries makes this a complex process. Second, there is the practical challenge of collecting the data; it is generally impractical to include all the institutions in a country in the measurement effort and thus a sample of institutions must be designed, selected, and recruited. Compliance with the data collection request will be neither complete nor uniform across countries and analytic procedures will be necessary to correct for nonresponse. Third, it is necessary to establish suitable metrics for the performance of an institution and to select metrics that are meaningful across countries. From these, summary measures must be produced across institutions within and among disciplines and countries; there is a special difficulty when the units for which the measures are to be accumulated vary in size.

The population/the universe

The first step in measurement is to define the population of interest. The population must be defined in terms of content, units, extent, and time. The content is the information being sought – here it is the information about dissemination practices and policies of research institutes (RIs) and their implementation. Different units may be formed and reported for the same survey: the data collection in MORE-PE is presented for RIs, research areas (using OECD classification), and countries. The extent of the population is defined by the set of countries included in the study and within the country by the available and implementable materials that constitute the sampling frame.

DOI: 10.4324/9781003027133-3

Table 2.1 Organizational system of research institutions

	Examples of categories

Top/central level
  Research Body (RB)
	University; Faculty; College; School;
	  Society; Association; Academy;
	  Consortium;

Meso level
  (the one intended in this study)
  Research Institute (RI)
	Research Institute; Research Centre;
	  Department; Research Foundation;
	  Research Laboratories

Bottom level	Individual researchers
Micro-level	Research, lab groups

For a cross-national study, it is important to decide on the conceptualization of the universe being measured. Is it a study of the collection of countries as a unit or a set of studies of individual countries? For MORE-PE, each country comprises a separate and distinct subpopulation for which estimates are produced. Each country produces its own independent set of estimates, and these estimates can then be compared across countries. No attempt is made to produce overall estimates for the aggregate of countries, though the array of estimates across countries provides valuable contextual information; the priority is to produce useful (reliable and relevant) estimates for each country and utilize these in a cross-country comparison.

Some common structures must be agreed in order to provide comparability of estimates across the different national populations, though there must be flexibility in tailoring the sample design to the characteristics of these populations. The first step is to define uniformly across countries the units in the population for and from whom data will be collected. The research team defined the middle level (see Table 2.1) of "RI" as both the unit to be sampled and the unit of measurement (i.e., the units for which information would be collected); the meso level of RIs refers to the places where research is carried out and scientists work, which will be a better place to communicate their research and engage with external audiences. Table 2.1 illustrates the process of unit definition, and examples of sampling designs can be found in the country-level sample descriptions (see Chapter 15).

Sample design

There are always compromises in determining the implementation of a sample design; Figure 2.1 is a simple representation of the initial challenge. The *target population* is the set of all RIs in a country; the *sampling frame* is the set of materials (lists, etc.) that is available to select the sample. There are a number of reasons why the two may not be aligned. There may be many very

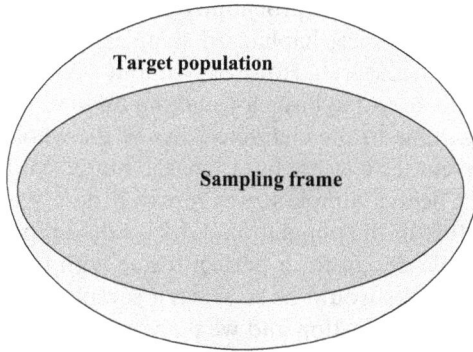

Figure 2.1 Diagram of sampling frame/target population.
Note: Adapted from the JPSM teaching course materials, created by Jim Lepkowski and the author.

small RIs with few researchers and little influence that, despite their numbers, account for a very small proportion of research activity; we may wish to exclude these from the sample because of cost. Some RIs may be new and may not have available contact information. On the other hand, some RIs that appear on the frame may no longer be functioning and are, therefore, not part of the population.

Some examples illustrate the variety of approaches. At one end of the scale, there are countries where, due to the relatively small size of the complete universe, all RIs were included; the Netherlands and Portugal fall in this category. In the center, we have countries with a predominance of state institutions of higher learning, such as the United Kingdom and Germany. In the United Kingdom, all 154 universities in the 2014 Research Excellence Framework were included; these comprise all the major research activities in the United Kingdom. In Germany, the only noteworthy exclusion was of private universities; the frame includes more than 90% of all academic staff in the German higher education system. At the other end of the scale is the United States, with many large public and private universities, where research activity is more dispersed. Here, the sampling frame was restricted to the 115 Carnegie Doctoral Universities with Highest Research Activity (out of about 3,000 four-year colleges in the United States); while this is a very small fraction of eligible institutions, it represents a substantial proportion of RI activity. For each university, all RIs were mapped from websites, giving a sampling frame of 10,308 RIs.

A "Representative" sample

The next step is to decide on how best to represent this population in the sample. A major decision is to decide whether to insist on a probability sample;

a probability sample is a sample produced by a design in which every unit in the frame has a known nonzero probability of being included in the sample. In many fields, expert choice, haphazard samples, or convenience samples form the basis of much research. However, without a probability basis for the selection, there is no scientific basis for making an inference from the characteristics of the sample to the characteristics of the whole population. Kiaer (1897), in a presentation to the International Statistical Institute in Berne which launched the field of sample survey research, described the objective as selecting a "miniature of the population" which would represent the whole.

The default sample design for a perfect frame with units that are essentially exchangeable units would be to select a sample where every unit has the same probability of selection and we place no structural constraints on the selection; this is simple random sampling. A simple random sample is a sample that (i) gives every unit in the population an equal chance of appearing in the sample and (ii) gives every combination of that number of elements the same chance of selection.

Figure 2.2 illustrates the conceptualization: We start with a frame that contains limited information about each of the units in the population (e.g., identifying information and field). From the frame, we select a sample of units for examination using a probability mechanism. For these units, we collect a

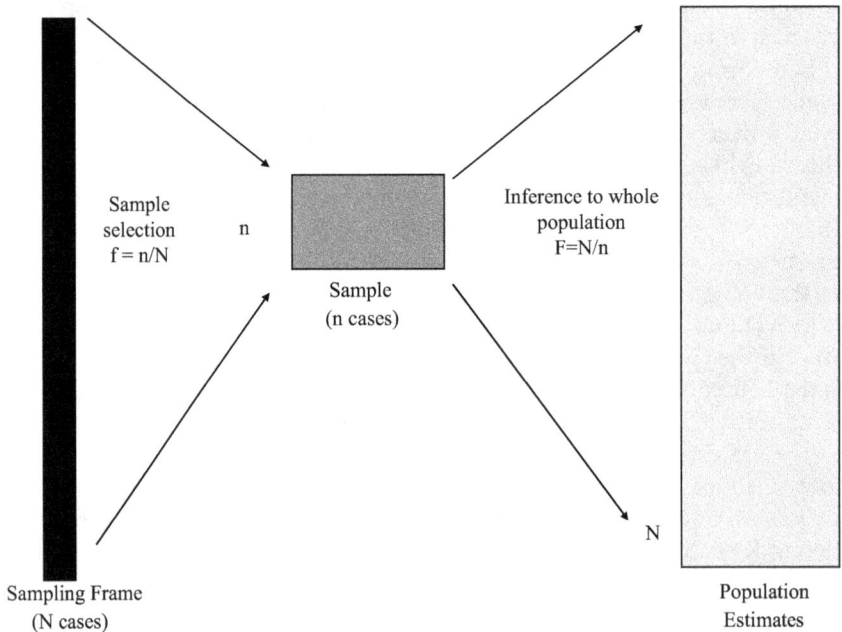

Figure 2.2 Conceptual scheme of the sampling frame.
Note: Adapted from the JPSM teaching course materials, created by Jim Lepkowski and the author.

broad range of information about the key characteristics under investigation; this is the survey data collection phase. Using the inverse of the probability mechanism, we then extrapolate this rich information about the sample to produce estimates of these characteristics of the whole population. It is the knowledge of how the sample relates to the population (the probability mechanism of selection) that provides the scientific warrant for this extrapolation step.

In practice, we do not see all the units in the population as being essentially equivalent, and we are not indifferent to the structure of the sample. In his seminal paper on applied sampling, Kiaer (1976) describes the objective of sampling as obtaining a "miniature of the population", the study of which permits description of the whole population. This means that in practical situations we structure the sample design so that important segments of the population are guaranteed appropriate representation and that we have a sufficient number of units in the sample for subpopulations for which we wish to obtain reliable estimates. We accomplish this through stratification and unequal probabilities of selection for different units (see the section below).

To deal with the heterogeneity of populations across countries, the sample design for each country may be specific for that country. To produce estimates that maximize the precision of comparisons between countries (for a given total sample size), the most efficient design is one that equalizes precision across countries.

Country samples

We use the term domain to indicate a (sub)population for which we wish to obtain an estimate. The estimates needed for MORE-PE are in the form of percentages (or proportions) of institutions that have a particular characteristic or level of activity. In order to produce an estimate of that percentage with a confidence interval of ± 7% (a standard error of 3.5%), a sample size of approximately n=200 is required. To provide acceptable estimates for MORE-PE, therefore, a sample size of about 200 is required for each domain (if a domain contains fewer than 200 units, all the units can be included).

Target domains for MORE-PE Stratum by research areas by country

MORE-PE aimed to produce (1) separate estimates for each participating country and (2) estimates for each of a set of research fields within each country. The team decided to consider research areas separately because they engage differently in public communication of research. This would avoid any bias by equalizing the representation of units across fields/disciplines that may have different likelihood of engaging in public communication (such as social sciences and natural sciences). The objective is to provide separate estimates for each of the OECD S&T classification's six research areas (Natural Science, Engineering and Technology, Medical and

Health Sciences, Agriculture, Social Sciences, Arts and Humanities). Consequently, each field within each country sample needs to provide a sufficient number of cases in the sample from each field to permit separate estimation of activity in the field and comparison across countries within areas.

Stratification

Stratification consists in dividing the population into distinct nonoverlapping subpopulations (the strata) and selecting a sample independently within each stratum. This permits us to tailor the sample design to each stratum to whatever extent we wish. At the basic level, it allows us to decide in advance how many units from the stratum will appear in the final sample. For MORE-PE, each country is a first-level stratum – a separate population from which an independent sample is drawn and for which independent (separate) estimates can be obtained. This allows flexibility in defining the units and the coverage of the sampling frame in each country. Within each country, each S&T research area is a second-level stratum, again permitting an independent sample of adequate size. The separate field estimates within a country can be combined to produce the overall country estimate. The within-country areas are the domains that determine the sample structure.

The objectives of MORE-PE specify that estimates of the parameters be produced for each field and each country and impose this common structural constraint on the sample designs. The simplest and most common form of stratified sample is the proportional stratified sample, where the probability of selection is the same within each stratum; this would result in a sample that mirrors the population both in structure and in extent. This would result in national samples that were very different in the mixture of fields for different countries.

As different research areas engage differently in public communication of research, the team was concerned that such samples would make comparisons across countries invidious, as the apparent performance of countries might be dependent more on the mix of fields in the country than on the comparative performance of scientists within disciplines. To avoid this, it was decided to equalize the number of RIs to be selected from each field in each country. This requires different sampling fractions (probabilities of selection) in different areas within a country; in statistical terms, this is a disproportional stratified sample. If the national samples are used without reweighting, it is important to realize that this decision has important implications for the interpretation of the results. The comparison is not a comparison of the actual behavior of scientists and RIs across countries, it is a comparison of the behavior of countries if the distribution of activity across fields was identical in all countries and consisted of equal numbers of RIs in each field in each country. This approach is analogous to the statistical technique of standardization used in statistical analysis to maximize precision (Kalton, 1968).

As MORE-PE aimed to produce separate estimates for each domain, the target sample size for each field within each country was 200; if the number of RIs in the country in a particular field was less than 200, all the RIs in that field in that country were included.

Response rates, nonresponse, and potential bias

A properly selected probability sample provides the most robust basis for scientific inference from a sample to a population. The challenges of implementing collection of the relevant data from all the members of such a sample are considerable in practice. Nonresponse, or the failure of a selected unit to provide data, could undermine the utility of the estimates derived from a survey. There has been considerable debate in the literature on the level of nonresponse that is acceptable, and the accepted level varies considerably across fields of investigation; an interesting assessment of the use of surveys in litigation, for instance, can be found in Diamond (2011). In general, response rates for surveys of institutions (businesses, schools, agencies) are lower than those for surveys of individuals. In business-to-business surveys, response rates of 20% or so are common. The question arises, of course, as to whether a probability sample with a low response rate is superior to a non-probability sample of the same size. This was vigorously debated in the preparations for the US National Children's Survey (Michael and O'Muircheartaigh, 2008). Among many of the epidemiologists and physicians who participated in the discussions about the sampling plan, there was a belief that a probability sample becomes essentially worthless if the retention rate or the initial response rate drops below some arbitrary level. It was contended that if the response rate fell below 50%, for example, then a purposive sample would be equally good if not preferred for its other attributes. A simple example was useful in countering this belief.

"For the Los Angeles metropolitan area (population 12 million), consider two possible implementations: (i) a probability sample with a response rate of 30% and an achieved sample size n = 3,000 and (ii) a volunteer sample of 3,000 patients. What is the appropriate inferential population for each? The probability sample, even with its low response rate, provides (without any additional assumptions) a basis for inference to the approximately 30% of the population that has demonstrated willingness to participate in the study. This gives a robust inference to about 3.6 million people. The volunteer sample provides an assumption-free inference to itself only that is to an inferential population of 3,000 people".

Sample surveys can be susceptible to bias in two ways. First, if the selected sample is not representative of the population, the results will provide evidence only for those given an opportunity to be selected. We have dealt with this by selecting, in every country, a probability sample with known probabilities of selection for the RIs; this allows us to make the inferential leap from the sample to the population. The second major threat of bias comes from nonresponse – failure of selected units to participate in the survey. Nonresponse leads to bias whenever nonrespondents differ from respondents in the characteristics being measured. If nonresponse is random, and there is no difference between respondents and nonrespondents, then nonresponse merely reduces sample size but does not introduce bias. However, it is often plausible to assume that willingness to respond to a particular survey is correlated with the target measures (in this case, the willingness of an RI to respond might be related to its communication strategies and practice).

The best protection against bias comes from maximizing the response rate. In MORE-PE, the standard protocol involved six approaches to the sampled RIs, the initial communication, and five reminders. This protocol led to response rates that were in the acceptable range of response rates for institutional surveys of this kind. For cost reasons, the survey was administered by mail and email. Other modes of data collection, such as direct phone calls and other targeted approaches, will typically obtain higher response rates, but the cost per unit of these approaches is much higher. Consequently, these enhanced procedures are typically too expensive to use for the whole sample. Cochran (1977) proposed a method to incorporate results from a (probability) subsample of nonrespondents to reduce the potential nonresponse bias while controlling costs. A subsample of nonrespondents to the initial approach is selected and the more expensive procedure employed with the subsample; this will typically lead to successful recruitment of a subset of this subsample, who are then used to estimate the characteristics for the whole class of nonrespondents.

The recruited sample now has two components: (i) the respondents obtained in the standard recruitment process – we can think of these as representing the "easier to recruit" units in the population and (ii) the respondents obtained through the more intensive recruitment process – we can think of these as representing the "hard to recruit" units in the population. In calculating the response rate overall, we combine the two rates so as to represent the "easy to recruit" and "hard to recruit" parts of the population appropriately. In incorporating them into the analysis, we weigh them with the inverse of the subsampling rates employed in selecting them.

As an illustration, Table 2.2 gives the calculation of the response rate for the Natural Sciences field in the United Kingdom, where this subsampling procedure was used effectively. An initial sample of n=185 RIs was selected, and these were approached using the standard protocol (an escalating sequence of five attempts by mail); 24 responses (13%) were received. Of the 161 nonrespondents, 32 (one in five) were followed up with repeated individual phone calls; seven of these (22%) responded. Each of these seven supplementary responses represents five of the initial nonrespondents in our estimate; thus in combining them with the initial responses, we give each of them a weight of five in the calculation, giving a weighted number of responses of 24=35=59. There is of course still potential for nonresponse bias, but it has been greatly reduced by obtaining this information.

Table 2.2 Response rate calculation for the Natural Sciences field in the UK

Initial sample	Initial responses		Subsample for Intensive follow-up		Responses to follow-up			Weighted response rate
Number	N	%	N	Sample fraction	N	Weight number	Response rate (%)	(%)
185	24	13%	32	f=1/5	7	7/f=35	22	(24+35)/185 = 32%

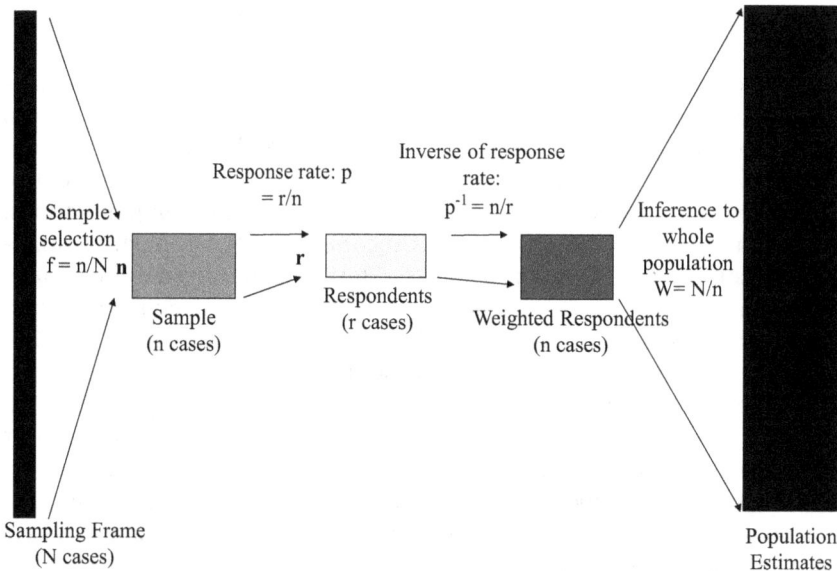

Figure 2.3 Enhanced inference schema, considering nonresponses.
Notes: Adapted from the JPSM teaching course materials, created by Jim Lepkowski and the author.

Figure 2.3 shows the enhanced inference schema taking nonresponse into account.

Calculating a country-level estimate of RI public engagement (combining estimates across fields within countries) and comparing estimates among countries (combining across countries)

In calculating a country-level measure, we need first to decide what our inferential population is. There are two approaches/possibilities. First, we could simply average across the sample observations. The general sample design targeted the same number of RIs in each area, regardless of the number of RIs in the area. Consequently, this unweighted sample estimate will in effect be giving each area the same weight in the overall estimate. This will bias the estimate toward the behavior of the least numerous fields in the country.

Second, we could weigh the observations so that each area is represented proportionally to its number of RIs in the country. This will produce an estimate that represents the distribution of RIs in the country appropriately. For comparison across countries, within-area comparisons are straightforward, as in general RIs within area within countries were selected with equal probabilities. If, however, we wish to compare national estimates across countries, we need to be aware that such estimates will have two components: (i) the level of PE in each area and (ii) the composition of the population of RIs across areas.

Many disciplines face this challenge in making comparisons across subpopulations, notably demography (in comparing mortality or birth rates) and epidemiology (in comparing infection rates and transmission rates). *Standardization* is a technique that is used in those disciplines to control for the compositional differences between populations and subpopulations. The method permits the separation of the impact of the population structure (or composition) from the estimate. Thus a standardized estimate uses a common composition and produces an estimate for each country that indicates what its average level of activity would be if the distribution of RIs in the country matched the standard structure.

If a particular area (for instance, TK) has typically much higher outreach efforts, then a country whose RIs are concentrated in this area will appear to have higher rates, even though the outreach levels by field might be very similar (TK would be nice to have an example). This standardization process is common in demography, in representing mortality rates by country. As age distributions across populations can differ greatly, mortality rates are usually presented as age-specific mortalities rather than as gross mortality rates to separate out compositional effects (reference here). Age-standardized mortality rates (using a common composition) may better represent the reality in the country. In general, it is worth bearing in mind that any aggregate summary that is used to compare countries with different structures or composition will both reveal and conceal complexities in the underlying reality. The conceptualization/meaning of the summary measure needs to be defined and justified.

Units of different sizes

There is one final complication that should be addressed in thinking about the design of studies such as this and the presentation of the results. When describing a population that comprises units of widely different sizes, it is important to distinguish different measures of average or total. Kish (1965) provides a striking illustration of the challenges of interpreting data about institutions. The launch of the first Sputnik satellite by the USSR in 1957 led to a search for an explanation of this failure of US science and US science education. Initially, the lack of science education was blamed; one report at the time indicated that almost half the high schools in the United States offered no physics, a quarter offered no chemistry, and a quarter no geometry. However, further investigation revealed that only about 2% of the population of high school students attended these schools. A second example comes in US Federal Government continuing surveys of banks, where the monitoring surveys rely on reports from a tiny fraction of all banks; these banks nevertheless account for more than 90% of the value of all banking transactions. If you were to describe the average performance of banks, giving each bank equal weight, your estimate would be dominated by the behavior of the small banks, which represent a tiny fraction of the financial sector. But if you weigh the estimates by the size of the bank, you essentially ignore the behavior of most of the banks.

Analysts must decide which is the appropriate measure for their analyses; the data collected in the MORE-PE survey provide a basis for calculating summary measures using the appropriate measures for different analyses.

References

Cochran, W. G. (1977). *Sampling techniques* (3rd ed.). New York and London: John Wiley and Sons.

Diamond, Shari Seidman (2011). Reference guide on survey research. In National Research Council (Ed.), *Reference manual on scientific evidence: Third edition*. Washington, DC: The National Academies Press. https://doi.org/10.17226/13163.

Kalton, G. (1968). Standardization: A technique to control for extraneous variables. *Journal of the Royal Statistical Society: Series C (Applied Statistics)*, 17(2), 118–136. https://doi.org/10.2307/2985676

Kiaer, A. N. (1897). *Den Representative Undersogelsemethode*. Oslo: Videnskapsselskapets Skrifter.

Kiaer, A. N. (1976). *The representative method of statistical surveys*. Oslo: Norwegian Central Bureau of Statistics.

Kish, L. (1965). Sampling organizations and groups of unequal sizes. *American Sociological Review*, 30(4), 564–572.

Michael, R. T., & O'Muircheartaigh, C. A. (2008). Design priorities and disciplinary perspectives: The case of the US National Children's Study. *Journal of the Royal Statistical Society: Series A (Statistics in Society)*, 171(2), 465–480.

Part II

Cross-National Comparisons

Chapter 3

Professionalizing the Communication of Research Institutes

Marta Entradas and João M. Santos

Introduction

Research universities and other research institutions have in recent decades experienced a paradigm shift in the ways they communicate and interact with publics. Seen before as "ivory towers", difficult to reach or understand nonpeers, the current and ever-increasing environment to promote dissemination of research and engage with broader audiences beyond the narrow scope of the academic community (European Commission, 2013, 2020) has pushed research institutions toward professionalizing their communication structures. This is seen, for instance, in the rise of press offices (PR) and press releases produced by scientific institutions (e.g., Vogler & Schäfer, 2020), or in the variety of communication activities for nonspecialists in many countries (Entradas et al., 2020; Mejlgaard et al., 2019), which are carried out by a growing community of communication professionals employed at universities (Elken et al., 2018).

This community has received increasing attention in the science communication literature in recent years, with particular focus on tasks and roles, and how the "profession" is developing within research organizations (Fähnrich et al., 2019). Most of this research has, however, focused on the central/PR officers of universities, while less is known about these professionals working at the research institutes hosted by these universities (Entradas, 2015; Entradas & Bauer, 2017). In this chapter, we add to the existing literature by examining the community of communication professionals at the institute level. We start with a brief account of the literature in the field of professionalized public communication, with a focus on scientific institutions. We then present empirical data on characteristics of communication professionals concerning their backgrounds, roles, and level of professionalization, and we identify patterns of professionalization across countries and areas of research.

DOI: 10.4324/9781003027133-5

Professionalizing public communication of science of research universities

Public communication between universities and publics has gained a new reputation over recent decades. From the 1980s on, we see an intensification in the number and in the level of activity of PR offices and increased pro-fessionalization of PR and marketing functions (Grunig & Grunig, 2008). PR officers have intensified their activity by acquiring a portfolio of me-dia contacts (Borchelt & Nielsen, 2014), and assisting researchers in media communication tasks such as preparing press releases and contacting the media and journalists (Marcinkowski et al., 2014; Neresini & Bucchi, 2011). Public information officers and the "'management profession" were seen as needed to create and maintain relationships between the university and so-ciety (Krücken & Meier, 2006) (see Chapter 1 of this book).

Research in the PUS and science communication literature that has ex-amined communication professionals is still emerging. It has mostly inves-tigated views and roles of these professionals toward public engagement, using small, convenience samples. This might be due to the fact that science communication, as a profession, is still relatively new (Trench & Bucchi, 2010). Despite this, the available evidence points to a growing community internationally and a move toward the development of a profession on its own (Evetts, 2003). Attempts of professionalization are evident in the proliferation of platforms, networks, and associations for public science communication (see, for example, the Australian Science Communicators organization, comprised of individuals who identify themselves with such a role, particularly those dedicated to communication tasks in institutions (Metcalfe & Gascoigne, 1995)), and in the training in science communica-tion on offer at universities in many countries (Trench, 2017).

Still at an early stage of professionalization, it is perhaps not surprising that formal training in science communication is not yet something seen among most communicators, with a significant number hailing from other more established professions, such as journalism, public relations, or science jobs, and more frequently trained in science (Brown & Scholl, 2014) than in communication (Miller, 2008). In terms of demographics, the available information is scarce, yet, it points to science communication being often connected to a women's activity (Miller, 2008; NCCPE, 2021). For instance, a recent study of public engagement professionals working at UK universi-ties during Covid-19 shows that 80% of the 128 professionals participating in the study were women (NCCPE, 2021).

The roles that these professionals perform are varied. The available re-search shows science communication tasks often clashing with others such as public relations and marketing, at the central PR offices of universities. For example, Peters et al. (2008) in an analysis of 40 interviews with PR professionals working in German universities found that "each year, PR offices commonly issue several hundred of press releases and respond to hundreds of journalistic enquiries" (p. 75), while also identifying as general goals maintaining a media presence, developing a brand and a positive image, and marketing products and services. Most recent research, describes the activities performed by public communication professionals as a "set of boundary spanning activities" (Schwetje et al., 2020). Their primary purpose is informational, but this requires managing a variety of stakeholders both in the organization and outside, all with different expectations. Communicators have self-described themselves as "gatekeepers", "service units", "bridge builders", "mediators", "counselors", "consultants", and even "court jesters" (Schwetje et al., 2020). Perhaps because of a lack of a defined professional identity, PR officers do not believe their work is appreciated by other organizational actors (Buhler et al. 2007; Wellcome Trust, 2015).

Yet, although science communication tasks seem limited, communication professionals associate the attainment of their university goals with contributing to the social role of universities and civic engagement (Elken et al., 2018; Koivumäki & Wilkinson, 2020), and identify their purpose as being social legitimization and public debate of science and technology, going beyond simple communication of scientific findings (Hvidtfelt Nielsen, 2010). But they also recognize challenges to the profession and their roles in facilitating public engagement. Public engagement professionals working at UK universities highlighted that public engagement is perceived as servicing more instrumental goals specific to research governance, while also facing issues of support in the organization (e.g., funding, training) (Watermeyer & Lewis, 2018).

With boundaries between the various roles often blurred, it has been proposed that an entire ecosystem of communicators exists, with varying roles and purposes (Fahy & Nisbet, 2011). This diversity may have some impact on institutional science communication. As some commentators have described it, it achieves neither science communication nor public relations goals, for example, by targeting stakeholders instead of general publics (Claessens, 2014) or not focusing on scientific findings *per se* but aimed toward promoting institutional image (Carver, 2014) and visibility of a few scientists (Marcinkowski et al., 2014).

While these studies have mainly focused on professionals working at central communication/PR offices, we know less about the personnel working at research institutes within these universities. In the current environment of increasing calls for public communication of research and competition between universities for students, funding and public visibility, it would not be surprising that professionalization would trickle down to the institute level: either as self-initiated activities by institutes that might feel that their communication initiatives and research are not given enough coverage by central offices or as part of the university communication strategy. This could perhaps be the case in larger universities or in countries with greater incentives for engaging in the activity (e.g., government and funding agencies' requirements). Also, and given the variety of "professionals" and backgrounds likely to be found, we were interested in examining the *level of professionalization* of this staff, i.e., how specialized in the field they are, and whether their level of professionalization would have any impact on the institute communication activity. The expectation is that institutes with dedicated, specialist staff will engage more intensely in public communication in its various formats. We investigate two main questions and one hypothesis as follows:

> RQ1: Who is the communications staff working at research institutes, i.e. what are their professional characteristics and roles?
> RQ2: How professionalised is this staff and how does the level of professionalisation vary across countries?
> H1: The level of professionalisation relates to the level of institutional public communication activity.

In the analyses that follow, we proceed in three steps. First, we describe the characteristics and roles of the communication staff working at research institutes (RQ1); second, we distinguish this staff according to their degree of professionalization and compare across countries and research areas (RQ2); and third, we examine the effects of level of professionalization on communication activity (H1). To address H1, we build on our previous analytical framework that examined public communication (P) as a function of the general organizational context – context factors (C) (e.g., country, discipline, size, research budget), and organizations' dispositions to communicate publicly – disposition factors (D) (e.g., available staff, funding and policies/guidelines, engagement of scientists) (see Chapter 16 for description of the framework for analysis). In the present study, we add the level of professionalization as a disposition factor and use hierarchical regression analysis to investigate the effects of professionalization on the communication activity of an institute.

Methods

Sample

For these analyses, we consider only the institutes that reported employing communication staff (N = 763 out of 2,030 institutes in the eight countries; corresponding to around 38% of the full sample). We defined communication staff in the study as anyone carrying out public communication tasks at the institute (excluding individual researchers). Overall, compared to the whole sample, these are larger institutes (M=159 researchers, SD = 303.4; average institute M=120; SD=414.7), with higher research budgets (most reporting budgets over 1M (41%) compared to the average in the full sample reporting <100,000 Euros; 26%), a communication policy in place (64%) (compared to 40% in the full sample).

Staff characteristics and level of professionalization variables

The level of professionalization of staff is assessed through an index of professionalization which was built using several variables related to characteristics of the communication staff. These included: (1) dedication to communication tasks (exclusively/partly dedicated; 2 variables); (2) type of contract (temporary contract for a research project, temporary contract with the institute, permanent contract; 3 variables); (3) full- or part-time employees; (4) previous experience (no previous experience, researcher, marketing/PR/comms, design/multimedia, journalist/science writer, administrative, project/finance/HR/managers; 7 variables); and (5) training in communication (post-graduate, undergraduate, workshops/short courses, no formal training; 4 variables). All questions were asked to measure the number of staff as we wanted to capture all employees working at institutes. For instance, the question on the type of contract asked "Please indicate what type of contracts the staff at your research institute have: Please count each person only once" (four options for background).

These variables were recoded into categorical variables to normalize the data. Using exploratory factor analysis with Varimax rotation (Hair et al., 2014), several variables were removed due to having a Measure of Sampling Adequacy score under 0.50 (Hair et al., 2014) (temporary contract for a specific research project, temporary contract with the RI, permanent contract with the RI, workshops/short courses) and inadequate communalities (Kolenikov & Bollen, 2012) (full-time employees, "previous experience – Administrative", and "previous experience – project/finance/HR/managers").

A two-factor solution was obtained, explaining 44.2% of total variance. Items loading in the first factor were related to specialized work, whereas the second factor was mostly loaded by items alluding to unspecialized work. This tentative structure was specified in Confirmatory Factorial analysis (CFA); the model fit was good (Barrett, 2007; Kline, 2016) ($\chi2$ = 51.01, CFI=0.98, RMSEA=0.05, TLI=0.97, df=19, p<0.001). We kept the two-factor structure and used the estimated scores obtained from CFA as two distinct measurements for level of professionalization, with "specialist" referring to professionalized work and "nonspecialist" referring to communication work performed by untrained, part-time, or staff responsible for a mix of tasks.[1] Table 3.1 reports the results of the CFA. Table 3.2 shows the descriptive statistics for the two indices.

Table 3.1 Factor loadings for a two-dimension CFA model

Latent factor	Indicator	SE	Z	Beta	Sig
Specialist	Exclusively dedicated to PE tasks	0.032	25.363	0.799	0.000
Specialist	Training in comms – Post-graduate degree	0.040	14.627	0.588	0.000
Specialist	Training in comms – Undergraduate degree	0.044	13.162	0.575	0.000
Specialist	Previous experience – Marketing/PR/comms	0.039	16.286	0.630	0.000
Specialist	Previous experience – Design/ multimedia	0.045	13.296	0.598	0.000
Specialist	Previous experience – Journalist/sci writer	0.042	14.050	0.595	0.000
Nonspecialist	Part-time employees	0.115	8.077	0.928	0.000
Nonspecialist	Previous experience – No previous experience	0.065	5.306	0.344	0.000

Table 3.2 Descriptive statistics for the professionalization indices

	Min	Max	Mean	Standard Deviation
Specialist	−0.948	2.471	0.06	0.77
Nonspecialist	−0.753	1.922	0.07	0.70

Notes: Values are presented as Z-scores. For specialist, higher scores mean higher degree of professionalization (staff more professionalized); for nonspecialist, higher scores mean higher degree of nonprofessionalization (i.e., staff more nonspecialized, part time with no training).

Context and disposition variables and hierarchical regressions

Three dependent variables were used in the regression analysis – *public events, traditional media channels,* and *new media channels.* These variables correspond to indices of intensity of an institute's participation/organization of public communication activities. They were generated through CFA from items of activities (for public events' index, we used 19 items; for traditional channels index, we used 13 items and six items for new media index) (see Chapter 16 of this book for description of variables). The final model yielded good fit ($\chi2 = 627.54$, CFI=0.96, RMSEA=0.04, TLI=0.95,

BIC=56474.49, df=142, p<0.001) and resulted in factorial estimates which were used as indexes of intensity.

Three sets of independent variables were introduced in steps. In step one, we added *context variables* including research area, country, size, and research budget. The second block adds *communication disposition variables* including communication staff, percentage of researchers participating in public communication of the institute (active researchers), communication policy, and funding for public communication activities. The third block is the *level of professionalization* that adds professionalization metrics – specialist and non-specialist indices. See Table 3.3 for description of these variables.

Table 3.3 Descriptive statistics of the control variables

Categorical variables	N	%	Categorical variables	N	%
Country			*Average research income*		
Germany	195	25%	< 100,000€	124	19%
Italy	102	13%	100,000€–250,000€	90	14%
Portugal	114	14%	250,000€–500,000€	83	12%
The Netherlands	50	6%	500,000€–1M €	82	12%
The United Kingdom	49	6%	> 1M €	263	41%
The United States	95	12%	*Annual budget on PE*		
Brazil	63	8%	None	17	3%
Japan	95	12%	< 1%	198	32%
Field of science			1–5%	255	42%
Natural Sciences	211	27%	5–10%	80	13%
Engineering & Technology	122	16%	>10%	62	10%
Medical & Health Sciences	133	17%	*Communication policy*		
Agricultural Sciences	44	5%	No	234	36%
Social Sciences	151	19%	Yes	421	64%
Humanities	102	13%			
% Researchers PE					
None	5	1%			
< 10%	77	10%			
10–20%	139	18%			
20–40%	154	20%			
40–60%	154	20%			
60–100%	144	19%			
Don't know	48	6%			
Continuous variables	M	SD			
Size of the institute	159.52	303.41			
Size of the PE staff	2.77	4.61			

Communications staff working at research institutes

Background, previous experience, and training

The data show that about a third of all institutes covered by the survey (38%; N = 763 institutes) reported having staff dedicated to public communication tasks. Yet, in only 50% of those is the staff exclusively dedicated to such tasks. Among the 763 institutes, the average institute has one person exclusively dedicated (M = 1.2, SD = 3.2) and three persons partly dedicated to public communication (M = 2.8, SD = 4.3). This would correspond to around 2.7 persons per team and a total of around 2.060 communication professionals working in these institutes. Few have a temporary contract to work on a specific research project (M = 0.5, SD = 1.6) or have temporary contracts with the institute (M = 0.7, SD = 1.5), yet most staff have permanent contracts (M = 2.4, SD = 6.0). This might be only an indication that these numbers correspond to permanent staff (administrative for instance) who also carry out communication tasks as part of their jobs. Also, on average, a communication team is composed of two full-time employees (M = 2.5, SD = 4.9) and one part-time employee (M = 1.4, SD = 4.5).

Regarding prior experience, individuals who were previously researchers (39%, M=0.87, SD=1.78) and Marketing/PR officers (37%, M=0.63, SD=1.23) seem to represent the bulk of this staff some reported administrative roles in the past (31%), and less journalism (20%), while others have no prior experience in communication at all (23%). This is opposed to what has been found among communicators working at the central level, whose previous experience is more likely related to journalism and PR.

When it comes to formal training in a communication-related field, most staff have no training (57%), with undergraduate (21%), and post-graduate degrees (29%) comparatively underrepresented among the staff who were more likely to have attended a workshop or a short course in communication (46%). Table 3.4 summarizes these findings.

Table 3.4 Characterization of communication staff. Percentages do not sum up to 100% because multiple options (one person could be counted in more than one category for each dimension) (N=773)

	N	%	M (number of staff)	SD
Dedication to communication tasks				
Exclusively dedicated	373	50%	1.46	3.29
Partly dedicated	620	83%	2.85	5.62

Type of contract

Temporary contract for a research project	171	23%	0.57	1.63
Temporary contract with the institute	293	40%	0.72	1.55
Permanent contract with the institute	537	72%	2.44	6.04

Time allocated

Full-time employees	598	79%	2.58	4.90
Part-time employees	387	52%	1.47	4.58

Previous experience

No previous experience	141	23%	0.49	1.44
Researcher	240	39%	0.87	1.78
Marketing/PR/comms	227	37%	0.63	1.23
Design/multimedia	129	21%	0.34	0.92
Journalist/science writer	122	20%	0.39	2.16
Administrative	191	31%	0.55	1.53
Project/finance/HR/managers	69	11%	0.18	0.67

Training in communication

Training – Post-graduate degree	209	29%	0.63	2.31
Training – Undergraduate degree	153	21%	0.36	0.94
Training – Workshops/short courses	341	46%	1.17	4.04
No formal training	415	57%	1.87	4.58

Roles

From a list of ten roles, we asked institutes "how frequently does the 'communication staff' at your research institute conduct the following" (Figure 3.1). The most common answers were, by decreasing order, "manage the website" (84%), "compose print materials" (73%), "organize public events" (53%), and "motivate researchers to get involved in public communications" (52%). Less frequently, these professionals "intervene in moments of institutional crisis" (65% never/rarely doing it), organize training for researchers (68% never/rarely), and assist researchers on research grants (53% never/rarely). Occasionally or rarely/never, they contribute to "creating action plans for communications" (59%) and "deciding on policies with the leadership" (59%). This seems to suggest that professionals working at the level of institutes are more likely to be involved in the institutes' activities, and have little influence on the institutional communication strategy, and institutional events.

Level of professionalization of staff

We examine the level of professionalization (RQ2), using the factor scores from CFA. As described earlier, factor analysis revealed two types of communication professionals at research institutes: (1) *specialist staff* – individuals more likely to be exclusively dedicated to communication tasks, have training in communication, either at a post-graduate or undegraduate level, and have previous work experience in a communication-related field and (2) *nonspecialist staff* – individuals more likely to be partly dedicated to communications, often part-time employees, without experience in communication-related fields. Figure 3.2 illustrates these findings, showing that nonspecialist scores are, in general, higher than specialist scores. This indicates a predominance of nonspecialist staff over specialist staff at institutes suggesting that communication tasks are more often carried out by nonspecialists.

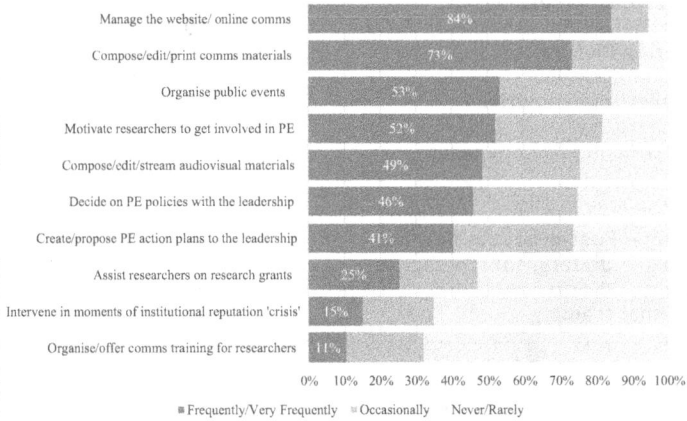

Figure 3.1 Frequency of the various public communication roles (N=763).

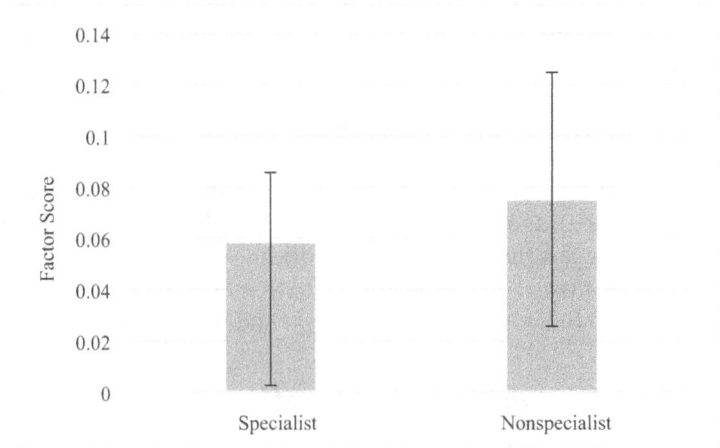

Figure 3.2 Comparison of specialist and nonspecialist scores across participant institutions.
Note: error bars indicate 95% confidence intervals.

Level of professionalization and associated roles

We then examined the relationships between the level of professionalization and the roles of public communication staff. We were interested in understanding whether differences existed between tasks carried out by specialist and nonspecialist. Self-reported roles were then regressed on the specialist and nonspecialist scores, and the results are shown in Table 3.5.

Overall, the data show that most roles are common among specialist and nonspecialist staff (Table 3.5). Yet, there are exceptions with roles such as "crisis management" (B = 0.077, p <.01), "compositing communication materials" (B = 0.077, p <.05), and "organizing training for researchers" being specific to specialist staff. "Composing audiovisual materials" shows a significant effect on both types of staff, but the effect is stronger for specialist, indicating that although this is a common task, it is more frequently performed by specialist staff. On the contrary, "assisting researchers in research grants writing" has a negative coefficient, indicating that this role is unlikely to be performed by communication staff, perhaps because this is a task more likely to be of the responsibility of research managers (Trindade & Agostinho, 2014).

These results show that while most of these professionals perform similar roles, which are often related to the institute activities, specialist staff are more likely to be involved in tasks that require knowledge and training in

Table 3.5 Regression of roles on specialist/nonspecialist scores (N=600)

Roles	Specialist	Nonspecialist
Decide on communication policies with the leadership	0.058 (0.034)	0.061 (0.035)
Create/propose communication action plans to the leadership	0.011 (0.038)	−0.026 (0.039)
Motivate researchers to get involved in public communication	0.061 (0.032)	0.034 (0.033)
Intervene in moments of institutional reputation "crisis"	0.077** (0.027)	0.022 (0.028)
Compose communication materials	0.077* (0.037)	0.035 (0.038)
Compose audiovisuals	0.115*** (0.030)	0.092** (0.031)
Manage the website/online comms	0.048 (0.031)	0.010 (0.032)
Organize public events	0.020 (0.029)	0.021 (0.030)
Organize/offer comms training for researchers	0.120*** (0.029)	0.024 (0.030)
Assist researchers with research grants	−0.125*** (0.022)	−0.070** (0.023)
R^2	0.284	0.093

Notes: *** p < 0.001; ** p < 0.01; * p < 0.05

the field and also to contribute to institutional communication strategy, unlike untrained, nonspecialists.

Level of professionalization across countries and areas of research

Finally, we look for any patterns in the level of professionalization across countries and areas of research. For these analyses, we use the indices of level of professionalization (specialist and nonspecialist). First, we note the distribution of institutes with communication staff across countries and areas of science. More institutes report employing communication staff within the institute in Germany (54%), Portugal (51%), and Brazil (37%), less in the United States (36%), the United Kingdom (26%), Italy (28%), and Japan (29%). As for areas of research, institutes in the natural sciences (41%), engineering and technology (39%), and agriculture (39%) have tendentially more staff dedicated to communication staff, while the medical sciences (40%), the humanities and arts (31%), and the social sciences (35%) have less communication staff.

Level of professionalization across countries

Of all countries, Brazil has the highest specialist and nonspecialist scores, indicating more specialised and fully dedicated staff, and higher degree of specialist (more and fully dedicated) and higher degree of nonspecialist (part time, with no experience). Lagging behind and with a similar pattern are institutes in Germany and in the Netherlands, with also more specialist and nonspecialist staff. Italy and the United States show degrees of professionalization around the average. Japan shows the lowest degree of professionalization (lower scores for specialist), followed by Portugal and the United Kingdom, indicating that institutes in these countries have overall less qualified communication staff.

Also important to note is the fact that Portugal, Italy, and the United States of America, by decreasing order, have overall lower scores on the nonspecialist index compared to the other countries. This suggests that many institutes in these countries are likely to rely on nonspecialists that carry out public communication tasks. It is also interesting to note, that despite Portugal being among the countries reporting more staff dedicated to public communication tasks (around 74% reporting staff within the institute both partly and fully dedicated, compared to 40% in Japan), the level of professionalization is low. This suggests that the fact that an institute has communication staff fully dedicated to communication does not necessarily mean that staff is professionalized (Figure 3.3).

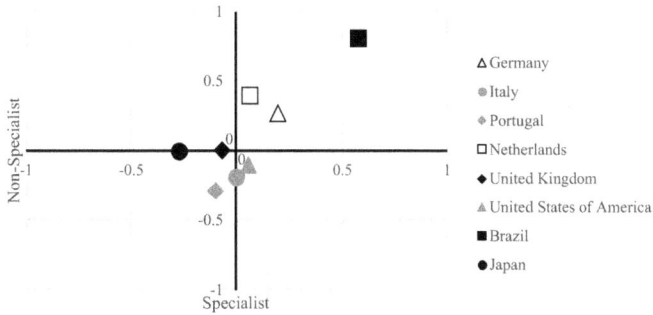

Figure 3.3 Specialist versus nonspecialist scores by country.

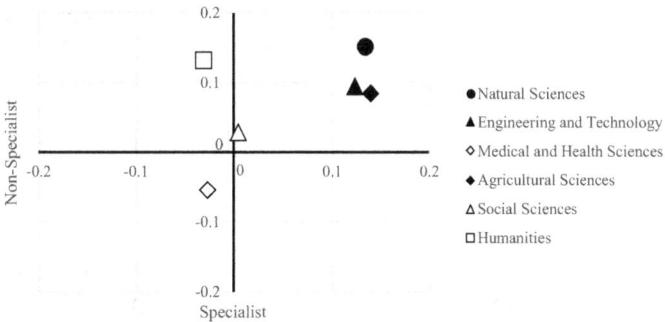

Figure 3.4 Specialist versus nonspecialist scores by research area.

Level of professionalization across research area

As for areas of research, two clusters can be observed. Higher scores in both specialist and nonspecialist indices are found among the "hard" sciences, in particular, the natural sciences, the engineering and technology, and the agricultural sciences. This indicates higher professionalization of staff among these sciences. On a lower level of professionalization, are the medical and health sciences, the humanities, and the social sciences. The humanities show a lower specialization score than the other sciences, and a high nonspecialist score, indicating that in this area, most communication tasks are carried out by nonspecialist staff. Social sciences show average scores on both indices, indicating some level of professionalization. The lowest scores are found among the medical and health sciences, for both specialist and nonspecialist, indicating overall a lower level of professionalization in these fields than the other sciences, which might also relate to the low activity found among these sciences (Entradas et al., 2020) (Figure 3.4).

Relationships between professionalization and institutional communication

To examine the relationships between levels of professionalization and levels of public communication activity of an institute (H1), we run regression analysis controlling for context and disposition variables. Model 1 (context variables), for all the three dependent variables (public events, traditional media, new media channels), shows important effects on specific fields of science and countries. Size has a consistent effect on all three outcome variables (p <.01), and research budget has a significant effect on traditional and new media channels (p <.05) further confirming findings of increased media relations among institutes receiving larger science budgets (Entradas & Santos, 2021). Model 2 shows significant effects for all variables; these are kept significant in Models 3. These results are no different from earlier work that showed associations between the level of engagement and the context and commitment of the institution to public communication (Entradas et al., 2020).

Model 3 introduces the professionalization variables. The model shows that *specialist* staff is positively associated with increased use of traditional media channels (B = 0.338, p <.001), new media (B = 0.398, p <.001), and public events (B = 0.229, p <.001), and the effects are stronger for media channels. On the contrary, *nonspecialist* shows no effects for any of the dependent variables. This suggests that specialist staff facilitates communication, while nonspecialists shows no evidence of contributing to increased public communication activity. Moreover, adding the professionalization variables significantly increases the model fit (p <.001), further confirming the importance of specialist staff of an institute. It is interesting to note that the "size of staff" ceases to be significant in Model 3 for all three dependent variables suggesting that the number of staff might be irrelevant when staff is dispersed among various roles and that trained staff (quality) is more important than the number of staff (quantity). Table 3.6 summarizes the results of the regressions.

Discussion and the "professionalization thesis"

In this chapter, we described levels of professionalization and roles of the communication staff working at research institutes and comparatively across countries and disciplines. We find that about 38% of all institutes surveyed

Table 3.6 Hierarchical regression models

Variables	Traditional media channels			Public events			New media channels		
	Model 1	Model 2	Model 3	Model 1	Model 2	Model 3	Model 1	Model 2	Model 3
Context variables									
Natural Sciences	0.019 (0.133)	0.145 (0.126)	0.123 (0.122)	0.125 (0.127)	0.261* (0.120)	0.244* (0.118)	-0.029 (0.130)	0.102 (0.123)	0.076 (0.116)
Engineering and Technology	-0.286 (0.152)	-0.135 (0.142)	-0.184 (0.138)	-0.209 (0.144)	-0.055 (0.135)	-0.089 (0.134)	-0.248 (0.148)	-0.094 (0.139)	-0.151 (0.132)
Medical and Health Sciences	-0.273 (0.150)	-0.078 (0.143)	-0.108 (0.139)	-0.325* (0.143)	-0.119 (0.136)	-0.142 (0.135)	-0.318* (0.146)	-0.116 (0.140)	-0.148 (0.133)
Agricultural Sciences	0.033 (0.196)	0.206 (0.185)	0.104 (0.180)	0.079 (0.187)	0.271 (0.176)	0.210 (0.175)	-0.149 (0.192)	0.036 (0.181)	-0.090 (0.172)
Social Sciences	-0.078 (0.133)	0.018 (0.124)	-0.017 (0.120)	-0.291* (0.126)	-0.197 (0.118)	-0.223 (0.117)	-0.094 (0.129)	0.004 (0.121)	-0.036 (0.115)
Germany	0.264 (0.143)	0.367** (0.134)	0.231 (0.132)	0.442*** (0.136)	0.552*** (0.128)	0.471*** (0.129)	0.319* (0.140)	0.424*** (0.131)	0.256* (0.126)
Italy	0.371* (0.147)	0.339* (0.139)	0.252 (0.136)	0.592*** (0.140)	0.577*** (0.132)	0.514*** (0.132)	0.417** (0.144)	0.393** (0.135)	0.292* (0.130)
Portugal	-0.028 (0.146)	-0.132 (0.136)	-0.191 (0.134)	0.208 (0.138)	0.110 (0.129)	0.062 (0.131)	0.119 (0.142)	0.018 (0.133)	-0.045 (0.128)
The Netherlands	0.511** (0.194)	0.661*** (0.182)	0.536** (0.178)	0.619*** (0.184)	0.777*** (0.173)	0.710*** (0.173)	0.672*** (0.189)	0.829*** (0.177)	0.670*** (0.170)
The United Kingdom	0.202 (0.196)	0.135 (0.183)	0.074 (0.177)	0.476* (0.186)	0.405* (0.174)	0.360* (0.172)	0.875*** (0.191)	0.804*** (0.178)	0.735*** (0.169)
The United States	0.218 (0.164)	0.229 (0.154)	0.073 (0.152)	0.351* (0.156)	0.346* (0.146)	0.239 (0.148)	0.640*** (0.160)	0.645*** (0.150)	0.463** (0.146)

(Continued)

	Traditional media channels			Public events			New media channels		
Variables	Model 1	Model 2	Model 3	Model 1	Model 2	Model 3	Model 1	Model 2	Model 3
Brazil	0.607***	0.694***	0.434**	0.716***	0.821***	0.673***	0.679***	0.774***	0.446**
	(0.169)	(0.159)	(0.163)	(0.161)	(0.151)	(0.159)	(0.165)	(0.155)	(0.156)
Size	0.000**	0.000**	0.000	0.000**	0.000**	0.000*	0.000*	0.000*	0.000
	(0.000)	(0.000)	(0.000)	(0.000)	(0.000)	(0.000)	(0.000)	(0.000)	(0.000)
Research income	0.065*	0.080**	0.065*	0.034	0.052*	0.042	0.072*	0.088***	0.070**
	(0.029)	(0.028)	(0.027)	(0.028)	(0.026)	(0.026)	(0.029)	(0.027)	(0.026)
Communication disposition									
Size of staff		0.020*	0.004		0.016*	0.007		0.020*	-0.000
		(0.008)	(0.008)		(0.008)	(0.008)		(0.008)	(0.008)
% Researchers participating		0.644***	0.574***		0.528***	0.488***		0.547***	0.459***
		(0.145)	(0.141)		(0.138)	(0.137)		(0.141)	(0.135)
Communications policy		0.365***	0.277***		0.332***	0.278***		0.357***	0.251***
		(0.081)	(0.080)		(0.077)	(0.077)		(0.079)	(0.076)
Annual budget on PE		0.116**	0.087*		0.152***	0.134***		0.138***	0.102*
		(0.043)	(0.042)		(0.041)	(0.041)		(0.042)	(0.040)
Professionalization									
Specialist			0.338***			0.229***			0.398***
			(0.072)			(0.070)			(0.069)
Nonspecialist			-0.034			-0.056			-0.016
			(0.078)			(0.076)			(0.075)
R^2	0.114	0.246	0.298	0.159	0.282	0.304	0.143	0.270	0.350
ΔR^2	0.114	0.132	0.053	0.159	0.123	0.022	0.143	0.127	0.079
ΔF	4.111***	19.338***	16.518***	6.047***	18.951***	6.892***	5.332***	19.333***	26.956***

reported having staff within the unit responsible for communication tasks; of these, only about a half is fully dedicated to such tasks; in the other cases, institutes have either access to the staff central communication offices of their institutions (50%) or have do not contact with communication staff at all (11%).

In terms of professionalization, when we analyze the level of professionalization of the staff working in the institutes that reported them, we find a slightly predominance of nonspecialist staff, with about less than half of this staff being considered "specialized", i.e., fully dedicated and with communication experience. Our results point also to important relationships between the level of professionalization and the level of communication of an institute. The fact that specialist staff is associated with more intense communication and nonspecialist staff is not, points to a positive contribution of specialization to the institutional communication activity; the same is not true for nonspecialist. This is not surprising; as fully dedicated, trained staff will know about their field and have a focus on their tasks, more than part-time employees with no experience, who might also perform a mix of tasks (administrative included), and thus be less available for communication tasks. Nevertheless, the growing activity of communication observed at this level (Entradas et al., 2020) as well as expectations for increased resources could lead to expectations that public communication will become professionally competent at the meso level. This thesis on the increasing professionalization is expressed in the decentralization hypothesis, which expects capacity building at the institute level for their own communication, of which communication staff is an important part of.

The findings also show that level of professionalization is a more important contributor to institutional level of communication than the size of an (untrained) team. This suggests that a smaller but dedicated staff with specific training might accomplish more than a more sizable team with less training and less specialized. We also found differences in levels of professionalization across areas of research and countries. For instance, professionalized staff is more in evidence among the natural sciences, agriculture, and engineering. The differences can be partly explained by the fact that "hard" sciences have longer traditions of public communicating – see physics, astronomy, and agriculture (the land universities in the United States). It could be that natural sciences have greater incentives to seek specialized communicators, such as public demand. As for countries, one possible explanation for the differences found might be the differing degrees of maturity of the profession and science communication cultures in these countries (Mejlgaard, Bloch, & Madsen, 2019), for example, the existence or not of professional associations, national and government policies (Entradas, Junqueira & Pinto, 2020). But overall, these findings suggest that professionalization of staff is likely to be happening at a different pace in different countries and disciplines, and, as such, it might take longer for professionalization to reach higher levels in certain disciplines and countries.

These findings also reinforce previous ones that have shown the importance of communication disposition variables for institutional communication (e.g., Entradas & Bauer, 2017) and add another indicator to the models – the degree of professionalization of staff. Our professionalization scale suggests that several metrics can be used to benchmark the level of professionalization including exclusive dedication, training in communication, and previous experience in related fields. This is in alignment with descriptions of professionalization in the field mentioned in the literature (Brown & Scholl, 2014; Trench, 2017; Trench & Bucchi, 2010). The robust indices indicate that these characteristics can be reliably used as measures of professionalization in science communication. In contrast, nonspecialization can also be measured through contrasting metrics including being a part-time employee and having no previous experience, but this seems to add no statistical value to the models.

Our study does not provide information on the content of this communication. As such, the fact that an institute hires specialist staff might not mean better communication. However, the stronger effects of specialist staff on the level of media communication (traditional and online) compared to public events suggest specialists are more likely to focus on media communication and relations. The question remains about whether institutions are on an arms race for visibility or communicating science substance?

Note

1 The FAs and professionalisation index were developed in collaboration with Ahmet Suerdem, under a consultant contract with the project.

References

Barrett, P. (2007). Structural equation modelling: Adjudging model fit. *Personality and Individual Differences*, *42*(5), 815–824.

Borchelt, R. E., & Nielsen, K. H. (2014). Public relations in science: Managing the trust portfolio. In *Routledge handbook of public communication of science and technology* (pp. 74–85). London: Routledge.

Brown, P., & Scholl, R. (2014). Expert interviews with science communicators: How perceptions of audience values influence science communication values and practices. *F1000Research*, *3*, 128. https://doi.org/10.12688/f1000research.4415.1

Buhler, H., Naderer, G., Koch, R., & Schuster, C. (2007). *Hochschul-PR in Deutschland. Ziele, Strategien und Perspektiven*. Wiesbaden: Deutscher Universitätsverlag.

Carver, R. B. (2014). Public communication from research institutes: Is it science communication or public relations? *Journal of Science Communication*, *13*(03), C01. https://doi.org/10.22323/2.13030301

Claessens, M. (2014). Research institutions: Neither doing science communication nor promoting 'public' relations. *Journal of Science Communication*, *13*(3), C03.

Elken, M., Stensaker, B., & Dedze, I. (2018). The painters behind the profile: The rise and functioning of communication departments in universities. *Higher Education*, *76*(6), 1109–1122.

Entradas, M. (2015). Envolvimento societal pelos centros de I&D. In *40 anos de Políticas de Ciência e Ensino Superior em Portugal* (pp. 503–516). Coimbra: Almedina.

Entradas, M., & Bauer, M. M. (2017). Mobilisation for public engagement: Benchmarking the practices of research institutes. *Public Understanding of Science*, *26*(7), 771–788. https://doi.org/10.1177/0963662516633834

Entradas, M., Bauer, M. W., O'Muircheartaigh, C., Marcinkowski, F., Okamura, A., Pellegrini, G., Besley, J., Massarani, L., Russo, P., Dudo, A., Saracino, B., Silva, C., Kano, K., Amorim, L., Bucchi, M., Suerdem, A., Oyama, T., & Li, Y.-Y. (2020). Public communication by research institutes compared across countries and sciences: Building capacity for engagement or competing for visibility? *PLOS ONE*, *15*(7), e0235191. https://doi.org/10.1371/journal.pone.0235191

Entradas, M., Junqueira, L., & Pinto, B. (2020). The late bloom of (modern) science communication. In *Communicating science: A global perspective* (pp. 693–714). Australia: ANU Press.

Entradas, M., & Santos, J. M. (2021). Returns of research funding are maximised in media visibility for excellent institutes. *Humanities and Social Sciences Communications*, *8*(1), 216. https://doi.org/10.1057/s41599-021-00884-w

European Commission. (2013). Options for strengthening responsible research and innovation: Report of the Expert Group on the State of Art in Europe on Responsible Research and Innovation. Publications Office of the European Union. https://data.europa.eu/doi/10.2777/46253

Evetts, J. (2003). The sociological analysis of professionalism: Occupational change in the modern world. *International Sociology*, *18*(2), 395–415. https://doi.org/10.1177/0268580903018002005

Fähnrich, B., Metag, J., Post, S., & Schäfer, M. (2019). Hochschulkommunikation aus kommunikationswissenschaftlicher Perspektive. In *Forschungsfeld Hochschulkommunikation* (pp. 1–21). Wiesbaden: Springer Fachmedien. https://doi.org/10.1007/978-3-658-22409-7_1

Fahy, D., & Nisbet, M. C. (2011). The science journalist online: Shifting roles and emerging practices. *Journalism*, *12*(7), 778–793. https://doi.org/10.1177/1464884911412697

Grunig, J. E., & Grunig, L. A. (2008). Excellence theory in public relations: Past, present, and future. In A. Zerfass, B. van Ruler, & K. Sriramesh (Eds.), *Public relations research: European and international perspectives and innovations* (pp. 327–347). Wiesbaden: VS Verlag für Sozialwissenschaften. https://doi.org/10.1007/978-3-531-90918-9_22

Hair, J. F., Black, W. C., Babin, B. J., & Anderson, R. E. (2014). *Multivariate data analysis*. London: Pearson Education Limited.

Hvidtfelt Nielsen, K. (2010). More than "mountain guides" of science: A questionnaire survey of professional science communicators in Denmark. *Journal of Science Communication*, *09*(02), A02. https://doi.org/10.22323/2.09020202

Kline, R. B. (2016). *Principles and practice of structural equation modeling*. New York: Guilford Press.

Koivumäki, K., & Wilkinson, C. (2020). Exploring the intersections: Researchers and communication professionals' perspectives on the organizational role of science communication. *Journal of Communication Management, 24*(3), 207–226. https://doi.org/10.1108/JCOM-05-2019-0072

Kolenikov, S., & Bollen, K. A. (2012). Testing negative error variances is a Heywood case a symptom of misspecification? *Sociological Methods & Research, 41*(1), 124–167.

Krücken, G., & Meier, F. (2006). *Globalization and organization world society and organizational change.* Oxford: Oxford University Press.

Marcinkowski, F., Kohring, M., Fürst, S., & Friedrichsmeier, A. (2014). Organizational influence on scientists' efforts to go public: An empirical investigation. *Science Communication, 36*(1), 56–80. https://doi.org/10.1177/1075547013494022

Mejlgaard, N., Bloch, C., & Madsen, E. B. (2019). Responsible research and innovation in Europe: A cross-country comparative analysis. *Science and Public Policy, 46*(2), 198–209.

Metcalfe, J., & Gascoigne, T. (1995). Science journalism in Australia. *Public Understanding of Science, 4*(4), 411.

Miller, S. (2008). So where's the theory? On the relationship between science communication practice and research. In *Communicating science in social contexts* (pp. 275–287). Berlin: Springer.

NCCPE. (2021). *PEP Insights Research report is out: The impact of COVID-19 on public engagement.* Bristol: National Co-ordinating Centre for Public Engagement.

Neresini, F., & Bucchi, M. (2011). Which indicators for the new public engagement activities? An exploratory study of European research institutions. *Public Understanding of Science, 20*(1), 64–79. https://doi.org/10.1177/0963662510388363

Peters, H. P., Heinrichs, H., Jung, A., Kallfass, M., & Petersen, I. (2008). Medialization of science as a prerequisite of its legitimization and political relevance. In D. Cheng, M. Claessens, T. Gascoigne, J. Metcalfe, B. Schiele, & S. Shi (Eds.), *Communicating science in social contexts: New models, new practices* (pp. 71–92). Dordrecht: Springer. https://doi.org/10.1007/978-1-4020-8598-7_5

Schwetje, T., Hauser, C., Böschen, S., & Leßmöllmann, A. (2020). Communicating science in higher education and research institutions: An organization communication perspective on science communication. *Journal of Communication Management, 24*(3), 189–205. https://doi.org/10.1108/JCOM-06-2019-0094

Trench, B. (2017). Universities, science communication and professionalism. *Journal of Science Communication, 16*(05), C02. https://doi.org/10.22323/2.16050302

Trench, B., & Bucchi, M. (2010). Science communication, an emerging discipline. *Journal of Science Communication, 9*(3), C03. https://doi.org/10.22323/2.09030303

Trindade, M., & Agostinho, M. (2014). Research management in Portugal: A quest for professional identity. *Research Management Review, 20*(1). https://fles.eric.ed.gov/fulltext/EJ1022036.pdf. Zugegriffen: 23 November 2017

Vogler, D., & Schäfer, M. S. (2020). Growing influence of university PR on science news coverage? A longitudinal automated content analysis of university media releases and newspaper coverage in Switzerland, 2003–2017. *International Journal of Communication, 14*, 22.

Watermeyer, R., & Lewis, J. (2018). Institutionalizing public engagement through research in UK universities: Perceptions, predictions and paradoxes concerning the state of the art. *Studies in Higher Education, 43*(9), 1612–1624.

Chapter 4

Public Duty or Self-Interest? Public Communication of University-Based Research Institutes after an Era of Governance Reforms in Europe

Frank Marcinkowski

Introduction

Existing research shows that scientists often view public engagement activities as a kind of duty they must fulfil (Davies, 2008; Martin-Sempere et al., 2008). Public communication on science and technology is perceived as part of the social responsibility of science – as a kind of *public service* that one performs for the benefit of society – and not as something one does for oneself (Loroño-Leturiondo & Davies, 2018). This seems especially consequential for scientists at publicly funded universities, who owe accountability to taxpayers for their use of public funds. No wonder, then, that publicly funded universities in all European countries are regularly expected to perform public engagement activities. However, the European higher education systems have undergone a profound reform process over the past three decades. One major goal of these reforms was to transform higher education institutions (HEIs) from typical public service organizations into so-called entrepreneurial universities. Inspired by the ideology of neoliberalism, measures were taken to organize the national higher education systems and the European higher education area as a whole into a market in which individual universities behave like typical contestants in a competitive environment. Under these conditions, rationally acting actors (universities, departments and chairs) are forced to optimize their own profile and reputation management and make the well-understood self-interest their top priority.

Thus, research institutes (RIs) at universities find themselves in a dilemma: while the public engagement paradigm obliges them to see themselves as public service institutions, the governance structure of the higher education sector nudges them to behave like self-interested market participants. This leads to an interesting research question that will be addressed in this chapter: Which of the two impulses shapes the structure and content of European HEIs' public communication, public engagement with lay audiences or reputational competition in the higher education market? This paper argues that the public engagement activities of rationally acting research institutions can only be expected under two conditions: if they are

DOI: 10.4324/9781003027133-6

induced by substantive incentives or if they are enforced by the sanctioned decisions of a hierarchically higher level (e.g., central management, national policies). If neither is available, I suspect it will be the case that RIs that have to assert themselves in a quasi-market with other players in the competition for scarce resources will act with self-interest. This also applies to the intentions they pursue with their public communication. In other words, what research institutions communicate to whom, and through which channels, is shaped by the governance structures in which they operate.

From deficit to dialogue: public engagement with science

The dialogical turn in science communication is often understood as a reaction to the failure of one-sided paternalistic approaches, which are associated with the term *deficit model* (Miller, 2001; Trench, 2008). The deficit model starts out from the assumption that science has the duty to provide the public with scientific information at its own discretion in order to compensate for existing knowledge deficits. In this respect, the model insinuates a one-sided, top-down communication process under the extensive control of science and with the help of the mass media. Even though both models of science communication (the deficit model and public engagement) exist side by side, a dialogical understanding of science communication in research and practice seems to dominate today.

Public engagement with science (PE) is based on the idea of dialogical communication between science and the public, in which the fixed distribution of roles is abandoned so that both sides might act as speakers as well as audiences. Not only science is encouraged to communicate its evidence-based view of the world – society also communicates its view of scientific-technological civilization with a claim to attention and serious consideration by the scientific community. Furthermore, the communication of science should reveal the limits of its own possibilities and the dangers of certain technical solutions so that the starting points for objection and criticism by society become visible. Ideally, this results in a mutual learning process in which not only the public is 'enlightened' about science in general and specific scientific topics, but the scientific community can also get to know the needs and demands of laypeople. In this way, the priorities of science might be shifted, research agendas adapted and the governance of scientific knowledge production expanded (Gregory & Miller, 1998; Irwin, 1995; Wynne & Irwin, 1996).

With this ideological foundation, PE appears to be an instrument for putting scientific institutions at the service of the public. In other words, the public service association is already established in the idea of public engagement. In contrast to science-based business, where PE easily mutates into an expanded form of market research and public relations (Gregory et al., 2007), the public sector seems to be an ideal breeding ground for the idea and practice of PE. This applies, in particular, to basic research in the publicly funded universities of European welfare states. If anywhere at all, the idea of science as a public good has its ancestral home in the public sector institutions of European higher education systems.

From government to governance: conditions and consequences of higher education reforms in Europe

Since the late 1990s, the governance structures of higher education systems have undergone enormous changes in many European countries. This was primarily a consequence of the triumph of neoliberal ideologies, which also inspired social democratic governments, starting with New Labour under Tony Blair and taken up by the red–green coalition under Gerhard Schröder in Germany (Newman, 2001). One of the core concerns of neoliberal politics is to wrest as much as possible from state control and make it available for the insatiable accumulation of capital: old-age and healthcare, broadcasting, energy and water supply, large parts of the state infrastructure and much more. While marketization finally gained ground in European higher education systems, this does not necessarily mean a complete withdrawal of the state from its responsibilities for the universities. Rather, state control has been replaced by state supervision enforced by complex meta-governance (de Boer & Jongbloed, 2012). Where a complete privatization of previously publicly operated institutions was politically unenforceable, as was the case in Germany for the higher education system, the second best solution was to make public institutions behave like private actors in a market. Various instruments were used for this purpose, first and foremost deregulation, the establishment of quasi-markets by changing the financing systems, the promotion of public–private partnerships and the admission of private providers (de Boer & Jongbloed, 2012). The idea behind this is that market participants always behave rationally and efficiently, while state institutions are notoriously inefficient and ineffective. The associated reform model is called New Public Management (NPM; OECD, 1995). Thus, the triumph of neoliberalism was accompanied by the rise of NPM as a new organizational approach for the public sector. According to this approach, universities were to be managed in a business-like manner – that is, by resorting to methods and instruments of corporate management (Braun & Merrien, 1999; Pollitt & Boukaert, 2000). Indeed, the European higher education and research sector has been a victim of NPM since the late 1990s, just as the healthcare system or public transport have been.

In general, NPM in higher education attaches great importance to the results achieved by the institutions instead of detailed regulations at the outset. As many decisions as possible should be made at the decentralized level, at the universities, because they are the first to know relevant steering information. Therefore, universities are given more responsibility for their financial management. In doing so, the output-oriented model of governance replaces the traditional input-oriented model of detailed governmental accounting. In addition, a large number of competences on behalf of staff, appointments, examinations and internal management are delegated to the universities. However, increased responsibility naturally goes hand in hand with the task of justifying oneself for one's actions to the relevant stakeholders, not at least the state as the main money source. Self-legitimation, thus, becomes a dominant – if not *the* dominant – task of the external communication of HEIs.

In addition, the NPM reforms were accompanied by far-reaching changes in financing between and within universities (de Boer et al., 2009). First, the proportion of basic funding is becoming smaller and smaller, and the proportion of project funding is increasing. By project funding, I mean funds distributed by the state in competitive processes: in other words, funds which the universities can apply for and which are then awarded on the basis of specific services to be provided and the quality of these services. A prime example of this is the so-called Excellence Initiative or the Higher Education Pact 2020 in Germany (Civera et al., 2020). More and more central funds are also being allocated within HEIs on the basis of applications and competition between the institutes. Second, basic funding is not fully guaranteed either; only a small proportion is measured by the number of employees at a university or one of its subdivisions or by the number of students. The greater part of the basic funding is, in turn, awarded on the basis of the services provided and their quality.

As with all artificially staged competitions (Le Grand & Bartlett, 1993), however, one major problem arises: The actual consumers of this service do not appear at all. Normally, demand indicates which provider offers the best product at the lowest price and is, therefore, successful on the market. This mechanism is switched off in artificial markets. Instead, other systems are needed that symbolize service quality and market success (Le Grand et al., 1998). Consequently, all kinds of so-called performance measures by key figures, often devised by weakly legitimized private stakeholders (Brandenburg & Federkeil, 2007), must replace consumer demand. This led to the development of an increasingly elaborate reporting system and controlling instruments, which until then had only been known in commercial enterprises. Key figures form an important basis for this. They are treated as a valid reflection of the university's own performance, used as a basis for decision-making and for internal and external communication. The importance of key figures is increased by the fact that external institutions (e.g., ministries of science, ranking agencies, university councils) measure universities exclusively on quantitative indicators. The result is a research and teaching enterprise that revolves around staged competition for measurable metrics: It is not oriented towards what would be scientifically necessary but rather towards producing impressive numbers. Consequently, science loses originality, creativity, quality and professional autonomy (Binswanger, 2015).

It is reasonable to assume that this competitive situation between and within universities has a formative influence on the communication policy of research institutions. Artificial competition is also attention-grabbing competition. It is the constant attempt to draw the scarce attention of the relevant stakeholders – that is, those who decide on the allocation of money – to one's own institution. Public visibility and media presence become a relevant addition to the system of indicators that represent the performance and quality of an institution. In addition, other indicators, such as placement in one of the many rankings, the pleasant title of a 'university of excellence' or the graduation and third-party funding rates, are only refined through public communication. Where once the scientific ideal was to find and prove new knowledge, today the priority is at least as much about 'proving we are the best'. This leads to a situation in which the politically intended competition for scientific excellence is inevitably accompanied by a self-reinforcing contest for public attention. As a result, all public communication activities of HEIs tend to have an advertising character that is actually incompatible with the ideal of public engagement as described above (Marcinkowski & Kohring, 2014).

This line of reasoning leads me to the following three expectations for the structures and contents of the public communication of RIs at universities under the NPM regime:

> H1: The communication activities of RIs are mainly motivated by the self-interest of the institutions rather than by a perceived obligation towards the lay public.
> H2: Accordingly, the communication of RIs is more often directed at relevant stakeholders in politics, business and funding institutions than at the lay public.
> H3: Consequently, the majority of activities involve traditional press work/public relations rather than public engagement.

From one country to the other: comparative perspectives

In the empirical section of this chapter, I create a comparative design to test the previously formulated assumptions, with a special focus on the European countries in the MORE-PE dataset, namely, Portugal, Germany, the

United Kingdom, the Netherlands and Italy. Although the basic idea and goals of university reforms are similar in all OECD countries – to increase efficiency, enhance effectiveness and emend performance – there are substantial differences in the intensity and content of NPM-based reforms between countries (Curaj et al., 2012; de Boer et al., 2008; Krüger et al., 2018; Paradeise et al., 2009). I argue that the following two features of higher education governance are particularly suited to negatively influencing the idea and practice of genuine public engagement activities: the depth and breadth of the decision-making powers given back to universities or their so-called regulatory autonomy. On the one hand, more autonomy encourages universities to develop specific profiles and strategies, but at the same time, it increases their accountability for success and failure. The greater the responsibility that must be assumed by the hierarchical leaders of the universities, the more their external communication must be geared towards blame avoidance and preventive self-justification. As a result, the newfound autonomy not only becomes a vehicle of indirect forms of state control (Enders et al., 2012) but also increases dependence on constant image management and positive publicity. In addition, I assume that the more market-based the financing of universities is – that is, the smaller the share of guaranteed basic financing and the larger the share of the budget allocated on the basis of performance measurement – the more the external communication of the organization should be committed to self-marketing and window dressing. With notoriously scarce resources for communication, both necessities go at the expense of elaborate public engagement activities.

In a comparative study, de Boer, Enders and Schimank (2008) measured the state of NPM reforms in the universities of four European countries by using five dimensions: state regulation, stakeholder guidance, academic self-governance, managerialism and competition. This approach is driven by the conviction that the so-called new governance of higher education is not a model for all cases but rather consists of a specific mix of these ingredients in each individual case. For the purpose of this chapter, two dimensions are of interest. First, the withdrawal of the state (state regulation) can be interpreted as a gain in formal autonomy in the sense meant above. In addition, the dimension of competition is important because it indicates the sharpness of the rivalry within and between universities for scarce resources (money, personnel, prestige).

To start with the latter, according to the study, the current British university system is fully designed in the sense of NPM insofar as weak performance in research can result in the complete withdrawal of institutional funding, which obviously intensifies the pressure to not only be good but also *look* good. The authors consider the Dutch system to be the second closest to the NPM benchmark in terms of the severity of resource competition. Germany follows, at some distance, in third place. The study shows the same order for the factor of 'state regulation'. The authors see the United Kingdom and the Netherlands as close to each other in this respect although both are some distance from the NPM benchmark of complete managerialized self-control. Germany is also in third place, by some distance, in this dimension. Overall, the higher education system in Germany, where the Humboldtian ideal of the unity of research and teaching still enjoys much support, proves to be comparatively traditional although there is some movement there, especially with regard to competitive pressure. The United Kingdom proves to be particularly close to the ideal of the NPM model, while the Dutch system holds a middle position between the two. Unfortunately, neither Italy nor Portugal is covered in that investigation.

In another comparative study, de Boer and Jongbloed (2012) used a set of eight indicators to measure the level of university autonomy in seven European countries in terms of financial discretion, dependency on state budget, student selection, staff recruitment, etc. A quantitative analysis of the data from this study (pp. 566–577) reveals that British universities enjoy by far the greatest formal autonomy to behave freely like market-type actors. All continental university systems follow the United Kingdom by a wide margin but have very similar levels of formal freedom in direct comparison, so no further gradation makes sense. Germany was, unfortunately, not part of this study. Another comparative study of the implementation of NPM reforms in the higher education systems of seven countries (Broucker et al., 2015) attests to the United Kingdom's very strong competition for scarce financial resources, particularly forced by the Research Excellence Framework, the budget cuts of the 1990s and the drastic increase in tuition fees. Although performance indicators and quality criteria are increasingly being used to allocate budget funds in Portugal as well, the actual redistributive effect of these measures – and, thus, the competitive pressure – is comparatively low. According to this study, the performance-based payment of universities is also on the rise in the Netherlands, with the country occupying a kind of middle position between the United Kingdom and Portugal in terms of the intensity of competition. As far as the degree of freedom of action is concerned, Dutch universities have gained autonomy since the 1990s, albeit within narrow limits set by the state, compliance with which is monitored by the constant evaluation of research and teaching. The assessment is very similar for Portugal: Here, universities develop their own profiles and make strategic plans, albeit within the guidelines of state policy, enforced by measures of accreditation and evaluation. The autonomy of universities in both countries is, thus, a relative one. In the United Kingdom, with its traditionally largely independent universities, the starting point is different in this respect: Compared to the high degree of autonomy once enjoyed by British universities, state influence has increased considerably so that today they enjoy only a little more freedom than universities on the continent.

With all due caution due to the diversity of comparative methodologies, the existing literature suggests that the United Kingdom comes closest to meeting the 'ideal' of the NPM benchmark among the countries included here. At the other end is Germany, with what is, by comparison, a traditionalist university system. Portugal and the Netherlands are somewhere in between, with the Netherlands closer to the United Kingdom and Portugal closer to the German model. The available literature does not allow the drawing of conclusions about Italy's position, so I do not make the country part of the hypothesis. Against this background, the final hypothesis reads as follows:

> H4: The tendencies towards a marketing-oriented communication policy, as described in the first three hypotheses, are most evident in British universities, followed by the Netherlands, Portugal and, finally, Germany. Where Italy places itself on this scale remains an open research question.

Data and measurements

All following analyses are calculated with the data set of the MORE-PE joint project. For details of the data, see Chapter 16. For the country comparison, I have selected only European countries since the arguments on the NPM reform process refer to European higher education systems. In detail, the following countries are the subject of the analysis: The United Kingdom (N = 188), Germany (N = 245), the Netherlands (N = 142), Italy (N = 366) and Portugal (N = 224). The total size of the data set comprises N = 1,165 RIs from five European countries. Due to missing values for some variables, the number of cases may fall short of these values in individual analyses.

Rationales for public communication

I use the question of reasons for public communication by RIs to operationalize the two central motivational dimensions from my theoretical argument, namely, 'self-interest' versus 'public responsibility'. To this end, in the first step, I tried to classify the eight given items based on conceptual considerations (note: a statistical analysis of the dimensions, for instance, by factor analysis, is not possible due to data coding). The categorization is based on the primary beneficiary of an activity (higher-level units, the research unit itself or the general public) and also distinguishes between imputable driving forces. I use this classification to group the institutes according to the motives they identified as 'most important'.

I then constructed a second variable to combine the motives that were cited as most important and second most important. Since I have no evidence to suggest that respondents consider the most important motive twice as important as the second most important motive, I give equal weight to both. That means the combination of public duty (rank 1) and external constraint (rank 2) ends up in the same value as the identical combination in reverse order. For the combination of the most important and second most important reason, I use the three combined motives, resulting in a variable with six values.

Audiences

We asked our respondents how frequently the institution's scientists have contact with nine different audiences (responses were given on a 3-point scale ranging from 1 = *never* to 3 = *frequently*). I combined the query into three groups, each with a mean index of contact frequency, namely, general public (general public, schools, students outside the classroom), stakeholders (representatives of cities, municipalities, associations, industry, government and state agencies, NGOs) and media (journalists and media).

Public perceptions

Drawing on an exploratory factor analysis of four statements about the general audience, I formed two variables. One variable represents the typical view of the public engagement approach as being about actively engaging the general public, while the second expresses the traditional view of the public understanding of science and humanities (PUSH) approach that lay audiences need to be educated by scientists, and as a result, public support and trust in science will increase. The question wording reads as follows: for the PUSH view, 'The public needs to be educated by those who are knowledgeable' and 'If the public knows more about our research, they will be more likely to support it'. For the PE view, the items read 'The public wants to contribute to science' and 'We would like the public to become more actively involved in decisions about the research conducted at our research unit'.

Public events

We asked the respondents how often scientists from their institution organized or participated in events in the past 12 months. Responses were given on a 5-point scale ranging from 1 = *never* to 5 = *weekly*. Based on an exploratory factor analysis, a total of eight of the 11 different event types were distinguished into two classes and a mean index was calculated for both, namely, *involving events* (citizen science projects, exhibitions, participatory events, etc.) and *traditional knowledge transfer* (public lectures, events in schools, etc.). Both indices are reasonably reliable, with Krippendorf's alpha > .63 and > .68, respectively.

Channels for communication

We asked respondents to indicate how frequently they used various communication channels to communicate with the nonscientific public in the past 12 months. Responses were given on a 5-point scale ranging from 1 = *never* to 5 = *weekly*. I divided the total of 13 different channels into two classes and calculated a mean index for both, namely, *mass media channels* (interviews, press releases, press conferences, etc.) and *target group media* (materials for schools, policy papers, brochures for special target groups, etc.). Both indices are sufficiently reliable, with Krippendorf's alpha > .70 and > .80, respectively. I sorted the frequency of use of online channels into two categories: *standard* and *advanced* online media. According to this, websites, Facebook and Twitter form the standard repertoire. These channels are used most frequently in all countries and by all institutions. I assumed that these channels primarily indicate their involvement in the attention competition. I calculated the frequency with which the institutes use these channels to a fairly reliable mean index (alpha = .713). This contrasts with three other channels that can be considered much more elaborate from the production side, namely, podcasts, YouTube videos and blogs. I assumed that these channels are used when one wants to involve the audience in one's own research. The three indicators are summed to produce a just-reliable mean index (alpha = .615).

Results

The following section is structured along the four hypotheses I formulated above. I use intertitles (instead of mere numbers) to allow a better orientation in the text.

Rationales for communication

To provide evidence for H1, I first present a simple frequency analysis of the institutes' claimed most important motives for their communication activities. When asked why they communicate with the nonscientific public, just under 46% of the institutes (N=1,019) claim that they do so in the public interest. Slightly more than a quarter of all institutes (27%) cite interest in themselves as their most important motive. The proportion of institutions that state that they are fulfilling political requirements is just as large (27%). A simple chi-square test confirms that the group sizes are significantly different; motivations are not equally distributed in the population, Chi² (2, 1.019) = 70,483, p < .001. According to this, the majority of all institutions are mainly driven by one of the motives I have assigned to the category of 'public duty'. However, if we look at the two individual items from Table 4.1 (left column) that are prototypical for instrumental communication in the

Table 4.1 Classification of reported rationale for communication with nonscientific audiences into three categories (question wording: 'What would you say is the most important, the second most important, and least important reason for your research unit to undertake communication with nonspecialist audiences?')

Item	For whom?	Driving force
1a We aim to respond to the policy/mission of our host institution/university		
1b We aim to respond to the policy of our funding bodies	Them	External constraint
1c We aim to respond to national policies of public engagement		
2a We want to raise our research profile		
2b We want to attract funding		
2c We want to get public support for the research we do	Us	Self-interest
2d We aim to recruit new generations of scientists		
3a We want to disseminate our research to the public		
3b We want to listen and involve the public in our research	Community	Public duty

self-interest (attract funding) and communication in the public interest (involve the public), we get a somewhat different impression: Just under a quarter of the institutes (23%) openly admit that the acquisition of third-party funding is the most important motive in their external communication, while less than a fifth (18%) claim that their first priority is to involve the nonscientific public.

Since the reasons for the communication of RIs are manifold, we looked at the combination of the two most important motives in a further evaluation step. For this purpose, I used the combination variable described above. As Table 4.2 shows, few institutes are motivated by either public or self-interest (roughly 23%), but a larger group of institutes (30.5%) report being motivated by both self-interest and communicating in the public interest. Within the group of either/or institutes, units whose most important motives can be interpreted as purely self-interest (12%) form a slightly larger group than those institutes that name public interest as the most important and second most important motive for their communication activities (11%).

However, the largest group of institutions states that their communication activities follow both motives simultaneously: self-interest and public obligation. Moreover, the numbers suggest a conjecture I expressed above:

Table 4.2 Combination of mentions as the most important and
second most important rationale for public
communication of research institutes (N = 1,005)

Combination: most important/second most important motivation	Frequency
Self/self-interest	12.1% (122)
Public/public interest	11.0% (111)
Political/political interest	5.5% (55)
Self/public interest	30.5% (307)
Political/public interest	23.2% (233)
Political/self-interest	17.6% (177)
Sum	100% (1,005)

Communication in the public interest is more likely when institutions follow policy directives (from their own organization or the state) that call for it. At least, just under a quarter of the institutes (23%) state that they communicate because it is politically required and, at the same time, in the public interest.

Overall, the evidence on H1 is mixed. A dominance of self-interest in image politics, profiling, fund-raising, etc., is not as clear as assumed. However, the data do not provide a clear rejection of the thesis either. Indeed, the idea of an obligation to the general public is still a strong motivation for many institutes to communicate with the nonscientific public.

Engaging whom?

The second hypothesis concerns the target groups of the external communication of RIs. Recall that I have assumed that relevant stakeholders from whom one expects immediate benefits are addressed more frequently than the general public, from whose goodwill one does not directly benefit. In this respect, the frequency of contact with specific target groups can be understood as a further proxy for self-interested communication.

As the mean values of contact frequency in Table 4.3 show, the majority of the RIs' communication offers are directed at the general public ('anyone who might be interested', the questionnaire states). Just as often, the institutes' communication is directed at students (outside of teaching) although the question must remain open as to whether this refers to students at their own institution, students at other universities or even future students. Without being able to clarify this question, it can nevertheless be clearly stated that stakeholders are addressed less frequently than the general public. For the sake of completeness, intermediaries (the news media) that can be used to address a wide variety of groups have been added to the table. A T-test for paired samples reveals that the mean values of contact frequency of the general audience and stakeholders are significantly different (t(1.099) = 17.535, p < .001). The effect size of .55 for Cohen's d indicates a moderately

Table 4.3 Frequency of contact with different publics: mean values (1 = never to 3 = frequently; N between 1,009 and 1,099)

Publics	M (SD)		M (SD)
General public	2.46 (.556)		
Schools	2.35 (.652)	General public	2.42 (.439)
Students outside teaching	2.46 (.580)		
News media/journalists	2.28 (.545)	Intermediaries	2.28 (.545)
Members of local municipalities/ councils/associations	2.22 (.595)		
Delegates from industry	2.24 (.661)	Stakeholders	2.13 (.476)
Governments/policymakers	2.10 (.642)		
Nongovernmental organizations (NGOs)	1.96 (.649)		

strong influence of the target group on contact frequency. Thus, the greater frequency of contact with the general public compared to the frequency of contact with specific stakeholders is not a statistical coincidence but can be interpreted as intentional.

In addition to the frequency of contact with target groups in the entire sample, it is also relevant to ask whether there is a relationship between the frequency of contact with individual target groups and the motivation for public communication discussed above. According to the theoretical argument of this chapter, the governance of universities shapes the rationality of action of institutes and chairs, and this rationality, in turn, influences communication activities. In order to examine a possible connection between the motivation for communication and the target group concept of the institutes, I calculated simple bivariate correlations between the most important motive for communication and the frequency of contact with the two most important target groups. The calculation reveals a negative correlation between self-interested motivation for public communication and frequency of contact with the general public ($r = -.15$, $p = .000$). In contrast, I find a weak but positive correlation between targeting the general public and the public duty rationale as the most important motivation ($r = .07$, $p = .028$). Thus, although the totality of all studied institutions more often address the general public as relevant stakeholders with their communication, the expression of a motivation strongly oriented towards self-interest tends to reduce communication efforts towards the lay public. This finding supports the general argument.

What to do with the general public?

Since the very broadly formulated category of 'general public' initially leaves open who or what exactly the respondents imagine by this, we additionally asked for images of the audience. For the purposes of this chapter, I created a variable that distinguishes a specific public engagement view of the public from another view of the public that is closer to the traditional public understanding of the science paradigm (see Methods section). This variable is intended to help assess whether it is, indeed, permissible to interpret the more frequent addressing of the general audience, automatically, as an indicator of closeness to the public engagement approach. In fact, this is not valid: The data show that only about 15% of the institutes interpret the general audience in terms of the PE paradigm, while about 45% of the institutes understand the concept of 'general audience' in terms of the deficit model. Furthermore, audience image is significantly correlated with motivation for communication: Institutes that feel motivated by the public interest are more likely to represent an image of the general audience that is close to the PE paradigm ($r = .11$, $p = .001$), while institutes whose most important motivation is self-interest are more likely to represent a paternalistic view of the general audience, as implied by the PUSH approach ($r = .14$, $p = .000$).

Overall, the data also provide mixed evidence on H_2. On the one hand, the majority of all communication activities are directed at the general public. On the other hand, however, this is not necessarily evidence of public engagement activities because the majority of institutes represent a concept of the general public that is not compatible with the PE paradigm. Finally, there is evidence to support the assumption that self-interest as the primary driver for communication leads to an increased focus on stakeholders as addressees.

Which activities and channels?

The third hypothesis deals with the intensity and nature of public communication efforts by university-affiliated RIs. The expectation is that classical top-down activities and dissemination dominate, while more involving and interactive forms of communication are rare because they are less conducive to the institutes' self-promotion and, at the same time, more costly. I start with some descriptive figures to give an impression of the proportions of individual activities. In doing so, I have compared two types of activities in pairs, each of which is intended to represent, ideally, a more marketing-oriented approach to communication and a more PE-oriented approach. This distinction is most plausible for events organized or staffed by the institutes. The results are clear and in the expected direction: Across all institutes, events for one-sided knowledge transfer ($M = 2.51$), such as public lectures, are more frequent than participatory events ($M = 2.01$), such as exhibitions. The mean frequency of both types of events differs significantly (p < .001). The same is true for communicating through mass media channels (M = 2.26), which are typically used to get one's message across, compared to the less frequent use of group-specific channels (M = 1.92) to engage a specific target group.

In contrast to these two pairwise comparisons, the use of online media by universities is difficult to interpret. Research on the use of online media and especially social networks by universities is still very patchy and fragmented. This is especially true for the higher education systems of continental European countries, while online communication by universities in the United States or the United Kingdom has been researched somewhat more intensively. Accordingly, the findings are inconsistent and fragmented. With all due caution, the limited insights can be generalized in the following way: Operating a website is, to a certain extent, part of the standard repertoire of HEIs today and does not index any particular engagement in the online sphere. The use of social media is widespread in Anglo-Saxon countries and is chosen there primarily for a marketing-oriented communication approach aimed primarily at students and prospective students. Universities and RIs in continental European countries use social media such as Twitter and Facebook less frequently, and even where corresponding profiles are maintained, communication via these channels is generally not very intensive. They mostly have social media accounts because that is part of the game today (the 'me too' approach), but not because they could be used to realize a specific form of communication that would not be possible on other channels. Accordingly, online media are primarily additional distribution channels for their usual forms of communication and traditional content (Metag & Schäfer, 2018).

Against this background, it is virtually impossible to conclude a specific communication strategy or intention from the mere frequency of use of various online media (other data are not available here). My classification of online media is based on the speculation that traditional channels (websites, Twitter, Facebook) are operated with comparatively limited effort and ambition; they are meant to symbolize timeliness but are assumed to have no special significance for public engagement. Where they are used intensively, they are aimed primarily at students, but not at the general public. In contrast, there are three other online channels that can be regarded as much more elaborate from the production side, namely, podcasts, YouTube videos and blogs. They provide space to present complex contexts in more detail and, thus, possibly index the claim of knowledge transfer. I assume that these channels are primarily used when one wants to involve the audience in one's own research. One example is the many new podcasts that – at least in Germany – were used by prominent virologists during the months of the Covid-19 pandemic to inform the general public about the latest knowledge about the virus.

To the extent that this interpretation is permissible, the figures indicate very clearly that online media are used by university institutions on average much less frequently for knowledge transfer and public engagement than as additional dissemination media for their usual content. Standard online channels (M = 2.78) are used much more frequently than more elaborate online offerings (M = 1.63). Again, the mean difference is highly significant at p < .001. All findings presented up to this point provide support for H3.

Does motivation make a difference?

To test whether there is a relationship between activity structure and institutional rationales for communication, I calculated a series of one-way analyses of variance in which the grouping variable described in Table 4.2 serves as the factor and the activities as the dependent variables. In Table 4.4, only the mean comparisons of the groups of particular interest are shown. The result of this analysis shows, first of all, that institutions that are primarily motivated to communicate by the perceived interest of the general public in science and research do more of everything when it comes to communication. With a somewhat daring formulation, one could say that orientation towards the public interest actually acts as a motivator for communication, while self-interest acts as an inhibitor. Institutions that claim to communicate in the public interest address traditional media channels more often than those that admit to following primarily self-interested motives, but they also use target-group media more often. They use elaborate digital formats more often, but they also communicate more via Twitter and Facebook. They organize not only more involving events but also more dissemination events. From a statistical point of view, none of the post hoc tests for mean difference are significant, so it cannot be ruled out that the measured differences are random or caused by other reasons than motivation (e.g., resources). However, the homogeneity of the findings suggests that motivation makes a difference.

Country comparison

Finally, I tested the comparative hypothesis that public engagement is least likely in countries where universities and their RIs are most closely aligned with the NPM model. On the contrary, under the conditions of NPM universities, the public communication of RIs should primarily serve self-promotion and stakeholder cultivation functions. To make this assumption empirically plausible, I have picked out a total of seven indicators that, according to the analyses reported so far, can be considered as attributes of a public relations approach of institutional communication. Table 4.5 shows the measured values for the individual countries and, in addition, a rank that indicates, for each indicator individually, where a country is located in comparison with the others. Here, rank 1 indicates the greatest closeness to

Table 4.4 One-way analyses of variance for types of activities by institutional motivation for communication (results of post hoc multiple comparisons according to Tamhane-T2)

Most important & second most important motivation combined

Frequency of:	Self-interest mean (SD)	Public interest mean (SD)	Self & public interest mean (SD)	p
Top-down mass media activities	2.1648 (.72841)	2.22774 (.76154)	2.2885 (.78218)	n.s.
Targeted media activities	1.9271 (.71800)	2.0631 (.66392)	1.8712 (.65108)	n.s.
'Standard' online media	2.6510 (1.16028)	3.0168 (1.12536)	2.8316 (1.12173)	n.s.
'Advanced' online media	1.6250 (.77837)	1.8474 (.85803)	1.6599 (.78869)	n.s.
Engaging events	2.0403 (.64731)	2.0826 (.74341)	1.9209 (.66906)	n.s.
Diffusion events	2.4621 (.74434)	2.5177 (.76853)	2.4828 (.75151)	n.s.

Table 4.5 Measured values and rankings of seven self-promotion indicators for five countries

	NL	ITA	UK	GER	POR
Most important motivation: self-interest (% of RIs)	41.3 (1)	25.8 (4)	33.1 (2)	33.5 (2)	7.8 (5)
Single most important motive: attract funding (% of RIs)	29.2 (2)	31.5 (1)	12.8 (5)	23.3 (3)	14.9 (4)
Targeting stakeholders (% 'frequently')	10.2 (1)	5.7 (3)	6.5 (2)	3.0 (5)	5.8 (3)
PUSH view of public (mean)	3.80 (1)	3.87 (1)	3.71 (1)	3.65 (4)	2.19 (5)
Diffusion events (mean)	2.64 (1)	2.80 (1)	2.39 (3)	2.42 (3)	2.19 (5)
Mass media (mean)	2.54 (1)	2.53 (1)	2.20 (2)	2.07 (2)	1.92 (3)
'Standard' online (mean)	2.80 (1)	2.82 (1)	3.07 (1)	2.35 (5)	2.90 (1)
Mean of ranks	1.142	1.714	2.285	3.428	3.714

public relations, while rank 5 accordingly indicates the least affinity with public relations among the countries to be compared. Note that different rankings are assigned only if the percentage value differences are greater than 1.0 and if the mean differences are statistically significant in the post hoc test.

I have calculated an average of the rankings in order to be able to assess the positions of the individual countries. According to this, two groups of countries stand out. On one side are the Netherlands, Italy and the United Kingdom, whose motivation and communication clearly point to self-interested, top-down activities when it comes to public communication. At the other end of this scale are Portugal and Germany, where such tendencies are comparatively less pronounced. Thus, the comparison countries correspond more or less closely to the expectations expressed above, based on the state of implementation of NPM reforms. Italy, which was not part of the hypothesis, is more similar to the Netherlands and the United Kingdom than Portugal or Germany in terms of motivation and communication structure. Otherwise, one might have expected the United Kingdom to take the top position in terms of instrumental communication. This is not the case. In fact, the gap between British and German universities is not much wider than with the Netherlands, which, not entirely unexpectedly, ranks very far 'left' on the scale. This may indicate that not all features of the NPM model (e.g., regulatory autonomy, financing) affect the communication activities of the institutes in the same strength and direction. Of course, this analysis assumes that all the indicators used here actually index what should be measured and that they all have the same weight. Both assumptions are debatable.

Moreover, this analysis says nothing about what PE activities can be expected from the institutes and how the countries relate to each other in this respect. In fact, 'engaging events' are by no means less frequent at British and Dutch institutes than in Germany. And the share of institutes that state 'public duty' as their most important motivation is even higher in the United Kingdom and the Netherlands than in Germany (but not higher than in Portugal). Such figures give an indication that instrumental and involving forms of communication certainly coexist and that motivations for communicating with the outside world tend to be diverse rather than one sided. With these limitations, H4 can nevertheless be considered confirmed.

Discussion and conclusions

My main argument in this chapter is that university RIs in European higher education systems have been induced by political intervention to behave like self-interested economic actors in an artificially created market for educational services. I suggest that this will also be reflected in the structures and content of their public communications with external audiences. In particular, I have suspected that public engagement activities that are committed to the common good will be deemed to be contrary to the system logic under these conditions. In this respect, the argument assumes a progressive dominance of instrumental and marketing-oriented communication of the RIs.

Of course, such a long-term development cannot be proven with the data available here from a cross-sectional study. In this respect, it is more a matter of describing the state of affairs at a given point in time. It is not easy to answer the title-giving initial question with this snapshot. On the one hand, the reported results leave the impression that commitment to the common good is still strongly anchored, at least in the self-image of European universities. On closer inspection, however, the question arises of the extent to which one may actually believe these assurances. Although a majority of all institutes claim that their communication is primarily motivated by the public interest in science and research, the interest in directly involving nonscientific audiences in their own research is a minority position. This is evidenced not only by self-reported motives but also by the nature of activities and events. In the majority of all cases, what is understood as communication in the public interest is traditionally one-way knowledge transfer in the sense of the PUSH paradigm. The view that the majority of institutes have of the general public also fits in with this. Here, the image of a public of laypeople in need of academic education by the better-knowing research personalities dominates. Finally, among the target groups that are addressed particularly frequently are students or future students (schools). Although I have included this group in the overarching category of 'general public' and interpreted communication with it in terms of public service, it can also be interpreted differently, namely, as an effort to keep one's own clientele satisfied and happy. As far as an institute's own students are addressed, it is a kind of customer service and, with it, instrumental communication in the interest of the organization. As far as the activities at schools are concerned, the communication efforts could also be interpreted as instrumental since they are aimed at gaining a competitive edge in the lucrative student market.

Thus, while it is safe to say that the communication of European universities' RIs is far from meeting the ideal of PE, the study also fails to demonstrate evidence of institutes' complete surrender to a marketing-oriented approach to public communication. The picture that becomes visible is multifaceted. It indicates a juxtaposition of different communication strategies, activities and motivations rather than the clear dominance of a single approach. This was probably to be expected and, moreover, makes it clear that the governance reforms of the last decades, described above as the NPM reforms of higher education, do not have a monocausal formative power for universities' public communication. In fact, a number of other factors (national traditions, resource endowments, political directives, etc.) play a role in this process, and the individual reform steps have different consequences in different countries. All of this could not be taken into account here and limits the conclusiveness of the study. In addition, it should be kept in mind that the present study covers the meso level of universities – that is, the level of RIs within a university. It is very possible that the assumptions formulated at the outset already determine strategic thinking and communication at the level of university management but still need some time to trickle down to the level of the institutes. It may be that the RIs are currently the battleground where the contest between a more public relations-oriented approach to communication, as advocated by university management, and a more public engagement-oriented thinking, as preferred at the more research-oriented institutes, will be decided. In this respect, the next logical step for research interested in this area would be to conduct multilevel studies on the motivation for public communication by universities.

However, the present study leaves no doubt that instrumental communication and self-serving intentions are a strong driver of public communication by university RIs. My strongest evidence for this is the self-reported motivations that the institutes claim for themselves to justify why they communicate the way they do. Moreover, it appears that these motivations make a difference in what institutes communicate and how they communicate it. Finally, I have provided evidence that self-interested motivation is stronger where governance reforms have increased the artificial competition and regulatory autonomy of universities. Accordingly, there can be no doubt that the governance reforms described at the beginning also have an impact on the structures and content of the public communication of university-based RIs, which this research set out to prove.

References

Binswanger, M. (2015). Sinnlose Wettbewerbe in der Wissenschaft. In P. Grimm & O. Zöllner (Eds.), *Ökonomisierung der Wertesysteme – Der Geist der Effizienz im mediatisierten Alltag* (pp. 73–87). Franz Steiner Verlag.

Brandenburg, U., & Federkeil, G. (2007). How to measure internationality and internationalisation of higher education institutions. Indicators and key figures. CHE-Working Paper No. 92. Gütersloh.

Braun, D., & Merrien, F. (1999). Governance of universities and modernisation of the state: Analytical aspects. In D. Braun & F. Merrien (Eds.), *Towards a new model of governance for universities? A comparative view* (pp. 9–33). Jessica Kingsley.

Broucker, B., de Wit, K., & Leisyte, L. (2015). *An evaluation of new public management in higher education: Same rationale, different implementation.* Paper presented at the 37th Annual EAIR Forum 2015, Krems, Austria.

Civera, A., Lehmann, E. E., Paleari, S., & Stockinger, S. A. (2020). Higher education policy: Why hope for quality when rewarding quantity? *Research Policy, 49*(8), 104083.

Curaj, A., Scott, P., Vlasceanu, L., & Wilson, L. (Eds.). (2012). *European higher education at the crossroads: Between the Bologna process and national reforms.* Springer Science & Business Media.

Davies, S. R. (2008). Constructing communication: Talking to scientists about talking to the public. *Science Communication, 29*(4), 413–434.

de Boer, H., Enders, J., & Jongbloed, B. (2009). Market governance in higher education. In B. M. Kehm, J. Huisman, & B. Stensaker (Eds.), *The European higher education area: Perspectives on a moving target* (pp. 61–78). Sense.

de Boer, H., & Jongbloed, J. (2012). A cross-national comparison of higher education markets in western Europe. In A. Curaj, P. Scott, L. Vlasceanu & L. Wilson (Eds.), *European higher education at the crossroads: Between the Bologna process and national reforms* (pp. 553–571). Springer.

de Boer, H. F., Enders, J., & Schimank, U. (2008). Comparing higher education governance systems in four European countries. In N. C. Soguel & Jaccard, P. (Eds.), *Governance and performance of education systems* (pp. 35–54). Springer.

Enders, J., De Boer, H., &Weyer, E. (2012). Regulatory autonomy and performance: The reform of higher education re-visited. *Higher Education, 65*, 5–23. DOI 10.1007/s10734-012-9578-4.

Gregory, J., Agar, J., Lock, S., & Harris, S. (2007). Public engagement of science in the private sector: A new form of PR? In M. W. Bauer & M. Bucchi (Eds.), *Journalism, science and society: Science communications between news and public relations* (pp. 203–213). Routledge, Taylor and Francis Group.

Gregory, J., & Miller, S. (1998). *Science in public: Communication, culture and credibility*. Plenum/Basic Books.

Irwin, A. (1995). *Citizen science: A study of people, expertise and sustainable development*. Routledge.

Krüger, K., Parellada, M., Samoilovich, D., & Sursock, A. (Eds.). (2018). *Governance reforms in European university systems. The case of Austria, Denmark, Finland, France, the Netherlands and Portugal*. Springer International Publishing.

Le Grand, J., & Bartlett W. (Eds.). (1993). *Quasi-markets and social policy*. Macmillan.

Le Grand, J., Bartlett, W., & Roberts, J. (1998). *A revolution in social policy: Quasi-market reforms*. The Policy Press.

Loroño-Leturiondo, M., & Davies, S. R. (2018). Responsibility and science communication: Scientists' experiences of and perspectives on public communication activities. *Journal of Responsible Innovation, 5*(2), 170–185.

Marcinkowski, F., & Kohring, M. (2014). The changing rationale of science communication: A challenge for scientific autonomy. *Journal of Science Communication, 13*(3), C04.

Martín-Sempere, M. J., Garzón-García, B., & Rey-Rocha, J. (2008). Scientists' motivation to communicate science and technology to the public: Surveying participants at the Madrid Science Fair. *Public Understanding of Science, 17*(3), 349–367.

Metag, J., & Schäfer, M. S. (2018). Hochschulkommunikation in Online-Medien und Social Media. In B. Fähnrich, J. Metag, S. Post, & M. S. Schäfer (Eds.), *Forschungsfeld Hochschulkommunikation* (pp. 363–391). Springer VS.

Miller, S. (2001). Public understanding of science at the crossroads. *Public Understanding of Science, 10*(1), 115–120.

Newman, J. (2001). *Modernising governance. New labour, policy and society*. Sage.

OECD. (1995). *Governance in transition: Public management reforms in OECD countries*. OECD.

Paradeise, C., Ferlie, E., Bleiklie, I., & Reale, E. (Eds.). (2009). *University governance. Western European comparative perspectives*. Springer.

Pollitt, C., & Bouckaert, G. (2000). *Public management reform: A comparative analysis*. Oxford University Press.

Trench, B. (2008). Towards an analytical framework of science communication models. In D. Cheng, M. Claessens, T. Gascoigne, J. Metcalfe, B. Schiele, & S. Shi (Eds.), *Communicating science in social contexts: New models, new practices* (pp. 119–135). Springer.

Wynne, B., & Irwin, A. (1996). *Misunderstanding science: The public reconstruction of science & technology*. Cambridge University Press.

Chapter 5

Perceived Successfulness of Public Engagement at Research Institutes

John C. Besley and Anthony Dudo

The question

One way to understand how members of the scientific community perceive the successfulness of their public engagement efforts is to ask them. In this chapter, we, therefore, explore a direct question asking about the perceived successfulness of engagement efforts included in the MORE-PE survey described in this edited volume. This chapter specifically assesses the degree to which we might better understand self-perceived public engagement successfulness as a function of some of the other questions built into the survey. The theoretical rationale for the study comes from a relatively loose interpretation of "Excellence Theory" from public relations scholarship and its focus on meso-level predictors (including manager background, organizational location, and other factors) of whether a public relations team will be viewed as successful by organizational leaders (Grunig & Grunig, 2008; Grunig, Grunig, & Dozier, 2006). The research is also informed by individual-scientist-level studies of public participation (Besley & Dudo, 2017; Besley, Dudo, Yuan, & Lawrence, 2018; Poliakoff & Webb, 2007). To our knowledge, this is the first study in the area of science communication to look at perceived public engagement successfulness using a quantitative approach. While the available data are imperfect, we hope that the argument and analyses presented provide an impetus for future work.

The ultimate rationale for looking at self-reported perceived successfulness is a belief that science communicators likely have some sense of whether the public engagement activities that they take part in are successful. These perceptions—like any self-perceptions—are certainly imperfect and future research might explore the nature of discrepancies between perceived and actual success. However, we also see substantial value in using data collected as part of the MORE-PE study to learn what we can about the statistical correlates of self-reported success. The practical hope is to identify factors that represent some of the choices that strategic communicators might make to increase the likelihood of engagement efforts. For example, if having a public engagement plan or access to a public engagement specialist

DOI: 10.4324/9781003027133-7

seems to increase perceived successfulness than the scientific community might wish to explore prioritizing the availability of such resources. This chapter can be understood as secondary data analysis in as much as the designers of the MORE-PE project were not specifically focused on building statistical models to predict perceived successfulness of scientific research centers' and institutes' public engagement efforts. Our view, however, is that there are enough potential questions that the data can help explore. These are presented below.

Past research

Research on Excellence Theory from public relations scholarship provides one of the few bodies of work to help quantitatively understand what factors might help organizations achieve success when communicating strategically. The name "Excellence Theory" comes from the theory's developers' efforts to speak to ideas that were circulating in the management literature at the time, but the core of the work is an effort to identify best practices within groups tasked by their organization with building long-term relationships with stakeholders (Grunig & Grunig, 2008). Among other things, key variables identified by researchers emphasize the value of having the person in charge of building relationships with stakeholders think at a strategic (versus tactical) level and having the trust of the organizations' overall leadership team. Strategic thinking, in this regard, can be understood as evidence-based, goal-oriented thinking that works backward from organizational goals to immediate communication objectives to the tactics needed to achieve these objectives (Hon, 1998). Further, communications teams benefited from different types of diversity and a recognition that the task of building relationships with stakeholders should generally be coordinated with other types of marketing efforts but still kept somewhat separate. A commitment to true dialogue was also seen to be helpful (Grunig & Dozier, 2002; Grunig & Grunig, 2008; Grunig, Grunig, & Toth, 2007).

As noted, enumerating these types of benchmarks is important because they all suggest potential choices (e.g., increasing diversity) that someone could make in an effort to improve communication.

The focus on building long-term relationships is also helpful to understanding contemporary public relations scholarship, including Excellence Theory. Relationships can be partly understood in terms of trust-related beliefs such as perceptions of integrity and caring (i.e., benevolence) as well as mutual benefit and overall satisfaction (Colquitt & Rodell, 2011; Hendriks, Kienhues, & Bromme, 2015; Hon & Grunig, 1999). The key is to recognize that public relations scholars do not typically focus on outcomes such as short-term affect or behaviors such as sales in the way of related fields such as advertising. Rather, at least at the academic level, the expectation is that people involved in public relations will think about building meaningful, long-term relationships based on substantive interactions between organizations and their stakeholders. In doing so, the hope is to build legitimacy or a "license to operate" for the organization and this legitimacy facilitates organizational goal achievement over the long term (Burke, 1999). In this regard, the focus on building trusting relationships clearly represents a substantive overlap between public engagement-oriented science communication and public relations thinking (Borchelt & Nielsen, 2014). This is not to say public relations activities related to science may not have problematic elements (Bauer & Bucchi, 2008; Weingart, 1998), but it should also be understood that ideas from public relations scholarship can help us understand public engagement efforts (VanDyke & Lee, 2020).

Looking specifically at the questions available to analyze with the MORE-PE survey, there are several variables that might speak to issues of experience and diversity. Specifically, one might expect that (H1) bigger organizations, (H2) organizations with more experience doing public engagement, (H3) organizations that budget for public engagement, and (H4) organizations with dedicated public engagement staff would all be more likely to be able to draw on the strategic thinking that Excellence Theory highlights as a core best practice. Size might also increase the likelihood of diverse thinking, experience, and access to stakeholders. Similarly, the MORE-PE survey instrument asks respondents whether their organization has a public engagement strategy or policy. This represents a more direct indicator of strategic thinking and should also, therefore, be associated with perceived successfulness (H5).

Recent scholarship on individual predictors of public-engagement-related choices may also provide insight into selecting potential drivers of perceived engagement successfulness. A central argument of this emerging literature is that we can understand science communicators' choices as behaviors. This means that we can use the types of models developed to understand individual-level behavior related to health or the environment choices to understand behavior related to communication choices. Specifically, building on the Theory of Planned Behavior (Fishbein & Ajzen, 2010), the Integrated Behavioral Model (Fishbein, 2009; Montano & Kasprzyk, 2015) argues that intention to enact planned behaviors occurs as a function of individuals' salient attitudes (i.e., evaluative beliefs), normative beliefs, and efficacy beliefs related to the behavior. Besley and his colleagues have used this logic to try to understand scientists' overall willingness to take part in public engagement (Besley, Dudo, Yuan, et al., 2018) as well as intent to prioritize specific communication tactics (Besley, O'Hara, & Dudo, 2019) and objectives (Besley, Dudo, & Yuan, 2018). In general, the results of this work suggest that attitudinal, normative, and efficacy beliefs are all statistically associated with communicative willingness (Besley et al., 2019).

In the current context, the MORE-PE survey includes a set of measures for perceived public interest in public engagement participation, trust in science and scientists, and the belief in the idea that filling public knowledge deficits will lead to greater support for research. The current study conceptualizes perceptions of (H6) low public interest and (H7) public trust as attitudes (i.e., evaluative beliefs) related to public engagement and, thus, potential predictors of overall evaluations of engagement willingness. Unfortunately, the data do not include measures assessing perceived normative beliefs, or self-efficacy or behavioral control beliefs. Also, it should be noted that the survey underlying the MORE-PE project asks respondents to answer on behalf of their organization, rather than themselves, whereas the TPB is focused on individual beliefs and behaviors. Nevertheless, the operating assumption underlying asking the questions of organizational respondents is that the responses provide meaningful information about how organizations think about public engagement. It thus makes sense to expect that such views might be associated with the type of potential outcome variable discussed here.

In this regard, the criterion variable in the current study is perceived successfulness of engagement. This seems conceptually similar to behavioral willingness inasmuch as those who feel successful seem likely to be more willing to do a behavior. In other words, it seems logical to expect that scientific research units who see their audiences as relatively less interested in science and relatively less trusting in science are going to feel less successful when it comes to engagement (and, thus, less willing to engage, perhaps). In making these statistical predictions, it should be noted that the evidence seems to suggest that few groups are trusted more than scientists and that many people express substantial interest in a variety of scientific topics (National Science Board, 2018).

Beyond interest and trust, the study includes consideration of the academic field as a predictor of success with no specific expected direction (RQ1). The MORE-PE survey puts substantial focus on understanding differences between academic fields in how they think about public engagement (Ecklund, James, & Lincoln, 2012) although some past research has found few differences between fields in how they think about engagement (Besley, Dudo, & Yuan, 2018; Besley, Dudo, Yuan, et al., 2018). This variable is, therefore, included as a research question.

A final research question addressed in the current study is whether scientists who believe that educating the public can lead to support are more likely to feel successful (RQ2). This belief in the value of scientific knowledge as a driver of attitudes and behavior can seem somewhat naïve given past research suggesting a very limited relationship between knowledge and relevant attitudes and behaviors (National Academies of Sciences, 2016; Simis, Madden, Cacciatore, & Yeo, 2016). Nevertheless, for our purposes, scientists who believe that filling public deficits in knowledge can lead to support might feel successful because they often feel especially confident in their ability to transfer knowledge (Besley, Dudo, & Yuan, 2018). On the other hand, holding the simplistic belief in the efficacy of knowledge transfer might also get in the way of making the types of strategic communication choices that help build relationships associated with feelings of successfulness. It also seems worth noting that belief about the relationship between education and support is not clearly an evaluative belief—it seems more like a cause/effect belief. Ultimately, the decision was made to include this belief in the analysis largely because it is such an important element of discussions about effective communication that it seems worth testing the degree to which holding this belief affects a key outcome as such perceived successfulness.

In summary, the current study involves an opportunistic effort to explore whether a range of different factors are associated with perceived public engagement successfulness. The limitations associated with this study will be described in the concluding section. For now, however, we see substantial value in making the best possible use of the available data. The specific measures and analyses used to test the proposed hypotheses are outlined in the following section.

The current study does not seek to address hypotheses or research questions related to the country. As noted below, data from five different countries are included, but these are analyzed separately. This means that they might, thus, be seen as attempted replications rather than an effort to build a single model although a composite model is provided (without interactions by country). The reason for doing so is that, with just five countries, it did not seem reasonable to conduct a true multilevel model with interaction terms by country for each variable, given the small number of countries and somewhat small sample sizes within countries. Parsimony instead suggested simply exploring differences between results in a largely qualitative way with country-level differences—if evident—left for future explorations.

Method

The respective "national story" chapters elsewhere in this volume provide the sampling for each country included in the analysis. One important thing to remember, however, is that the unit of analysis is a research unit in the form of a research center or institute (rather than an academic department or college). Here, we describe the specific questions used from the MORE-PE survey for this chapter as well as the analysis plan. We describe the questions in English, but respondents saw surveys in their own language. Five countries from the MORE-PE project are used in the current study because of their availability at the time of the analysis and because these represent countries with somewhat overlapping research cultures. These include Italy, Germany, the Netherlands, the United Kingdom, and the United States.

Questions

The MORE-PE survey included a single question focused on perceived successfulness that serves as the criterion variable for the current analyses. Specifically, respondents were asked:

Perceived Successfulness: "Overall, how successful do you think your public engagement efforts have been in enhancing the mission of your unit."

Respondents could answer using a 5-point scale going from "very successful" (coded as "5" for our analyses) to "very unsuccessful" (coded as "1") with "neither successful, nor unsuccessful" as the middle category (coded "3"). As can be seen, respondents in three of the countries included in the analysis (Germany, Italy, and the United Kingdom) tended to select a response somewhat above the mid-point of the scale, whereas respondents in the remaining two (the Netherlands and the United States) tended toward the middle of the scale (Table 5.1).

The predictor variables used in the analyses include measures of research area, unit size, the unit's engagement experience and budget, the availability of at least one dedicated engagement staff member, and the availability of an engagement plan or strategy. These were measured in the following way.

Research area (RQ1): Consistent with the overall MORE-PE project's emphasis on academic field, dichotomous variables were used in the analysis. Respondents selected a primary area for their unit based on categories provided by the Organization for Economic Cooperation and Development. They included the "natural sciences," "engineering and technology," "medical and health sciences," "agricultural sciences," the "social sciences," and the "humanities." For these analyses, the humanities group was used as the reference category. The sampling plan had intended to have approximately equal size groups, but relatively few organizations seem to exist in the "agricultural sciences" and the natural science respondents seem to have been more likely to respond in many countries (see relevant chapters on national stories).

Size (H1): "How many people work at your research unit? Please consider all researchers, post-doctoral fellows, and Ph.D. students and technicians; do not count administrative staff." (Respondents were asked to enter a number; these were recoded so that 0=1, 1 1–5 = 2, 6–15 = 3, 16–50 = 4, 51–150 = 5, 151–300 = 6, 301–400=7; 32 larger units were removed from the analysis as they seemed too different to warrant inclusion.)

Past public engagement (H2): Twenty-nine questions asking respondents to estimate "roughly, how many times in the past 12 months" that their unit had taken part in various past public engagement activities were used to create a single engagement experience variable (alpha = .91 across countries). The original questions included items asking about face-to-face, mediated, and online engagement (see national chapters for specific details on items). This variable was also rescaled so that it retained its original 5-point scale where 1 = never, 2 = annual (once a year or less), 3 = quarterly (2–6 times a

Table 5.1 Descriptive statistics and correlations with perceived engagement successfulness efforts, by country

	Germany			Italy			The Netherlands			United Kingdom			United States		
	r	M	SD	r	M	SD	r	M	SD	r	M	SD	r	M	SD
Perceived successfulness (1–5)		3.58	0.78		3.80	0.74		3.44	0.76		3.70	0.80		3.49	0.85
Natural Sciences (Dich.)		35%			26%			22%			18%			26%	
Engineering/ Tech. (Dich.)		18%			19%			11%			13%			10%	
Medical/Health (Dich.)		11%			18%			18%			16%			16%	
Agriculture (Dich.)		5%			5%			2%			6%			11%	
Social science (Dich.)		17%			17%			31%			26%			21%	
Humanities (reference)		15%			16%			17%			21%			16%	

(Continued)

	Germany			Italy			The Netherlands			United Kingdom			United States		
	r	M	SD	r	M	SD	r	M	SD	r	M	SD	r	M	SD
Size (1–7)	.32*	4.15	1.22	-.09	4.35	1.25	.03	4.28	1.22	.31*	3.70	1.24	.12	3.16	1.20
Past PE (1–5)	.43*	1.95	0.58	.27*	2.14	0.62	.37*	2.15	0.58	.27*	2.05	0.44	.30*	1.96	0.55
5+ yrs PE Experience (Dich.)	.30*	66%		.20*	53%		.14	64%		.18*	61%		.11	69%	
Communication funding (1–4)	.25*	2.33	0.78	.05	2.47	0.89	-.01	2.33	0.61	.21*	2.78	1.09	.19*	3.05	1.17
Specialist Staff (Dich.)	.41*	55%		.14*	28%		.17*	35%		.19*	26%		.28*	36%	
Communication Policy (Dich.)	.31*	41%		.26*	59%		.21*	40%		.37*	33%		.19*	28%	
Perceived Low Public Interest (1–5)	-.34*	2.40	0.76	-.20*	2.50	0.74	-.22*	2.36	0.66	-.34*	2.24	0.67	-.25*	2.38	0.70
Perceived Public Trust (1–5)	.12*	3.19	0.78	-.02	3.07	0.90	.12	3.32	0.83	.16*	3.10	0.78	.08	3.09	0.85
Perceived Public Deficit (1–5)	.15*	3.78	0.81	.11*	4.00	0.80	-.05	3.79	0.77	.16*	3.89	0.74	.17*	3.95	0.77
DV valid/ Listwise	356	325		358	219		140	105		187	129		261	157	

Notes: PE = Public engagement; *p < .05 (one-tailed); listwise deletion.

year), 4 = monthly (7–20 times a year), and 5 = weekly (> 20 times a year). The means of 2 out of the 5 countries reported in Table 5.1 do about 1 of each type of activity per year. However, as noted in the national reports, some units do some activities far more often than other activities, so the mean might better be understood as suggesting that the average unit does about 30 engagement events per year.

5+ Years Engagement Experience (H3): Respondents were asked "how long" their unit had been "carrying out public engagement activities for nonspecialist audiences." They could indicate "less than 1 year," "between 1–5 years," and "more than 5 years." Only a small number of units (n = 43) across all countries indicated that they had been doing engagement efforts for less than a year (no countries were doing nothing), so this variable was recoded into a dichotomous variable where 1 = more than five years. Table 5.1 indicates that most groups had been doing some sort of engagement for at least five years.

Communication funding (H4): Respondents were asked to estimate "the percentage of the annual budget spent in the last 12 months on the public engagement efforts" of the unit. The question text further noted that "this can include actions such as the maintenance of the website, printing of brochures, organization of public events, etc." and also asked respondents "not to consider salaries of 'communication staff'." Response categories include none (coded as "1" for the analysis), < 1% (coded as "2"), 1–5% (coded as "3"), 6–10% (coded as "4"), and > 10% (coded as "5"). Respondents who indicated that they did not know how much budget their group devoted to engagement were coded as "none" (coded as 1 based on the logic that these groups likely were not spending a substantial amount of their budget on engagement activities). The means reported in Table 5.1 suggest that the average unit was spending a limited amount of their budget on engagement activities, not accounting for staff.

Specialist staff (H5): The availability of communication or public engagement staff was included in the analysis using a simple dichotomous question asking respondents whether their unit has "specialist staff responsible for public engagement activities." Such staff were defined as

> all employees who carry out public engagement tasks as part of their day-to-day responsibilities. This can include staff responsible for maintaining the website, organizing public events, supporting researchers in their public engagement work, producing the newsletter, responding to journalists, etc.

Table 5.1 indicates that the availability of such staff members varied fairly substantially by country, with Germany units being especially likely to have engagement staff.

Communication policy (H6): Respondents were asked directly whether their unit had a "public engagement policy" at the time of the survey (coded as a dichotomous variable). Table 5.1 again suggests variation by country, with Italian units seeming especially likely to have such a policy.

Perceived low public interest (H7): Low public interest perceptions were assessed using a set of four questions where respondents were given a statement and asked to indicate their agreement or disagreement on a 5-point Likert-type scale. This scale was anchored by strongly disagree agree (coded "1" for the analysis) and strongly agree (coded "5") with neither agree nor disagree as the middle category (coded "3"). Those who indicated that they did not know were also recoded into the middle category. The statements included "the public is not interested in the research conducted at our unit," "the public is not eager to learn about science," "the public is interested in a limited range of research such as dinosaurs, dolphins, and disasters," and "we cannot expect a large public to take interest in the research we do" (alpha = .71). The fact that the average score in all countries was below the scale midpoint suggests that most units tended to reject the idea of a disinterested public.

Perceived public trust (H8): Trust in science was measured using a single statement—"The public trusts science and scientists"—for which respondents used the same Likert-type scale described for the questions about interest. The average scores in Table 5.1 suggest that many respondents tended not to believe that the public trusts the scientific community.

Perceived public deficit (RQ1): Scientists' belief that the public would support science if they better understood it was also assessed with a single statement and a Likert-type response scale. Specifically, the survey presented respondents with the statement: "If the public knows more about our research, they will be more likely to support it." Table 5.1 reports means somewhat about the mid-point for all countries, suggesting that respondents tended to agree with this statement although not strongly.

Analysis

The analysis presented uses simple linear regression modeling using the "General Linear Model" (Type 3) function in SPSS 25. The decision to do so was for the sake of parsimony and because visual inspection of the key relationship suggests that, with the recoding described, a focus on linear relationships could adequately describe the data. Unstandardized regression coefficients along with overall effect size for each variable are reported. The primary analysis focuses on individual models of the five countries, but a final combined data run is also included. A hierarchical model suitable for nested data is not included because the small number of countries in the analysis would not provide substantial power to detect country-level differences. Interaction terms using the country dummy codes were tested sequentially with all the variables to see if they would improve the amount of variance explained by the models. However, the decision was ultimately made that they provided little additional insight such that it made more sense to present a relatively parsimonious model in Table 5.3. Correlations between perceived successfulness and the variables included in the model are also provided (Table 5.1). With five different countries and multiple variables per country included in the analyses, numerical data are included only in tables and are not repeated in the text for parsimony.

Results

A few patterns are evident in the reported models. First, neither research area (RQ1) nor belief in the deficit model (RQ2) appears to be associated with perceived successfulness. This is evident in both the models for individual countries (Table 5.2) and the summary model (Table 5.3). In contrast, variables associated with organizational size and experience were sometimes meaningful statistical predictors of perceived successfulness.

First, institute size appears to be weakly correlated with perceived successfulness in two countries (Germany and the United Kingdom), but not elsewhere (H1). Similarly, once controlling for other variables in the model, it actually appears that organization size is associated with slightly more perceived success in Germany, but less in both Italy and the Netherlands. The overall summary model does not show any relationship between perceived successfulness and organizational size (Table 5.3).

Variables associated with experience also sometimes seem to matter. Perceived engagement successfulness is associated with engagement experience in the past 12 months (H2) in three countries (Italy, the Netherlands, and the United States; Table 5.2) as well as the overall model (Table 5.3). The partial-eta square suggests that this variable is moderately important relative to the other variables in the model. Table 5.1 also highlights relatively large simple correlations between past engagement and perceived successfulness. However, having five or more years of engagement experience (H3) is a significant predictor only in Italy and, to a relatively small degree, the summary model. This pattern is also evident in the univariate correlations.

Variables associated with resources also varied in their connection to perceived successfulness. Communication funding (H4) appears not to be substantially associated with perceived engagement success, whereas access to specialist staff (H5) was associated with perceived successfulness in two countries (Germany and the United States; Table 5.2), as well as the overall model (Table 5.3), although the relative effect size seems similar to the other variables that were also significant predictors of success. Finally, having a communication policy seemed especially important in the United Kingdom

Table 5.2 Univariate General Linear Model (Type III) regression coefficients (unstandardized), standard errors, and partial-eta^2 for perceived successfulness of public engagement efforts, by country

	Germany			Italy			The Netherlands			United Kingdom			United States		
	B	SE	Prt-E^2	B	SE	Prt-E^2	B	SE	Prt-E^2	B	SE	Prt-E^2	B	SE	Prt-E^2
Intercept	2.92	.47	.16	3.90	.49	.24	3.83	.84	.18	4.11	.80	.19	3.06	.63	.14
Natural Sciences (Dich.)	−0.06	.14	.00	0.28	.15	.02	0.20	.25	.01	0.15	.22	.00	−0.07	.21	.00
Engineering/ Tech. (Dich.)	0.13	.15	.00	0.08	.17	.00	0.13	.32	.00	−0.32	.23	.02	−0.58	.25	.04
Medical/Health (Dich.)	0.19	.17	.01	0.11	.17	.00	−0.10	.25	.00	−0.17	.24	.00	−0.36	.22	.02
Agriculture (Dich.)$^+$	0.15	.22	.00	−0.15	.27	.00				0.19	.34	.00	−0.43	.24	.02
Social science (Dich.)	0.01	.15	.00	0.20	.18	.01	0.01	.22	.00	−0.13	.20	.00	0.06	.20	.00
Humanities (reference)															
Size (1–7)	0.10	.05	.02*	−0.09	.04	.02*	−0.14	.07	.04*	0.06	.07	.01	0.02	.06	.00
Past PE (1–5)	0.15	.09	.01	0.19	.09	.02*	0.50	.19	.07*	0.06	.18	.00	0.31	.13	.04*

	B	SE	p	B	SE	p	B	SE	p	B	SE	p	B	SE	p
5+ yrs PE Experience	0.15	.10	.01	0.24	.10	.03*	0.08	.16	.00	0.05	.14	.00	0.06	.13	.00
Communication Funding (1–4)	0.01	.06	.00	-0.01	.06	.00	-0.18	.14	.02	0.07	.06	.01	-0.01	.06	.00
Specialist Staff	0.27	.10	.03*	0.08	.11	.00	0.10	.18	.00	0.05	.19	.00	0.33	.13	.04*
Communication Policy (Dich.)	0.15	.10	.01	0.14	.11	.01	0.13	.17	.01	0.49	.16	.08*	0.02	.14	.00
Perceived Low Public Interest (1–5)	-0.21	.06	.06*	-0.12	.07	.01	-0.15	.12	.02	-0.36	.11	.09*	-0.18	.09	.03*
Perceived Public Trust (1–5)	0.11	.05	.02*	-0.06	.06	.01	0.07	.09	.01	0.10	.08	.01	0.04	.07	.00
Perceived Public Deficit (1–5)	0.05	.06	.00	0.07	.06	.01	-0.06	.10	.00	-0.03	.09	.00	0.12	.08	.02
r²/Adjusted r²/	.36	.32		.20	.14		.20	.08		.32	.23		.30	.23	
Final n	229			219			105			129			157		

Notes: PE = Public engagement; *p < . 05 (one-tailed); listwise deletion. +Not included for The Netherlands because of missing data.

Table 5.3 Univariate General Linear Model (Type III) regression
coefficients (unstandardized), standard errors, and partial-eta^2
for perceived successfulness of public engagement efforts, all
countries

	B	SE	Part-E^2
Intercept	3.44	.27	.17
Germany	0.00	.08	.00
Italy	0.22	.08	.01*
The Netherlands	−0.16	.10	.00
The United Kingdom	0.13	.09	.00
The United States (Reference)			
Natural Sciences	0.12	.08	.00
Engineering/Tech.	−0.06	.09	.00
Medical/Health	−0.02	.09	.00
Agriculture	−0.08	.12	.00
Social science	0.03	.08	.00
Humanities (reference)			
Size (1–7)	0.00	.02	.00
Past PE (1–5)	0.22	.05	.02*
5+ years PE experience	0.12	.05	.01*
Communication funding (1–4)	0.03	.03	.00
Specialist staff	0.18	.06	.01*
Communication policy (Dich.)	0.21	.06	.02*
Perceived low public interest (1–5)	−0.22	.04	.04*
Perceived public trust (1–5)	0.04	.03	.00
Perceived public deficit (1–5)	0.04	.03	.00
r^2/Adjusted r^2/Final n	.23	.21	839

Notes: PE = Public engagement; *p < . 05 (one-tailed); Interactions of variables
by country non-significant and, thus, removed from final model. Listwise deletion

and it is somewhat important across countries (H6). This pattern can also
be seen in the correlations reported in Table 5.1.

Although perceptions about public trust in science were not associated
with perceived engagement successfulness (H7) in the bulk of the models
presented in Table 5.2, perception of public scientific interest appears to have
been the most consistent statistical predictor in the analysis. In general, it
appears that units that believe the public is interested in science report per-
ceiving more success (H6). This pattern can be seen in the country-specific
models (Table 5.2), the summary model (Table 5.3), and the simple correlations (Table 5.1).

Discussion and limitations

The analysis presented is consistent with the idea that recent engagement
experience (H2), a history of engagement activity (H3), access to specialist
staff (H5), and a communication policy (H6) are all small but potentially

useful predictors of self-reported public engagement successfulness. Similarly, research units where the respondent indicated that they did not think the public was interested in engagement reported relatively less public engagement successfulness. These findings are consistent with the idea from excellence theory that emphasizes the importance of organizational factors in fostering engagement quality (Grunig et al., 2006). There were few consistent differences by country or area and also few differences as a function of perceived trustworthiness or deficit model beliefs. This lack of difference by audience-specific attitudes is not consistent with behavior change theories that typically show an important role for attitudes (Montano & Kasprzyk, 2015), but the nonrelationship might be a function of the nature of either the survey implementation (i.e., the survey was completed by one individual on behalf of an organization) or the fact that the available questions are not specifically focused on the relevant behavior. On the other hand, the lack of substantive difference by field seems to be consistent with the overall pattern found in studies of scientists; any such differences seem to be relatively minor (Bennett, Dudo, Yuan, & Besley, 2019). And the lack of difference by country might make sense inasmuch as we have no specific theory to suggest why the pattern of correlation between the types of variables included would be different in one country or another. On the contrary, it might be expected that positive views about one's audience or structural factors such an engagement plan would be positive predictors of success in any context.

Indeed, the three most important limitations that need discussion involve measurement, causation, and generalizability. With regard to measurement, an ideal study would involve getting more direct, multi-item measures of the types of key constructs suggested by past work on Excellence Theory, including better assessment of the nature of the people doing public engagement (training, role in the organization, etc.), as well as their evaluative beliefs/attitudes about engagement strategy. The MORE-PE project has some of this data but not in a way that would have allowed for that type of analysis to be conducted here.

The causal direction of the statistical relationships discussed is an especially important consideration for any study using cross-sectional data, but it seems especially relevant here. For example, the fact that two measures of engagement experience were associated with perceived successfulness might suggest that engagement experience might lead to a sense of success and this would allow us to argue that part of success is just getting started. On the other hand, it might be that people who feel successful would tend to engage more because they feel their investments in engagement are having an impact. Similarly, it might be that having public engagement staff and policies lead to success, but the counter argument is that those who feel successful might hire staff and create a policy. Finally, one might argue that groups who feel that the public lacks interest might be less likely to engage, but it might also be that groups who feel less successful have concluded that the public lacks interest. In all cases, our expectation is that reciprocal relationships are in play. However, the factors that we have included as predictors in our model are the types of things that someone interested in fostering engagement activity could attempt to change. For example, those interested in engagement could make the choice to begin trying out activities, hiring staff, and developing policies or strategies. Those seeking to enhance engagement activity might also seek to reassure potential communicators that interested audiences often exist (Nadkarni et al., 2019) and that part of strategy is identifying the audiences that would help achieve organizational goals and finding ways to spark their interest.

Finally, the current analysis was done only with five countries and the difficult nature of sampling research units makes it difficult to know if the current findings would reoccur in other samples. There were obviously some differences when looking at the individual countries (Table 5.2) rather than the summary model (Table 5.3), but the overall patterns seem clear and the individual country-level samples are fairly small, especially after listwise deletion. The issues with weak measurement might also affect generalizability and it is our hope that future studies might use more precise measures; the current data suggest that such research might prove fruitful.

Our ultimate sense is that the MORE-PE dataset provides some initial clues about potential ways to help science communicators feel more successful. Specifically, consistent with Excellence Theory, the data highlight the potential value of putting dedicated human resources into public engagement and ensuring that these resources develop a plan for success. It also suggests some value in "learning by doing" inasmuch as units with more engagement experience are the ones feeling the most successful. Future research should find ways to explore these possibilities.

References

Bauer, M. W., & Bucchi, M. (2008). *Journalism, science and society: Science communication between news and public relations.* London: Routledge.

Bennett, N., Dudo, A., Yuan, S., & Besley, J. C. (2019). Chapter 1: Scientists, trainers, and the strategic communication of science. *Theory and best practices in science communication training* (pp. 9–31). New York: Routledge.

Besley, J. C., & Dudo, A. (2017). Scientists' views about public engagement and science communication in the context of climate change. *The Oxford encyclopedia of climate change communication.* Oxford: Oxford University Press. doi: 10.1093/acrefore/9780190228620.013.380

Besley, J. C., Dudo, A., & Yuan, S. (2018). Scientists' views about communication objectives. *Public Understanding of Science, 27,* 708–730. doi: 10.1177/0963662517728478

Besley, J. C., Dudo, A., Yuan, S., & Lawrence, F. (2018). Understanding scientists' willingness to engage. *Science Communication, 40,* 559–590. doi: 10.1177/1075547018786561

Besley, J. C., O'Hara, K., & Dudo, A. (2019). Strategic science communication as planned behavior: Understanding scientists' willingness to choose specific tactics. *PLoS ONE, 14,* e0224039. doi: 10.1371/journal.pone.0224039

Borchelt, R. E., & Nielsen, K. H. (2014). Public relations in science: Managing the trust portfolio *Routledge handbook of public communication of science and technology* (pp. 74–85): London: Routledge.

Burke, E. M. (1999). *Corporate community relations: The principle of the neighbor of choice.* Westport, CT: Quorum Books.

Colquitt, J. A., & Rodell, J. B. (2011). Justice, trust, and trustworthiness: A longitudinal analysis integrating three theoretical perspectives. *Academy of Management Journal, 54*, 1183–1206. doi: 10.5465/amj.2007.0572

Ecklund, E. H., James, S. A., & Lincoln, A. E. (2012). How academic biologists and physicists view science outreach. *PLoS ONE, 7*, e36240. doi: 10.1371/journal.pone.0036240

Fishbein, M. (2009). An integrative model for behavioral prediction and its application to health promotion. In R. J. DiClemente, R. A. Crosby & M. C. Kegler (Eds.), *Emerging theories in health promotion practice and research* (2nd ed.). San Francisco, CA: Jossey-Bass.

Fishbein, M., & Ajzen, I. (2010). *Predicting and changing behavior: The reasoned action approach.* New York: Psychology Press.

Grunig, J. E., & Dozier, D. M. (2002). *Excellent public relations and effective organizations: A study of communication management in three countries.* Mahwah, NJ: Lawrence Erlbaum Associates.

Grunig, J. E., & Grunig, L. A. (2008). Excellence theory in public relations: Past, present, and future. In A. Zerfass, B. Ruler & K. Sriramesh (Eds.), *Public relations research* (pp. 327–347). Wiesbaden: VS Verlag für Sozialwissenschaften.

Grunig, J. E., Grunig, L. A., & Dozier, D. M. (2006). The excellence theory. In C. H. Botan & V. Hazleton (Eds.), *Public relations theory II* (pp. 21–62). Mahwah, NJ: Lawrence Erlbaum Assoicates.

Grunig, J. E., Grunig, L. A., & Toth, E. L. (2007). *The future of excellence in public relations and communication management: Challenges for the next generation.* Mahwah, NJ: Lawrence Erlbaum.

Hendriks, F., Kienhues, D., & Bromme, R. (2015). Measuring laypeople's trust in experts in a digital age: The Muenster Epistemic Trustworthiness Inventory (METI). *PLoS ONE, 10*, e0139309. doi: 10.1371/journal.pone.0139309

Hon, L. C. (1998). Demonstrating effectiveness in public relations: Goals, objectives, and evaluation. *Journal of Public Relations Research, 10*, 103–135. doi: 10.1207/s1532754xjprr1002_02

Hon, L. C., & Grunig, J. E. (1999). Guidelines for measuring relationships in public relations. *Institute for Public Relations.* Retrieved from http://www.instituteforpr.org/topics/measuring-relationships/

Montano, D. E., & Kasprzyk, D. (2015). Theory of reasoned action, theory of planned behavior, and the integrated behavioral model. In K. Glanz (Ed.), *Health behavior: Theory, research and practice* (5th ed.). Hoboken, NJ: Wiley-Blackwell.

Nadkarni, N. M., Weber, C. Q., Goldman, S. V., Schatz, D. L., Allen, S., & Menlove, R. (2019). Beyond the deficit model: The ambassador approach to public engagement. *BioScience, 69*, 305–313. doi: 10.1093/biosci/biz018

National Academies of Sciences, E., and Medicine. (2016). *Science literacy: Concepts, contexts, and consequences.* Washington, DC: The National Academies Press.

National Science Board. (2018). Chapter 7, Science and technology: Public attitudes and public understanding. *Science and Engineering Indicators*, from https://www.nsf.gov/statistics/2018/nsb20181/report/sections/science-and-technology-public-attitudes-and-understanding/highlights

Poliakoff, E., & Webb, T. L. (2007). What factors predict scientists' intentions to participate in public engagement of science activities? *Science Communication, 29*, 242–263. doi: 10.1177/1075547007308009

Simis, M. J., Madden, H., Cacciatore, M. A., & Yeo, S. K. (2016). The lure of rationality: Why does the deficit model persist in science communication? *Public Understanding of Science, 25*, 400–414. doi: 10.1177/0963662516629749

VanDyke, M. S., & Lee, N. M. (2020). Science public relations: The parallel, interwoven, and contrasting trajectories of public relations and science communication theory and practice. *Public Relations Review, 46*, 101953. doi: https://doi.org/10.1016/j.pubrev.2020.101953

Weingart, P. (1998). Science and the media. *Research Policy, 27*, 869–879. doi: https://doi.org/10.1016/S0048-7333(98)00096-1

Chapter 6

An emerging "Arms Race"

Resourcing the Public Communication Effort

Martin W. Bauer and Marta Entradas

Competition for attention seeking and the communication function of universities

In many advanced and advancing countries, the system of higher education and research-based universities has been expanding massively over the past half of a century, and this expansion has put a spotlight onto performance for purposes of accountability and reputation ranking. Some observers argue that higher education in many countries is now reaching a turning point of uncertainty for the future, after 100 years of predicted expansion towards ever bigger and better, and ever larger participation rates among the national youth (Mandler, 2020).

Universities and other research institutions internalise this changing context in the form of public–private management (NPM) ideas, moving in the direction of running the university as if it were a utility corporation providing a service, e.g. energy or public transport, for profit, and being supervised by a regulator who maintains a market to avoid monopolistic profiteering.

In this changing ecosystem, the communication function of universities has acquired a visible and central role. In part as an effect of NPM reforms to make public organisations more business-like with a focus on 'consumers' (e.g. Hemsley-Brown & Oplatka, 2006), communications structures are expanding and professionalised. This is visible in the proliferation of public relations (PR)/communications offices and managers responsible for managing relationships between the university and society (e.g. Krücken and Meier, 2006). National and international audit systems evaluate the performance of higher education in terms of research output, teaching quality, and impact on economy and society. These rankings, based on a multitude of indicators, are rendering the relative positions of institutions widely visible and create a game of competition for reputation. Higher education institutions respond to this challenge with the expansion of communication at the central level, at the level of research institutes where the action is (Entradas et al., 2020), and also among individual scientists (Bauer & Jensen, 2011; Entradas & Bauer, 2019) who cultivate their personal profile. This expansion of the communication function is our present concern.

DOI: 10.4324/9781003027133-8

Two other trends are linked and parallel this proficiency of communication. Science news and science reportage have equally seen massive expansion in the public spheres of most countries. People are able to read, listen, and watch ever more science news and science programmes in diverse formats, though this might have hit a ceiling by the mid-2000s; the Covid-19 years of 2020/2021 most likely being above this stabilising trend (Pansegrau & Bauer, 2019).

The mass media system is rattled on the back of new social media and Internet platforms. In the 21st century, traditional media based on large scale, capital intensive central printing, radio and TV broadcasting operations, are increasingly replaced by platforms with global reach that distribute information de-centrally produced on a different business model. The entry costs to media attention are very low, the costs of circulation are close to zero, and any costly gatekeeping and quality control is lost or delegated to censoring systems post hoc (i.e. AI systems which automatically recognise and edit out 'undesirable content' however defined). This broadens the access to mass media for many more actors and confuses the key function of a public sphere: to create a joint focus of attention out of which agenda setting and joint intentionality for society can emerge. Thus, global social media interfere with local public spheres by creating polarisation and encouraging the celebratory expression of segmented tribal mentalities and group thinking.

Some implications of these trends for science communication are condensed into the medialisation hypothesis. The medialisation hypothesis (Peters 2012; Weingart, 2012) expects a rising presence of science in the mass media, on the one hand, leading to increased public attention, and an over-adaptation on the side of scientists and scientific institutions to the logic of attention seeking, on the other hand. This requires the recognition of news values in addition to truth value that feed this rising appetite for public attention, which creates new dilemmas and tests for unintended consequences of good intentions. More science news is a good idea because it demonstrates the relevance of science to wider society; on the other hand, this search for more attention is risky for science: it changes the way science operates. Science communication does not only communicate the risks of anthropogenic climate change, genetic engineering, artificial intelligence, or nuclear power; it is itself a risky activity.

Empirical research with a focus on science medialisation at universities shows an intensification of attention seeking in the increase of media contacts (e.g. Koso, 2021; Vogler & Schäfer, 2020) and increased use of online means (Metag & Schäfer, 2019). However, the impact of this growing activity is less understood. One question that remains unanswered is whether there has been a refocusing of resources from an orientation on truth values towards public attention seeking.

Several trends point to such a potential pay-back risk for the conduct of science: Firstly, as hyperbole, or simply hype, becomes part of normal science operations, it loses its stigma of 'sensationalism' and defines common science rhetoric. The re-orientation on news values, in addition to truth values, selects scientific careers also on public visibility rather than research reputation. We understand 'visibility–publicity–popularity' as the prestige hierarchy of the public sphere and 'reputation' as the prestige hierarchy of the scientific community of peers. Secondly, when researchers decide what to study next, they might increasingly be lured towards 'systems' that afford attractive visuals because these will attract public attention; hence one studies butterflies rather than nematodes. Thirdly, the craving for public attention can create incentives for cutting corners and engaging dubious research conduct including fraud. Finally, more focus on attention seeking does not come cost-free, it will shift limited resources from ground research towards communication efforts, from the primary task of research to the supporting act of public communication. Note, we consider peer communication part of the primary task, the securing of truth values. It is the case that public communication enhances peer communication under conditions of extreme specialisation as demonstrated by Phillips et al. (1991) in the case of medicine, which makes the unambiguous separation between primary and secondary tasks of research more difficult. The primary task is increasingly unviable without the secondary task.

In this chapter, we will examine the latter of these trends and assemble evidence for a transfer of resources from research oriented on truth values towards attention seeking. We call it the **Arms Race for Public Communication (ARPC) hypothesis**. With increased competition among research institutes for limited funds, increasing proportions of funds and efforts are re-allocated from the primary task of research to the secondary task of celebrating this research. The arms race metaphor suggests that this competition creates niches for hypertrophic forms of communication with functional relevance mainly for market signalling (i.e. a kind of new Baroque culture). Because these niches are costly, they could, at some point, tip over into dysfunctionality. Our international project tries to establish baseline data to see where this trend is at, is moving towards and its tipping point might be.

Noam (1996) predicted radical changes that would happen to higher education universities with the roll-out of networked information technology. The worldwide web and the Internet lead to a fundamental redirection of the flow of information. Universities change from a place where information is centrally stored and students and researchers physically flock towards, to a place that seeks out and reaches out to people wherever they are. Research places become hubs in a global network, and they seek to attract people to operate that network via that hub wherever they are. That implies that many more locations can become hubs, seeking attention in competition with many other hubs; independently of whether these are of private or public or mixed legal status.

Real arms races are a historical reality. When military powers get locked into a race, they spend much resources on upgrading their weapon systems which become ever more destructive. To maintain a balance of potential forces, this escalation of procurement absorbs more and more state expenditures until becoming totally militarised and unsustainable. By analogy, in biology, 'arms races' are known by the **Red Queen effect** of increasing efforts only to keep pace with competitors. In Lewis Carroll's 'Through the Looking Glass', Alice complains that running in the looking-glass does not take her anywhere; the Queen replies: 'here, you see, it takes all the running you can do, to keep in the same place'. Efforts increase without real progress. Another well-known example is the **Peacock effect:** developing such elaborate and beautiful feathers to impress potential mates has costs; the peacock is a bird that no longer flies. The secondary function of feathers (attention of mates) dislodges entirely the primary one (flying). This analogy is cited by Kucharski (2020) in analysing the race for 'viral messaging' of seeking attention on social media (p175). The viral analogy only means 'spreading fast and widely'. Any criteria for successful versus unsuccessful content features are constantly invalidated as the user interests are shifting. On any topic, machine learning will show that being more dramatic, more evocative, and more surprising than anybody else requires ever more extreme contents, i.e. arming up for attention seeking (forget truth value, who cares?). This race shifts any focus from content creation to attention management. For this, the tricks of the trade (p176ff) are several and known as: broadcasting (one influencer serves many readers), user-to-user (many read and recommend; high reproduction rates), start with big initial seeding (many start-ups), honing in by peeking (monitor roll-out and respond quickly) and increase exposure through infinite scrolling (automatic page turning, video-music continuity). These tactics make us, on social media, messaging ever faster, ever broader, and ever more extreme, while our concern for truth value suffers from irrelevance.

Medialisation as ARPC hypothesis

The medialisation of science has two elements: science has ever more news space available, but this deflects scientific activity from the logic of research to that of managing public attention. This over-adaptation is consequential. We are further specifying this hypothesis with a potential for an 'arms race' for communication between institutions that conduct research: public communication (PC) needs to be intensified and professionalised without gaining any visible advantages, just to remain in the game (the potential Red Queen effect); public engagement creates new rituals of communication with unclear functionality for research (the potential Peacock effect).

Box I: a case study of academic arms race

An example of such an arms race is the annual ranking of German-speaking Economists, which the Swiss Neue Zuricher Zeitung (NZZ) is organising as the 'economist impact ranking'. For several years, the newspaper published an annual list of economists and their affiliations audited on three weighted criteria: scientific citations (2x), media references (in previous year; 1x) and policy influence (i.e. being mentioned by parliamentarians; 1x). To enter into competition, a name must have at least five citations for the past five years; comparing across years allows to show whether entries have moved up or down, or are treading water (NZZ, 2018). To reveal the arms race, we will have to observe how the criteria audited for this ranking are accelerating on increasing thresholds to reach the same rank position. This will take a few years; it might be too early to tell, but it is worthwhile to keep an eye on it.[1]

What might be a benchmark for the resourcing of science communication? We might compare the resourcing of communication across other sectors. In the pharma industry, the ratio of drug R&D to drug marketing is reported to be four units of production on six units of advertising (40:60); these figures are often highly controversial because what activities count as 'marketing' is contested (Matheson, 2008; Norris et al., 2005). In the film industry, the ratio is even more skewed towards the communication side and away from production; figures mentioned are 90:10. For every Hollywood blockbuster, $1 spent on making the movie sees $9 for getting it into cinemas and into the daily conversation, whatever this involves.[2]

Formally, an arms race model as proposed here has five elements to consider (Dawkins & Krebs, 1979):

1 (Competition) – A competes with B, either within or between types (e.g. economists compete among themselves; but economics might compete with other social sciences across types); institutes compete with others within the university, and with others outside their host institution.

2 (Primary and secondary tasks) – This competition entails the primary task of behaving and the secondary task of support for behaving in particular manner (e.g. running and running faster); the latter requires ever larger resources; for our purposes, we identify as 'research' as the primary task, and 'public communication about that research' as the secondary task.

3 (Attention seeking) – The secondary task is often evaluated on different criteria, e.g. beauty, attention, or threat alarm; 'exaptation' is the name

for a hypertrophic trait that shifts its function (e.g. 'walking faster' in the Red Queen effect and 'colourful, large feathers' in the Peacock effect). A bird's feathers primarily regulate temperature and allow for locomotion by flying. However, the peacock's feathers became co-opted by the mating game; they become long, heavy, and beautiful to the effect that the bird is unable to fly. In the same fashion, science communication expands beyond peer conversations, broadens its formats, and takes on a different function: attention seeking.

4 (Professionalisation) – Novel structures emerge which support this secondary task; the exaptation creates related features that sustain the secondary tasks. Science communication, once the hobby of scientists, is increasingly a professional and specialised activity with an independent career structure. The wider public co-opts the communication efforts of research that affords a logic of attention seeking above and beyond truth values.

5 (Limitation) – The communication system is potentially growing to a point of exhaustion, when in an extreme scenario, there is only communication activity and very little research left. Therefore, there must be a 'natural' limit for the secondary task in order not to overwhelm the primary task. However, this limit is difficult to determine because the point of inflection might depend on context. But the intuition remains: more is not necessarily better.

Three hypotheses derived from ARPC

From the arms race for public communication, our ARPC model, we derive three initial hypotheses which we want to put to the empirical test.

HI: There is variation in institutes' resource allocation to communication across countries

For there to be a competition, there has to be some trait variation. We will compare research institutes across several countries with regard to communication activities. For this, we use several indicators of resources they absorb. We have asked institutes in eight countries (1) whether they employ communications staff, either full time or part time (*decentral staff*); (2) whether they have access to communication support at central level (*central staff*), and (3) whether they maintain their own list of media contacts (*media contacts*). We also derive three indicators of resourcing of communication from the percentage of annual budget that goes on communications: (1) current funding for communications (*declared funding:* M=0.031, $0 < X < 0.13$, n=1610) and (2) expected funding in the future (*expected funding:* M=0.051, $0 < X < 0.13$, n=1614), and on the basis of staffing levels, we calculate (3) communications resourcing in percentage of the institute's annual budget, considering salary levels in different countries. This yields our own estimates of resources (*actual funding:* weighted for salaries PPP: M=0.026, 0 < X < 0.84; n=1487). *Declared funding* and *expected funding* are estimates provided by the units themselves in percentage of annual budgets, i.e. actor 'subjective' estimates. *Actual funding* is calculated based on staffing numbers, i.e. observer 'objective' estimates, in percentage of annual budgets. All data are from the MORE 2018 database of research units (see Entradas et al., 2020; also see Part 3 of this volume). We are considering eight countries in our comparison: Brazil, the United States, Japan, the United Kingdom, the Netherlands, Portugal, Italy, and Germany; research institutes are representative of six OECD disciplines which are Natural Sciences, Engineering & Technology, Medicine and Health Sciences, Agriculture, Social Sciences, and Humanities (N=2,030).

H2: Resourcing the communication lines up with level of competition

In a second step, we align the trait variation across research units with context information on level of competition. For this, we need indicators of competitive strive, of either national or international, or within research disciplines, between research units. We are using two aggregate indicators of competition: (a) for national competition, we measure the density of research population, i.e. the number of full-time researchers in higher education per country. The more researchers working in higher education in a country, the larger is the competition for national funding sources. We call this indicator *Researchers in HE*, and it refers to the total number of full-time researchers employed in higher education institutions in a country and discipline (OECD, UNESCO sources; scores 1137 < x < 65469). And for the international game, we consider b) the list of universities that are ranked on 2020 QS World University Ranking, and we count the number of universities listed for each country within the first 200 or 500 (https://www. topuniversities.com/qs-world-university-rankings). We call this indicator *international ranking*. The eight countries in our study receive scores 5 < x < 87. We are testing a monotonous relationship: the higher the competition (between countries or among disciplines), the more resources are going on average into communications (in our units). We expect a consistent pattern of increasing resourcing of communication activities across levels of competitions, for different indicators, and across different disciplines.

H3: Communications staff is aware of the competitiveness among units

Finally, we ask whether communications staff in research units are aware of this competition, and whether they consider engaging this competition to be part of their task description. Conscious engagement in competition is an additional factor in this escalating cycle.

For this, we consider the question *what are your rationales for doing PC*; from a list of eight rationales, informants pick the two most important ones (see Chapter 16 of this volume). We created an index by counting mentions, after having classified responses into either civic 'public engagement' (*PE*) *Rationales* (i.e. responding to policy, disseminate research, listen to publics), or the more corporate *PR Rationales* (i.e. raising profile, attracting funding, attracting support, recruiting students). About 48% of institutes mention PE rationales, and 52% mention PR rationales for their activities. By relating one to the other, we create a ratio score of PE over PR rationale (PR/PE = 1 means 'balanced'; < 1 means 'PE dominates'; > 1 means 'PR dominates').

Evidence of alignment of communication effort and level of competition

H1: There is variation in institutes' resource allocation to communication across countries

We can clearly identify variation in resource allocation per research unit across eight countries as shown in Figure 6.1. Staffing levels vary: 32% of research units in Germany report employing specialist PC staff; 40% of units report part-time PE staff, some of them in addition to full time. Overall with 50% of units with PC competences, Germany and Portugal are the countries with most PC staffing; at the other end is Japan with 10% of units with full-time PE

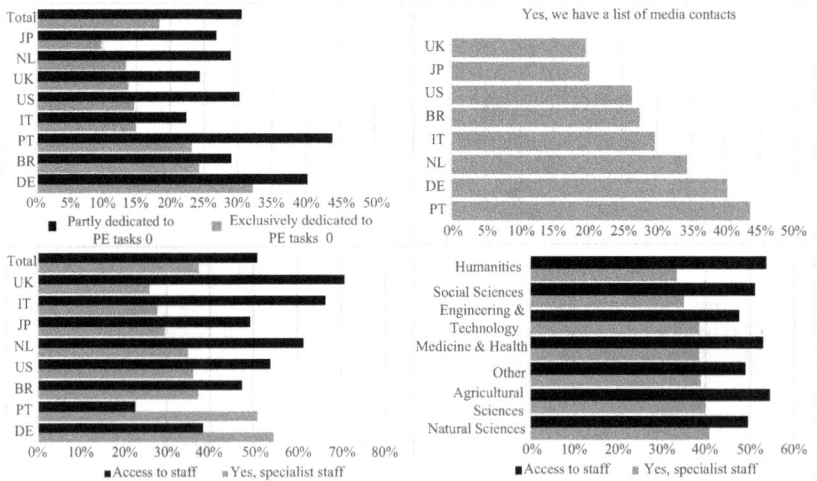

Figure 6.1 Percent of research units that (a) report their own full- or part-time communication staff, (b) maintain a media contacts list locally, and (c) employ decentral specialist staff or have access to specialists at the central level and for different countries and disciplines.

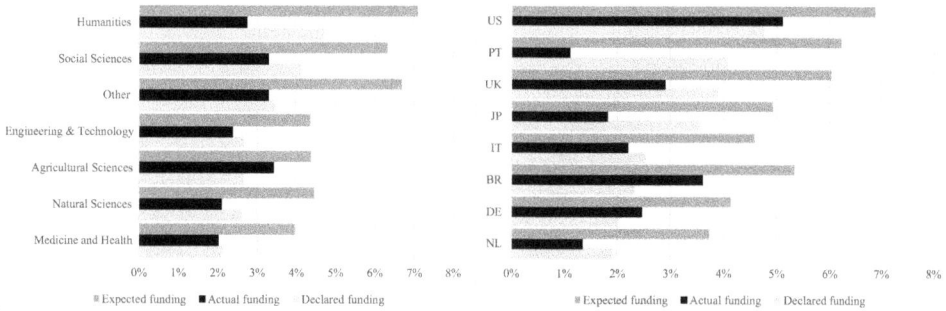

Figure 6.2 Estimates of annual resources allocated to PC activities in % of annual unit budget. Declared funding is the subjective estimates of current expenditure; actual funding is a calculation of PC expenditure including staff salaries in comparable currency; and expected funding is a subjective estimate of future communication expenditure.

staff. All in all about 50% of all units in Germany and Portugal report access to their own PE staff, while 26% of UK units have the same. Many universities also offer support for communication at the central level; this is reported by most units in the United Kingdom, the Netherlands and Italy, all with above 60% accessing central support. This is less the case in Portugal. Not all research units are equally media savvy to the extent that they keep their own media contact lists. We find that 44% of Portuguese units report such lists, but only 20% of the United Kingdom and Japanese units keep them.

Staffing is, among other things, a cost factor. We estimate how much of the annual research budgets go into communications staffing at each unit, as shown in Figure 6.1. Our estimates show that this 'declared funding' varies between 2% and 5% on average, be that across nations or across disciplines. We also observe that most units expect this commitment to rise in the next few years. *Expected funding* is generally higher and varies between 4% and 7%. The social sciences and humanities invest more on communication than the others. Across countries we find a clear ordering of average expenditure; however, there is some discrepancy between declared subjective and actual objective estimates of that commitment (see Figure 6.2).

The richer a unit is in terms of annual overall budget, the smaller is their declared 'subjective' commitment to public communication. However, if we consider the 'objective' actual funding, the relationship is direct: the richer a unit, the larger is their PC expenditure. It, thus, would appear that richer units tend to underestimate their PC effort, while poorer ones overestimate it.

H2: Resourcing the communication lines up with level of competition

Secondly, having examined the variation in PC commitment across countries and disciplines, we ask the second question: how this variation is lining up if

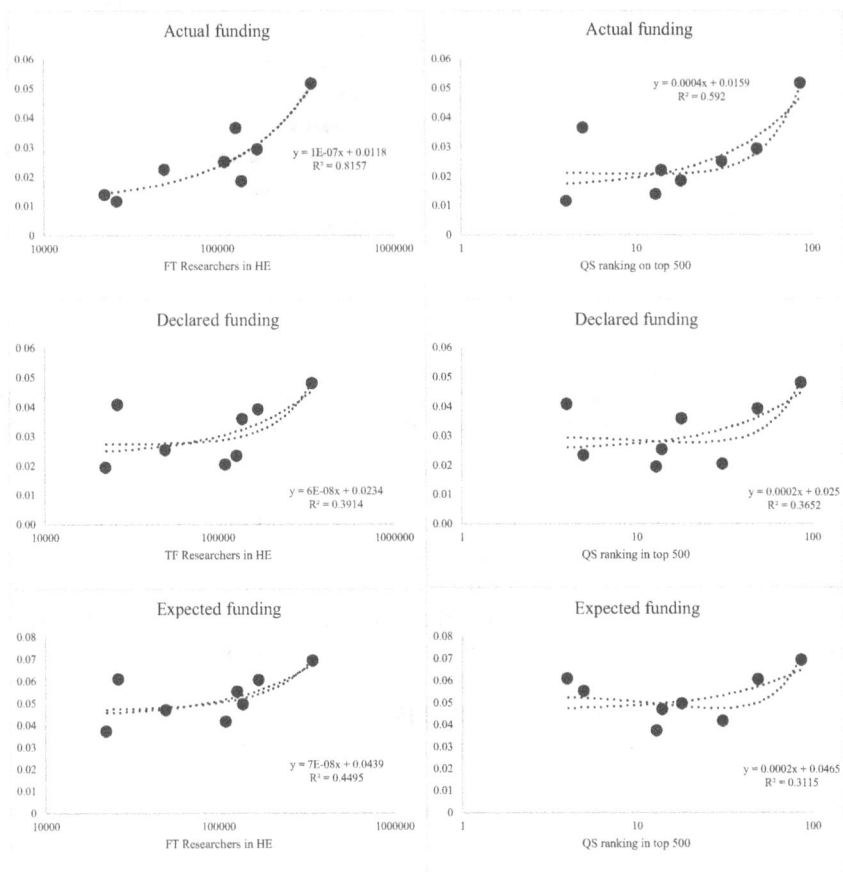

Figure 6.3 The graphics show the correlation of our three resource indicators (declared, future, and actual funding) with two competition indica-tors for different countries: (a) on the left, FT researchers in HE stands for the national population density chasing grants and (b) on the right, QS ranking stands for international competition. The dots represent the different countries.

we consider indicators of competition – both national and international? We do this in three steps: (i) first, we compare different indicators of resource allocation and different indicators of competition; (ii) second, we compare re-source allocation for different disciplines again competition; and (iii) finally, we summarise these relationships which give a clearer notion of a potential arms race over public communication among research institutes.

i   *Resource allocation and competition.* The results for three different in-dicators are shown in Figure 6.3. We are lining up actual, declared, and future PC funding with the density of the researcher population and with the number of universities in the top-500 list (both on log scales). Curve fitting shows that, for all countries, all relationships are linear with correlations between r=0.90 ($R^2 = 0.81$) and r=0.56 ($R^2 = 0.31$). This means the stiffer the competition between research institutes, the larger the percentage of PE expenditure. This alignment is the strongest on both indicators of competition when we consider the actual funding of the communication effort.

ii   *Effects of competition on resource allocation in different disciplines.* If we consider the effect of competition on the different OECD disciplines, we find that these are not equally affected by this intensifying pressure. We are using international competition – number of universities in top 500 – as the predictors and the 'declared' communications funding estimates as criterion. The linear curve fit is best for Engineering (r= 0.91) and the Natural Sciences (r=0.70), less for the others, and least for the Social Sciences (r=0.26) and Agriculture (r=0.23). Overall, the direction of fit is consistently linear: the stronger the international competition among universities, the larger is the declared PC expenditures; this trend is in evidence across all disciplines, but less so for social science and for agricultural research institutes.

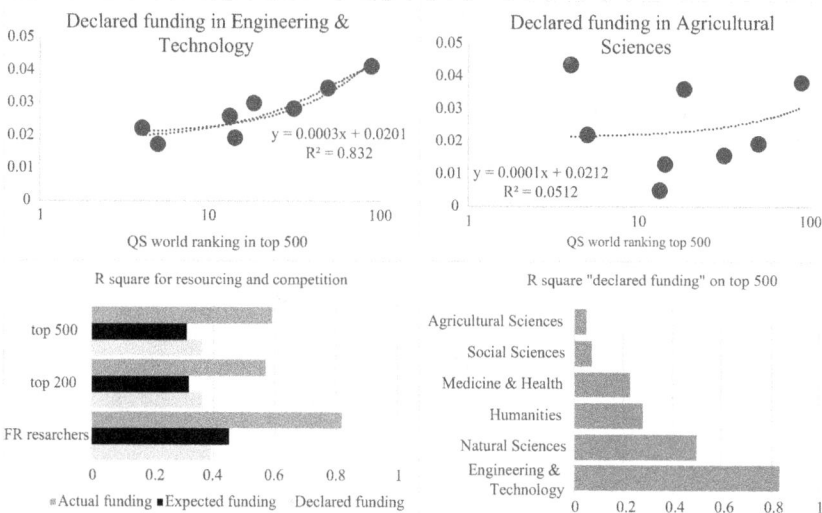

Figure 6.4 The top two graphics show the line-up of resources across countries for the disciplines at the extremes of the scale: the correlation between declared expenditure for 'engineering' and 'agriculture'. The bottom graph provides a summary overview of these alignments. We considered three indicators of resourcing (actual, declared, and future). Their alignment is the strongest with the density of the research population across the eight countries. The alignment is less strong when competition is indicated by university rankings. For the actual expenditure, this alignment is the strongest across all competition indicators.

Across disciplines, the alignment – best fitting with 'declared funding' – of communication effort and level of competition varies considerably (Figure 6.4). The curve fit best for Engineering with 83% variance explained, to lesser extent for Agricultural Sciences with only 5%. Let us also recall the above observation: expenditure for communication is overestimated among poorer units and underestimated among richer institutes.

H3: Communications staff is aware of the competitiveness among research institutes

Thirdly, we consider the awareness of competition. We did not directly ask officers/managers of research institutes whether they were in a competitive arms race. For this purpose, we need indirect evidence of entanglement. We asked institutes to identify the two most important rationales for engaging in public communication, which we grouped as PE Rationales and PR rationales, and from which we defined the PR/PE ratio index, for each country and each discipline.

Overall, 77% of research units report PE motives, and 52% also report PR motives as most important. This suggests a PR/PE ratio of 0.62, meaning that in 2018, PE motives remain more important than PR motives. In terms of avowed motives, competitiveness remains in the background and not foreground of communication managers' rationales.

Figure 6.5 points to these rationales in aggregate across countries and disciplines. There is little variation for PE motives, but more variation for espoused PR motives. About 74% of Dutch research institutes report PE rationales and 70% report PR rationales; the Dutch PR/PE index is, therefore, the highest at 0.95; PE and PR motives are more balanced, but PE still trumps PR. The countries order on a sliding scale, with Portugal being the leader in PE motives (84%), and a bare 30% PR motives; the competitiveness index is, therefore, the lowest with 0.36. A similarly sliding scale we find for disciplines: awareness of competitiveness is higher among the Engineers and lower among the Social Sciences and Humanities (index 82 vs. 61 or 57); the other disciplines are mid-range. It seems that for these two contexts, country and disciplines, PE remains dominant and PR does not trump PE rationales.

Looking at a possible alignment between actual competition and perceived competitiveness across countries, we find a potentially non-linear relation: the fitted curve is slightly inverted u-shaped ($R^2 = 0.23$). This suggests that countries at the lower end of international competition are also less aware of this competitive context and remain focused on the PE efforts, the same seems true of countries at the higher end of competition. It seems that the middle ground shows a more acute sense of competitiveness between 'blessed ignorance', on the one hand, and 'smug oblivion',

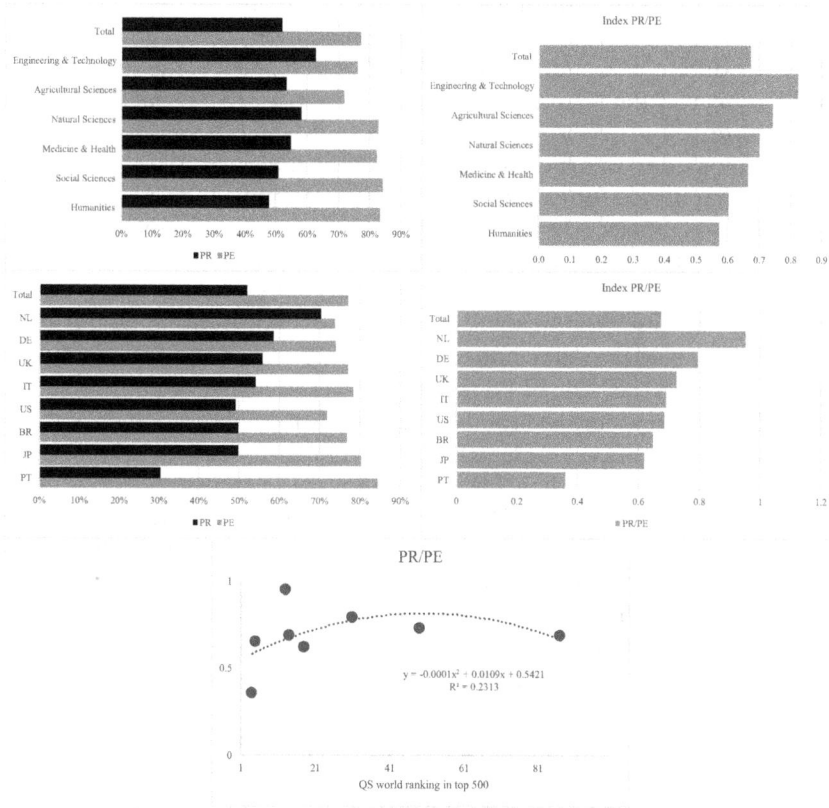

Figure 6.5 Awareness of competition in research units expressed by the rationale for communication PR rationales (funding, support, attention, recruitment) in relation to PE rationales (policy compliance, dissemination, involvement). The PR/PE index indicates the balance between these two: R > 1 means 'PR dominates'; R < 1 means 'PE dominates' the communication activities of units. Note, percentages add to more than 100% because of multiple mentions.

on the other hand. This could be another hypothesis that deserves further examination.

A potential arms race for public communication of science

In this chapter, we provided evidence of an arms race; the data are, indeed, consistent with such a hypothetical prediction. Research institutes are entangled in national and international competition for public reputation, and this affords increasing re-allocation of funding for public communication activities. This trend is demonstrated cross-sectionally in the linear alignment of average resources and levels of competition, across countries and across six different scientific disciplines. This alignment is the strongest for Engineering and the Natural Sciences and less so for the Social Sciences or Agriculture.

We demonstrate this alignment for different indicators of 'expenditure': actual, declared, and future funding of public communication efforts. Future expenditure is expected to increase for all contexts, which indicates further expansion is likely in the coming years. We also observed, when comparing declared and actual expenditure, that smaller research units overestimate and larger units underestimate their actual communication commitments.

Finally, we observe that communication staff are aware of this competitive context. But the logic of PE still dominates over that of public relation (PR) when we consider the espoused rationales for public communication. Our data line up different research institutes in cross-sectional comparison and provides compelling benchmark data for the alignment of public communication and competition. However, this does not yet suffice to make the full case for the ARPC hypothesis. For that we will need longitudinal data to verify, in each unit, the escalation of PC expenditure in the coming years, and this escalation needs to be aligned with the level of competition. Observation of this arms race should also distinguish more carefully between different races: e.g. between central and decentral communication within the university; between research units in any one country; and between research institutes and universities in the global marketplace for ideas, impact, and staff. What is the relative contribution of any of these types of competition to the expanding secondary task, the professional field of public communication of science?

An arms race for public communication will create parties who are interested in maintaining, expanding, and supporting this secondary task, which is consistent with developing a professionalism that is full of good intentions to do a 'better job'. The boom in science communication in recent years, as observed by the 'medialisation hypothesis', is already consistent with this professionalisation of the field. An increased focus on attention seeking away from truth orientation, which is often deplored in polemics about 'declining quality' of science communication or increasing mis- and disinformation about science, is a likely consequence of this development.

Our present diagnosis of an arms race, thus, specifies another risk to be included in the medialisation hypothesis: the reallocation of resources to 'non-essential' activities potentially 'starves' the core activity of research, leading to a Red Queen effect of intensifying efforts without progress, and in extremis, ending in a Peacock Effect, offering marvellous public communication but little substance of research. Let us end with a wild speculation in order to further elaborate such potential trends.

A concluding speculation: the dawning 'Baroque of science communication'

A brief glance into the history of art might allow us to appreciate what is happening with slightly different eyes and ears. A more positive note on the arms race arises from an historical analogy with an earlier arms race which created the 'Baroque style' of artistic expression. In other words, good things might arise from an arms race in communicating for 'better science', in the same way as the Baroque style emerged from the competition over the 'true religion' in the 17th century.

For modernist, ornament is a crime

The history of art recognises the 'Modernist Movement' at the turn of the 19th into the 20th century – with Vienna as one of its centres – for which 'ornamental art' is a 'crime against humanity' because it wastes effort and strains limited human resources and, thus, endangers health. This stripped-down aesthetics is morally loaded with the cultivation of the ascetic virtues of simplicity, precision, and clarity, leading to the avoidance or an outright ban of everything unnecessary or unessential, i.e. all ornamentation or embellishment of artefacts. This moral norm eschews the tastes of the wider public as Barbarity and proceeds unashamedly on the elitist path: only the selected few can fully achieve and appreciate this ascetic art form. This attitude was consequential for music (12-tone music), literature (Ulysses), design (art nouveau), and philosophy (Neo-Positivism), culminating in polemics against ornament and the misuse of language in 'nonsensical metaphysics' (Pfafigan, 1985) and contempt for the masses (Sammut & Bauer, 2021; Sloterdijk, 2000). We can recognise a similar attitude also among scientists or scientific institutions for whom any public communication of science is a step too far into the barbarism of the masses; though this attitude has become rare and less pronounced over the past 30 years.

An opposite attitude to art is the earlier Baroque lifestyle which exalted ornamentations in an expansive festival of celebrations. Until the early 20th century, 'baroque' was a term of insult; its meaning pejorative and referring to artistic work that was supposed to be confused, weird, unnatural, dissonant, forced, unnecessarily difficult, grotesque, corrupt, and overloaded with pomp and splendour, and only an inferior manifestation of art. Only, the 20th century brought the re-appreciation of Baroque as a style sui generis that combines general artistic features in a recurrent manner which we must appreciate as a 'way of being and seeing' rather than as an aberration of taste from an arbitrary standard (Woelfflin, 1932). In this new light, 'baroque' came to be defined by recurrent features, which we can even recognise in the 20th-century Jazz, such as the baseline over a progression of chords (basso continuo), the big band (concerto grosso), above all the cultivation of free improvisation over a given theme and a base line (basso continuo) and the Musical genre such as 'West Side Story' or 'Hair' as a large theatrical event involving music and acting (opera).

Recognising the creative achievement of Baroque ornamental art

The origin of Baroque is commonly dated in the period of 1600–1750. This is also the period of competition between the religious 'Reform' and 'Counter-Reform' in Europe and on the Global Stage for maritime expansion (Vaubel, 2005). In this context, artefacts were designed that exalted multi-sensory experiences involving large-scale visual, sound, and movement events. It extended to architecture, painting, sculpture, literature, and music and brought about new formats, including large installations (e.g. large paintings, churches, opera, and concerto). Many new developments emerged, for example, in music: the combination of polyphony with harmony (counterpoint, basso continuo), a rhetoric of codified thematic expressions (the Suite) and a catalogue of ornamentations. This historical arms race pitched musical composers of the Catholic European South like Monteverdi and Vivaldi against the Reformed North of Bach, Schuetz, or Buxtehude.

Baroque ecstatic exuberance cultivated creative communication of seemingly exaggerated and functionless manifestations (similar to biological exaptation as in the Peacock's feathers); brought flourishing artistic creativity to places where there was hitherto little (e.g. New World Baroque) and created a new tonal order, the well-tempered system of 12 minor and major scales, which is considered an early example of cultural rationalisation. Some recurrent features of Baroque chime with current concerns of science and science communication. One could start with Law's (2011) recognition of baroque features in the 'performative turn' of science. He calls for science studies to recover the 'rationality in flesh and body' (i.e. embodiment of cognition and emotions) including the features of theatricality, undoing of boundaries, heterogeneity, folding many in one, movement and self-consciousness, and apprehending otherness also in modern science.

In other words, the lesson of the historic Baroque might be the following: the ecstatic exuberance of Baroque expression is a creative achievement arising from a competitive arms race (of religion and theology). This race brought forth new formats of communication (Baroque art), a way of seeing the world with an aura of celebration and festivity, and all this in the service of a higher power (represented by the Church or the 'absolutist' King). This is far from misplaced investment and wasteful effort because 'Baroque' is the innovation. This might find an analogy in performative science and the current flourishment of science communication and its buzzling proliferation of formats and events of different size and shape. Once more, some of this event making might be considered superfluous and wasteful investment by some. However, positive consequences might arise when access to science is broadened by these formats; yet, there are risks along the way.

Baroque concerns about the misuse of creative communication

Even on these risks, there might be a useful analogy in the discussions of Baroque music at the time. Similar to present day discussions about science communication, we find voiced concerns about the misuse of the art and the appeal to authorities to prevent such aberrations.

Three types of misuses were identified in the critical commentary on Baroque music at the time: the overuse of dissonances (against the natural order of proportions), music playing in inappropriate situations (against beer happiness of music; 'devilish noise ... for pleasure rather than edification'), and poor-quality performances (Eggebrecht, 1996, 261ff). For all three, the authorities are called upon to police the performances and censor improper conduct. Similar appeals can be heard about present-day science communication when misinformation, poor quality, sensationalism and polarisation, particularly in new media and social media networks, are invoked. We are witnessing, at the same time, calls for policing these abuses on the basis of what seem arbitrary norms (fact checking, closure of websites). Maybe also the concern of the 17th century that 'music should praise the greater glory *of God*', echoes in current mantras that '*science must contribute to economic growth*'. However, to expect that censorship and regulation might put this right is most probably as misplaced in the 21st century as it was in the 17th century. More likely, it is the underlying arms race that drives the escalating performances. The final lesson of this analogy might be: If we want to stop the misuse of communication and the proliferation of 'wasteful Baroque science communication', we'd better focus on de-escalating this arms race.

Note

1 A recent request to the NZZ authors of that ranking to share the raw data was not successful; apparently no systematic data record is kept of the annual rankings and previous years' data seem to be lost [personal communication, 2nd Nov 2021].
2 Personal communication in 2018 with a neighbour of the first author, who works as a film producer in the British film industry.

References

Bauer, M. W., & Jensen, P. (2011). The mobilization of scientists for public engagement. *Public Understanding of Science*, 20(1), 3–11.
Dawkins, R., & Krebs, J. R. (1979). Arms races within and between species. *Proceedings of the Royal Society of London Series B: Biological Sciences*, 205(1161), 489–511.

Eggebrecht, H. H. (1996). *Musik im Abendland – Prozesse und Stationen vom Mittelalter bis zur Gegenwart*, Hamburg: Piper Verlag [2nd revised edition; chapter on Baroque, pp. 315–470].

Entradas, M., & Bauer, M. W. (2018). Die Kommunikationsfunktion im Mehrebenensystem Hochschule [the multi-level communication function of universities – trends and tensions]. In M. Schaefer et al. (Eds.), *Forschungfeld Hochschulkommunikation*. Wiesbaden: Springer Verlag, pp. 97–122 [original in German].

Entradas, M., Bauer, M.W. (2019). Bustling public communication by astronomers around the world driven by personal and contextual factors. *Nature Astronomy* 3, 183–187. https://doi.org/10.1038/s41550-018-0633-7

Entradas, M., Bauer, M. W., O'Muirchearteigh, C. et al. (2020). Public communication by research institutes compared across countries and sciences: Building capacity for engagement of competing for visibility. *Plos ONE*, 15(7). https://doi.org/10.1371/journal.pone.0235191

Hemsley-Brown, J., & Oplatka, I. (2006). Universities in a competitive global marketplace: A systematic review of the literature on higher education marketing. *International Journal of Public Sector Management*, 19(4), 316–338. https://doi.org/10.1108/09513550610669176

Koso, A. (2021). The press club as indicator of science medialization: How Japanese research organizations adapt to domestic media conventions. *Public Understanding of Science*, 30(2), 139–152. https://doi.org/10.1177/0963662520972269

Krücken, G., & Meier, F. (2006). Turning the university into an organizational actor. In G. S. Drori, J. W. Meyer, & H. Hwang (Eds.), *Globalization and Organization* (pp. 241–257). Oxford: Oxford University Press.

Kucharski, A. (2020). *The rules of contagion – why things spread and why they stop*. London: Profile books.

Law, J. (2011). Assembling the Baroque. *CRESC working paper series no 109*.

Mandler, P. (2020). *The expansion of universities and the crisis of meritocracy*. Oxford: OUP.

Matheson, A. (2008). Corporate science and the husbandry of scientific and medical knowledge by the pharmaceutical industry. *BioSocieties*, 3, 355–382.

Metag, J., & Schäfer, M. S. (2019). Hochschulkommunikation in Online-Medien und Social Media. In *Forschungsfeld Hochschulkommunikation* (pp. 363–391). Wiesbaden: Springer VS.

Noam, E. M. (1996). Electronics and the dim future of the university. *Bulletin of the American Society for Information Science*, Jun/Jul, 6–9 [reprint from *Science*, 1995, 270, 247–249].

Norris, P., Herxheimer, A., Lexchin, J., & Mansfield, P. (2005). *Drug promotion – what we know, what we have yet to learn*. Netherlands: WHO & Health Action International.

NZZ (2018). Diese Oekonomen praegen die Debatte. *Fucus der Wirtschaft*, 1 September, p. 29.

Pansegrau, P., & Bauer, M. W. (2019). The intensity of media attention as an index of authority of science. In M. W. Bauer, P. Pansegrau & R. Shukla (Eds.), *The cultural authority of science – Comparing across Europe, Asia, Africa and the Americas* (Vol. 40, pp. 86–104). New York: Routledge Studies of Science, Technology and Society.

Peters, H. P. (2012). Scientific sources and the mass media: Forms and consequences of medialization. In S. Rödder, M. Franzen & P. Weingart (Eds.), *The sciences' media connection – public communication and its repercussions*, Sociology of the Sciences Yearbook (Vol. 28, pp. 217–239). Dortrecht: Springer.

Pfafigan, A. (1985). *Ornament und Askese - im Zeitgeist des Wien der Jahrhundertwende*. Vienna: C Brandstaetter Verlag.

Phillips, D. P., Kanter, E. J., Bednarczyk, B., & Tastad, P. I. (1991). Importance of lay press in the transmission of medical knowledge to the scientific community. *The New England Journal of Medicine*, 325(16), 1180–1183.

Sammut, G., & Bauer, M. W. (2021). *The psychology of social influence – modes and modalities of shifting common sense*. Cambridge: CUP [see chapter 2 on 'crowding', pp. 27–50].

Sloterdijk, P. (2000). *Die Verachtung der Massen – Versuch ueber die Kulturkaempfe in der modernen Gesellschaft*. Frankfurt: Suhrkamp.

Vaubel, R. (2005). The role of competition in the rise of Baroque and Renaissance music. *Journal of Cultural Economics*, 29, 277–297.

Vogler, D., & Schäfer, M. S. (2020). Growing influence of university PR on science news coverage? A longitudinal automated content analysis of university media releases and newspaper coverage in Switzerland, 2003–2017. *International Journal of Communication*, *14*, 22.

Weingart, P. (2012). The lure of mass media and its repercussions on science. In S. Roedder, M. Franzen & P. Weingart (Eds.), *The Sciences' media connection – public communication and its repercussions*, Sociology of Sciences Yearbook (Vol. 28, pp. 17–33). Dortrecht: Springer.

Wölfflin, H. (1932). *Principles of art history*. New York: Henry Holt & Co.

Chapter 7

Public Engagement Profiles and Types of Research Institutes

Giuseppe Pellegrini and Barbara Saracino

Introduction

In recent years, governments and civil society organizations have been exerting considerable pressure on research institutions, which include universities and research institutes, to recognize the significant contributions they make to society (Urdari et al., 2017). This trend has also been partially responsible for causing research institutions to focus on developing so-called third mission communication and public engagement (PE) initiatives. The 'third mission' has almost always been interpreted as a technology transfer process in the quadruple helix framework, describing innovation as the result of interactions among stakeholders from different helixes (e.g., universities, industries, governments) and the environment, and/or of collaborations developed with the industrial sector (Miller et al., 2018). Alongside research and education, the third mission is a priority to which universities are particularly committed. Analyses of studies about the third mission show that it is difficult to find a more comprehensive vision which also involves aspects of public communication and engagement (Compagnucci & Spigarelli, 2020). Such initiatives generally aim to make research known and try to bring students and the public closer to the world of tertiary education. Our careful analysis of public communication and engagement activities highlights the various reasons behind research institutions' strategies and activities to foster dialog with various members of society in order to create a favorable environment for collaboration (Vakkuri, 2004).

Transmitting knowledge to be transformed into practical application is a process that has been widely developed since the 1980s (Etzkowitz, 1998, 2001) and constitutes another important objective of public communication. Universities, particularly in the United States, have taken up an entrepreneurial role by playing an important part in economic markets. To meet the needs of society and achieve production goals that involve the business world, research institutions often want to tear down the walls that turn them into ivory towers (Kapetaniou & Lee, 2017). However, sometimes, this approach hides their real intentions, which are to attract funding and increase

DOI: 10.4324/9781003027133-9

enrolment – a type of academic marketing – rather than to disseminate knowledge and generate involvement (Entradas et al., 2020). The means of engagement and the style of stakeholder involvement reveal universities' perspectives to varying degrees. Developing forms of engagement borrowed from marketing, for example, reveals a market-driven propensity to trade knowledge and attract resources in various ways. This vision of the commercialization of knowledge is also evident in the search for strategic partnerships and in the influence universities exercise in the local areas where they operate (Kotosz et al., 2016).

In other respects, communication aimed at the cultural and social growth of a local area favors a vision linked to community development, social inclusion and innovation (Laursen & Salter, 2004; Backs et al., 2019). Alongside this perspective, a strictly institutional type of communication and involvement is also present, in which universities have a duty to respond to government mandates by returning what they have received to society, thereby contributing to the development of their surrounding areas (Mora et al., 2015).

Studies on how academics communicate research highlight the difficulty of using appropriate language with stakeholders who do not operate within the university. The use of overly technical language and methods of dissemination, which are often focused on research projects entirely conceived within the scientific community, do not allow for effective encounters with nonacademic society (De La Torre et al., 2018). For this reason, on various occasions, effective meetings between scientists and other members of society are not conducted successfully.

Despite these difficulties, a process of reflection has developed, which has led the world of research to question the reasons behind the need to engage with nonexperts. Training courses, master's degrees and communication initiatives have been undertaken thanks to this process of reflection in order to foster greater awareness and develop communication skills (Giuri et al., 2019). These efforts have attempted to favor diverse methods of communication with various stakeholders and effectively involve them in the conception, activation and dissemination of research.

In this chapter, we describe research institutes' types based on their propensity for public communication and engagement. We identified said propensity by applying two combined analysis techniques to the entire dataset of the MORE-PE project: reducing some variables by constructing indices and reducing cases through a cluster analysis. While conducting this process, we paid careful attention to the quality of the data collected, to defining the aggregations of variables and to the possible systemic effects of the indices by using a highly transparent procedure (Hicks et al., 2015). The aim of the chapter is to define homogeneous groups based on the answers provided to our questionnaire and construct a typology using an exploratory data analysis approach (Tukey, 1977; Di Franco, 2015).

Methods

In this chapter, we refer to three levels of intellectual tools: concepts and conceptual structures, statements and explanations (Marradi, 2007). Conceptual structures encompass classifications, taxonomies and typologies. A typology is based on two or more *fundamenta divisionis* used simultaneously to identify a set of types, i.e., a set of groups characterized by n-dimensions. To obtain a typology, we used a Two-Step Cluster Analysis, considering nine active variables and six illustrative variables.

The many techniques defined as cluster analysis constitute a broad and varied whole. They all share the objective of assigning single cases to a limited number of groups, minimizing the heterogeneity among cases within the groups as much as possible and maximizing the heterogeneity among the different groups. It is useful to perform a cluster analysis after reducing the set of available variables with techniques such as principal component analysis or multiple correspondence analysis, or after having constructed indices. In this case, we obtained a double reduction of the dataset by first reducing the variables and then the cases.

The six illustrative variables are:

1 Country (Germany, Italy, Portugal, the Netherlands, the United Kingdom, the United States, Brazil, Japan)
2 Communications staff (Yes, No, Within the host institution (outside the research unit))
3 Communications funding (None, <1%, 1–5%, 5–10%, >10%)
4 Active researchers (None, <10%, 10–20%, 20–40%, 40–60%, 60–100%)
5 Time split between research and teaching in the research unit (Only research, More research than teaching, More teaching than research, Research and teaching equally balanced)
6 Research unit's main research area (Natural Sciences, Engineering and Technology, Medical and Health Sciences, Agricultural Sciences, Social Sciences, Humanities)

The nine active variables that we used in the Two-Step Cluster Analysis, and which contributed to the construction of the groups, are seven indices constructed from batteries of items and two single questions:

1 Public events
2 Traditional media channels
3 New media channels
4 Audiences
5 Positive views of the public
6 Negative views of the public
7 Evaluation of activities (Never or rarely, Occasionally, Frequently or always)

8 Rationales for engaging
9 Rationales for not engaging (Respond to policies, Attract, Disseminate to public, Other)

The *public events index* was constructed from a battery of 11 items (public lectures; exhibitions; open days, workshops, guided visits and similar event formats; science festivals/science fairs; National Science Week and similar formats of national events; science cafés and similar formats of public discussion events; UNESCO International Year, Fame Lab, European Researchers' Night and similar formats of international events; deliberative and participatory events on policymaking; events organized by private institutes; talks and workshops at primary/secondary schools; science projects by locals). Respondents answered each item by choosing from five alternatives (never, annually, quarterly, monthly, weekly). Therefore, the summation index is expressed in values from 11 to 55.

Another index, the *traditional media channels index*, was constructed from a battery of 13 items (newspaper interviews; radio interviews; TV interviews; other TV; press conferences; press releases; newsletters; brochures, leaflets, publications for the nonspecialist public; articles in magazines and newspapers for the nonspecialist public; multimedia, videos, films, podcasts; popular books; policy papers, briefings on policy issues for industry, politicians, policymakers; materials for schools). Respondents answered each item by choosing from five alternatives (never, annually, quarterly, monthly, weekly). Thus, the summation index values are from 13 to 65.

The *new media channels index* was constructed from a battery of eight items (website updates; blogs; Facebook; Twitter; Google+; Instagram; YouTube; podcasts). Respondents answered each item by choosing among five alternatives (never, quarterly, monthly, weekly, daily). The summation index is expressed in values from 8 to 40.

The *public perceptions index* was constructed from a battery of eight items (general public; schools; students outside teaching; members of local municipalities, councils, associations; industry delegates; governments, politicians, policymakers; nongovernmental organizations; media and journalists). Also, in this case, respondents answered each item by choosing from five alternatives (never, rarely, occasionally, frequently, very frequently), and the summation index values are 8–40.

The *positive views of the audience* and *negative views of the audience* indices were constructed from two batteries concerning public opinion. For each item, respondents chose from five alternatives (strongly disagree, disagree, neither agree nor disagree, agree, strongly agree). For the first index, the scores on seven positive items were summed and the values vary between 7 and 35. For the second index, the scores on six negative items were summed and the values vary between 6 and 30.

The *rationales for not engaging index* were constructed from a battery of nine items on researchers' reasons for not engaging in public communication activities. Respondents answered each item by choosing from four alternatives (definitely not true, unlikely true, likely true, very likely true). The summation index values are from 9 to 36.

For the Two-Step Cluster Analysis, seven continuous variables were obtained by summing the scores of the items and were recoded as seven ordinal variables with three equal categories (low, medium and high) using percentiles (Table 7.1).

In the results section, we will discuss the cluster analysis to explain the different characteristics of the groups identified via the statistical analysis. We will propose interpretations of empirical results through the use of the main theoretical references concerning the public communication of research and studies on the role of academic institutes.

Results

The Two-Step Cluster Analysis enabled us to identify three groups of countries from among those that participated in the More-PE survey, whose particular characteristics explain the different strategies their institutes have employed in recent years to communicate their research. This threefold division was made possible thanks to the use of the nine active variables described above, which allowed for an adequate reduction of complexity so as to carefully verify the differences among all the institutes involved.

Table 7.1 Index construction: reliability statistics and scale statistics

	Cronbach's Alpha	No. of items	Mean	Std. deviation	Minimum	Maximum	Percentiles 33.3	66.6	N
Public events	0.814	11	21.63	6.49	11	47	19	24	2030
Traditional media channels	0.885	13	26.19	9.01	13	61	22	29	2030
New media channels	0.797	8	14.49	5.58	8	40	11	16	2030
Audience	0.820	8	17.44	3.57	8	24	16	19	2030
Positive views of the public	0.820	7	19.68	5.94	7	35	19	23	2008
Negative views of the public	0.768	6	13.53	4.48	6	30	12	16	2008
Rationales for not engaging	0.881	9	19.68	6.83	9	36	17	24	2030

In IBM SPSS Statistics©, the Two-Step Cluster Analysis procedure is an exploratory tool designed to reveal natural groupings (or clusters), which would, otherwise, not be apparent, within a dataset. The algorithm employed by this procedure has several desirable features that differentiate it from traditional clustering techniques (Chiu et al., 2001; Bacher et al., 2004; Gelbard et al., 2007). Two-Step Cluster Analysis identifies groupings by running preclustering first and then by running hierarchical methods. In our analysis, we selected log-likelihood as the distance measure and had the number of groups determined automatically.

The selection of the Distance Measure determines how the similarity between two clusters is computed. The likelihood measure places a probability distribution on the variables. Categorical variables – which we employed – are assumed to be multinomial and independent. The procedure automatically determines the 'best' number of clusters by using the criterion specified in the Clustering Criterion group. In our case, the procedure left us with three groups, using both the Bayesian Information Criterion and the Akaike Information Criterion.

The following table and graphs show the differences among the three groups that emerged: both the variables that contributed to the construction of the clusters and the illustrative variables. The central cells in Table 7.2 show only the categories of the variables with the highest percentage for each group.

Taking up the reasons that favor public engagement, we can define the first group of institutes as the **diffusionist** group: the set of research units that wants to facilitate a significant two-way exchange of information with society to promote scientific and cultural contamination. These institutes not only seek to communicate the results of their work by enhancing the knowledge produced but also to enter into relations with various members of social through activities, initiatives and channels that go beyond the traditional circuits of the scientific community.

Analyzing the data, it emerges that the diffusionist group is very present among the Italian, Brazilian and Dutch institutes. These organizations use traditional and new media channels to a large extent to dialog, transfer knowledge and develop contacts with various areas of society. They have a more positive view of the public and also assess the effectiveness of their communication activities. They have a good budget for engagement activities, and more than 40% of their researchers participate in public engagement activities. In terms of core functions, these organizations carry out both research and teaching activities.

The strategies and actions adopted by these research institutes are oriented to build and maintain over time a capital of knowledge that is produced and maintained thanks to a favorable ecosystem (Frondizi et al., 2019). These institutes seek to promote and develop two-way communication initiatives to contribute to developing the knowledge society in which they are a key player.

In this perspective, research institutes, and universities in particular, have changed the traditional way in which they fit into the innovation process being

Table 7.2 Research institutes and three different communication strategies

| Active variables | Strategies | | | |
	Group 1: diffusionist	Group 2: institutional	Group 3: market-oriented	
Public events	High	Low	Medium	Chi-square=1167.144; Sig.=0.000
Traditional media channels	High	Low	Medium	Chi-square=1413.341; Sig.=0.000
New media channels	High	Low	Medium	Chi-square=736.679; Sig.=0.000
Audience	High	Low	Medium	Chi-square=1000.985; Sig.=0.000
Positive views of the public	High	Low	Medium	Chi-square=252.292; Sig.=0.000
Negative views of the public	Low	High	Medium	Chi-square=131.228; Sig.=0.000
Evaluation of activities	Frequently or always	Never or rarely	Occasionally	Chi-square=322.928; Sig.=0.000
Rationales for engaging	Disseminate to public	Respond to policies	Attract	Chi-square=278.284; Sig.=0.000
Rationales for not engaging	Medium	Low	High	Chi-square=253.300; Sig.=0.000

Strategies

	Group 1: diffusionist	Group 2: institutional	Group 3: market-oriented	
Illustrative variables				
Countries	Italy, Brazil, the Netherlands	Germany, Portugal, Japan	The United Kingdom, the United States	Chi-square=184.681; Sig.=0.000
Communications staff	Yes	No	Within the host institution (outside the research unit)	Chi-square=238.234; Sig.=0.000
Communications funding	>5%	None	1–5%	Chi-square=196.408; Sig.=0.000
Active researchers	>40%	<10%	10–40%	Chi-square=203.692; Sig.=0.000
Time split between research and teaching	Research and teaching are equally balanced	More teaching than research	More research than teaching	Chi-square=16.995; Sig.=0.009
Research area	Natural Sciences, Social Sciences, Agricultural Sciences	Medical and Health Sciences, Humanities, Engineering and Technology	Natural Sciences, Social Sciences	Chi-square=9.526; Sig.=0.483
%	**45.3**	**18.3**	**36.4**	

the subject of the research and second focusing on intellectual property or translating intellectual capital practice inside universities (Esposito et al., 2013; Vagnoni and Oppi, 2015). They have changed their role by placing themselves at the so-called fourth stage of Intellectual Capital theory: the development of exchange between institutes, businesses and civil society for the development of knowledge areas spread at local, regional and national level. For this type of action, communication becomes an important strategic asset, which explains the use of adequate resources we found in our data, the development of new activities and the open-up of listening spaces and dialogue with civil society.

Even clearer as an explanatory line of the trends presented is the model of the five helices. This model proposes the articulation of relations for research and innovation among governments, universities, businesses, civil society in a communicative context in which the media are increasingly relevant. In this way, it is even clearer how universities and research institutes need to position themselves to manage the ecological complexity of new ways of presenting and sharing research (Carayannis et al., 2012) (Figure 7.1).

The second group of institutes can be defined as **Institutional**, as they are strongly focused on a cognitive-academic mission that emphasizes the production of knowledge, technological advancement and the relationship with some specific economic sectors. Given that university departments and non-academic research institutes were included in our sample, this type of institute could also be defined as post-Humboldtian, as they are centered not only on scientific production but also on the context in which they operate, albeit with particular attention paid to productivity.

The research institutes in this group operate mainly in Germany, Portugal and Japan. Their commitment to communicating via both traditional and social media is below average, and they tend not to have a positive view of the public. These organizations are mainly driven by an institutional duty and very often by a government obligation that imposes certain standard methods of communication upon them. These institutes do not have staff dedicated to communication, nor a communication budget, and few of their researchers are engaged in public engagement activities. These research institutes mainly carry out teaching activities.

This type of orientation is largely recognized as the strategy of directing mass education toward the country's vocational training system and a relevant role of the State to shape national regulation on research and innovation policies. In this typology, institutes that enjoy good autonomy are recognized, even if they are largely financed from public resources. Of course, these institutes, like the other presenting types, have also been influenced by the third mission process, but they have probably not interpreted it in full and, in any case, have activated few channels for listening to the components of civil society. Recalling a well-known quotation from Wilhelm Von Humboldt, these institutes consider that, once their mission has been fulfilled, it should nevertheless be regarded as a public function of general interest and, therefore, without a strong need to develop more elaborate communication

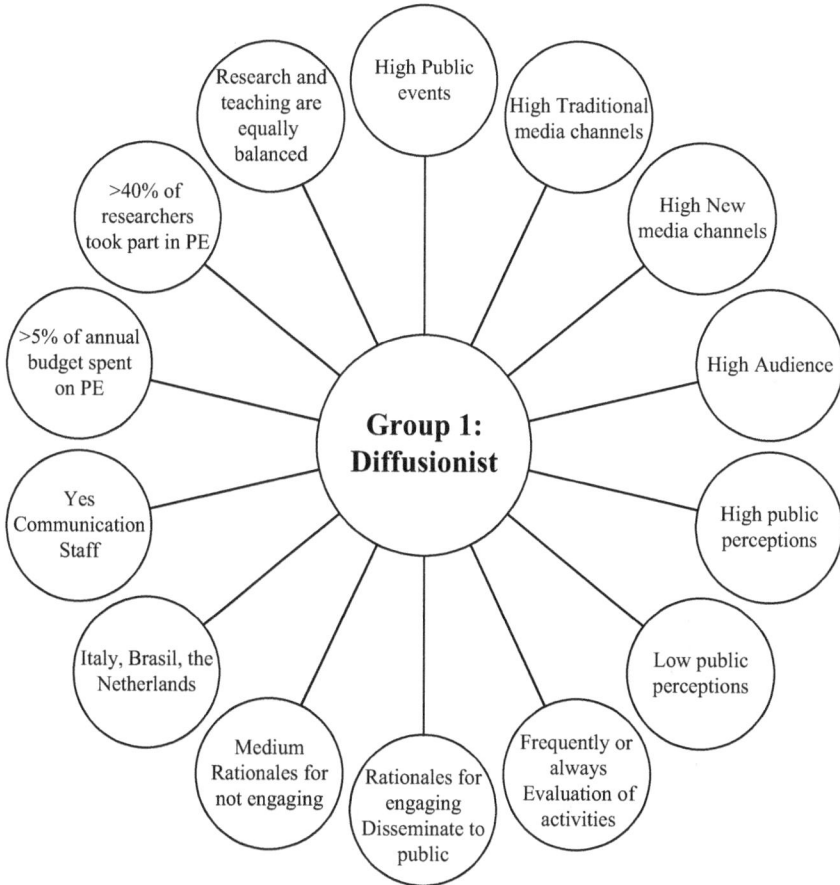

Figure 7.1 Diffusionist institutes main features.

strategies (Von Humboldt, 1854). In short, it is, therefore, possible to argue that these institutes, with their autonomy and educational orientation, to a large extent play a institutional role not only from the point of view of perfect research and teaching but also from a perspective of effective contribution to the public capital of knowledge (Frondizi et al., 2019) (Figure 7.2).

The third group of institutes can be defined as **market-oriented,** as they very often aim to develop activities that generate revenue and the sort of academic capitalism that has been described by various scholars since the 1990s (Slaughter and Leslie, 1997; Hoffman, 2011). These organizations often enjoy extensive private funding and, at times, have raised controversies on the issues of control and accountability in university–industry relations (Kirp, 2003, 208). Institutes of this type are found mainly in the United States and the United Kingdom. On average, they use both traditional and new channels to

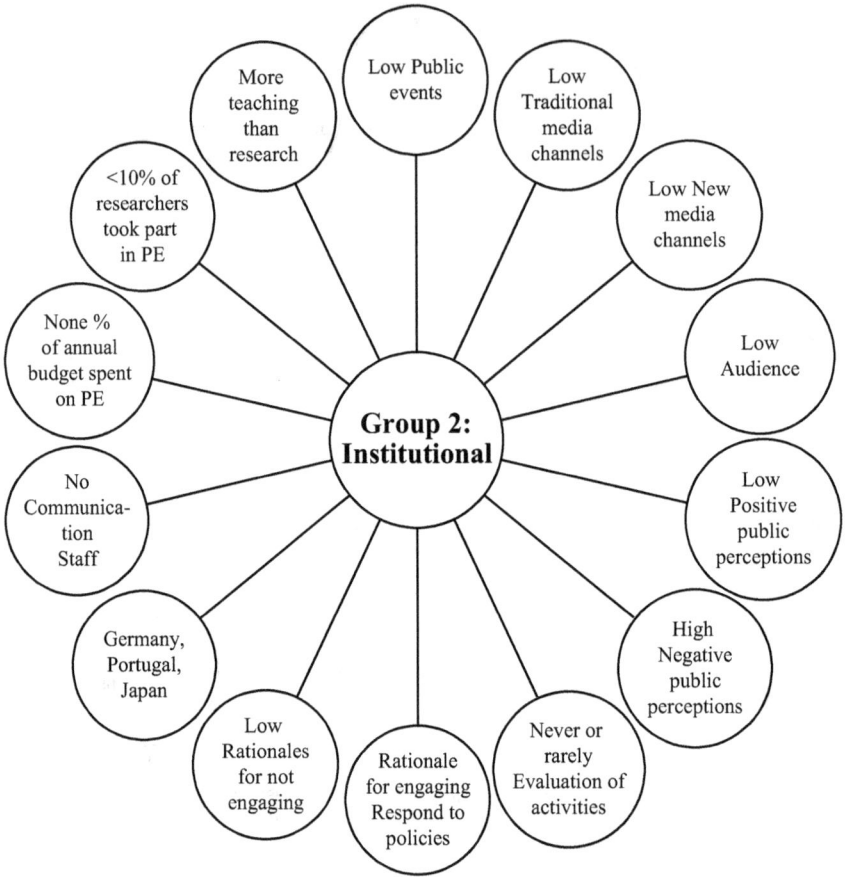

Figure 7.2 Institutional group institutes main features.

communicate their research work, and their communication strategy is mainly aimed at attracting funds. Between 1% and 5% of their budget is available for communication activities, and up to 10% of their researchers take part in public engagement activities. These institutes carry out mainly research activities.

It must be considered that, historically, in recent years in the United Kingdom higher education has moved quickly from near full public funding to a mixed publicly/privately funded tuition market. At the same time, costs for students have risen, particularly for full-time students (Marginson, 2018). As in the United States, the rules of institutional autonomy and academic freedom are deeply rooted in the United Kingdom. Furthermore, university autonomy is understood in terms of the university model as a company in its own right, led by a strategically minded president and executive, aware

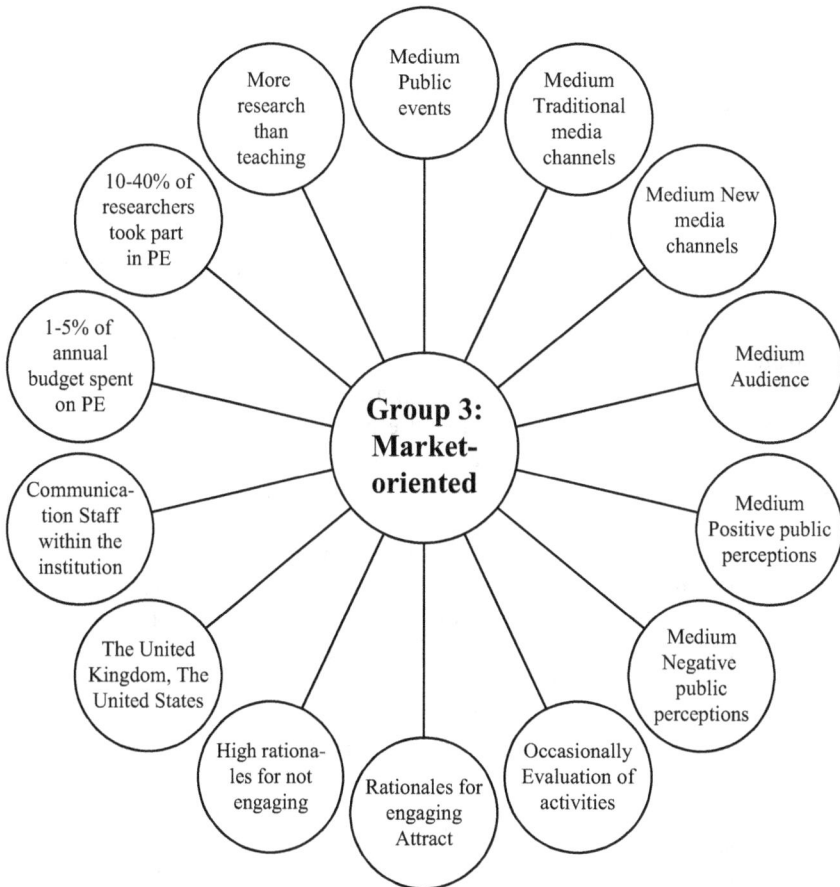

Figure 7.3 Market-oriented institutes main features.

of the need for revenue and prestige. In various ways, in fundraising and organizing activities, British and American universities operate with great financial autonomy, yet they receive public funds and compete with other higher education institutions for students, money and good academic staff.

In the light of these considerations, it is understandable that communication strategies and actions are strongly conditioned by these economic guidelines. Therefore, public communication initiatives also have the purpose of attracting resources by trying to maintain high public confidence in universities which, in the case of the United Kingdom, is very high (Marginson, 2018). The US higher education landscape has always been accompanied by a strong orientation toward the market, free competition and the autonomy of private research institutions. This means that universities have supported and still support some key market principles: self-reliance, rivalry and decentralized decision-making (Berman, 2012; Bok, 2003). In this light, the data collected confirm this trend with particular relevance for the effort to communicate to the public the research work and scientific progress achieved (Figure 7.3).

Conclusion

In this chapter, we have analyzed the characteristics of the various research institutes that participated in the MORE-PE survey to build a typology to identify three strong trends. The academic structure of the Humboldtian type – a vision of universities and research institutes as places of knowledge separate from the rest of society and deliberately little-influenced by the world of politics, and which influenced many universities and research institutes up to the second half of the twentieth century – is generally outdated. While we have witnessed research institutes all over the world, especially in Anglo-Saxon countries, entering into broader relationships with society, especially to collect resources and better orient their studies, there is still a strong propensity to preserve an academic mentality of self-referentiality and strong centralization of power. Recently, research institutes have increased their commitment to communicating their work to the public and engaging the public through various initiatives (Entradas and Bauer, 2017; Leshner, 2003; Marcinkowski et al., 2014). However, as described by Entradas and colleagues (Entradas et al., 2020), it is not clear whether a cultural change has actually occurred, or whether this effort is motivated, above all, by a need for visibility and a search for greater resources.

The typology presented in this chapter identifies these manifested ideals and approaches, which at times are not coherent. The institutes that follow a diffusionist model, for example, seem to want to exploit both the great potential of all means of communication and the strong of some legislation to facilitate institutes' third mission to enter into a closer relationship with the public by encouraging forms of dialog and participation to contribute to the development of the knowledge society. This is an important effort that creates space for social involvement and could enrich what research institutes produce with internal means and human resources. On the other hand, institutes that follow a more institutional approach try to direct communication more selectively, tempering their institutional obligations with the need to enjoy the public's favor in order to have the necessary funding for its activities. Finally, market-oriented institutes play their role and engage the public by accepting competition and innovation mechanisms typical of the business world and the economy principles of supply and demand.

These three types of institutes attempt to communicate their research to the public in order to play a significant role in society. It will be opportune to study how these types of public communication strategy develop in the near future, so we can understand which direction research institutes are taking to promote the social, economic and productive growth of their respective countries as well as to ensure democratic participation.

References

Bacher, J., Wenzig, K., & Vogler, M. (2004) SPSS twostep cluster – a first evaluation. *Univ. Erlangennürnb.*, 1, 1–20.

Backs, S., Günther, M., & Stummer, C., (2019) Stimulating academic patenting in a university ecosystem: An agent-based simulation approach. *J. Technol. Transf.*, 44 (2), 434–461. https://doi.org/10.1007/s10100-020-00716-3

Berman, E. P. (2012) *Creating the market university: How academic science became an economic engine*. Princeton, NJ: Princeton University Press.

Bok, D. (2003) *Universities in the marketplace: The commercialization of higher education*. Princeton, NJ: Princeton University Press.

Carayannis, E., Barth, T., & Campbell, D. (2012) The quintuple helix innovation model: Global warming as a challenge and driver for innovation. *J. Innov. Entrep.*, 1, 2.

Chiu, T., Fang, D., Chen, J., Wang, Y., & Jeris, C. (2001) A robust and scalable clustering algorithm for mixed type attributes in large database environment, in *Proceedings of the seventh ACM SIGKDD international conference on knowledge discovery and data mining – KDD '01*, New York, NY: ACM Press, 263–268.

Compagnucci, L., & Spigarelli, F. (2020) The Third Mission of the university: A systematic literature review on potentials and constraints. *Technological Forecasting and Social Change*, 161, 120284. https://doi.org/10.1016/j.techfore.2020.120284

De La Torre, E.M., Pérez-Esparrells, C., Casani, F., 2018. The policy approach for the Third Mission of Universities: the Spanish Case (1983 – 2018). *Regional and Sectoral Economic Studies*, 18, 13–33.

Di Franco, G. (2015) *EDS: Esplorare, Descrivere e Sintetizzare i dati*. Guida pratica all'analisi dei dati nella ricerca sociale, 3rd edition, Milano, Franco Angeli.

Entradas, M., & Bauer, M. W. (2017) Mobilisation for public engagement: Benchmarking the practices of research institutes. *Public Understanding of Science*, 26 (7), 771–788. https://doi.org/10.1177/0963662516633834

Entradas, M., Bauer, M.W., O'Muircheartaigh, C., Marcinkowski, F., Okamura, A., Pellegrini, G., et al. (2020) Public communication by research institutes compared across countries and sciences: Building capacity for engagement or competing for visibility? 15 (7): e0235191. https://doi.org/10.1371/journal.pone.0235191

Esposito, V., De Nito, E., Pezzillo Iacono, M., & Silvestri, L. (2013) Dealing with knowledge in the Italian public universities: The role of performance management systems. *J. Intellect. Cap*, 14, 431–450.

Etzkowitz, H. (1998) The norms of entrepreneurial science: Cognitive effects of the new university–industry linkages. *Res. Policy*, 27, 823–833. https://EconPapers.repec.org/RePEc:eee:respol:v:27:y:1998:i:8:p:823-833

Etzkowitz, H. (2001) The second academic revolution and the rise of entrepreneurial science. *IEEE Technol. Soc. Mag.*, 20 (2), 18–29. https://doi.10.1109/44.948843

Frondizi, R., Fantauzzi, C., Colasanti, N., & Fiorani, G. (2019) The evaluation of universities' third mission and intellectual capital: Theoretical analysis and application to Italy. *Sustainability*, 11 (12), 3455. https://doi.org/10.3390/su11123455.

Gelbard, R., Goldman, O., & Spiegler, I. (2007) Investigating diversity of clustering methods: An empirical comparison. *Data Knowl. Eng.*, 63, 155–166. https://doi.10.1016/j.datak.2007.01.002

Giuri, P., Munari, F., Scandura, A., & Toschi, L. (2019) The strategic orientation of universities in knowledge transfer activities. *Technol. Forecast. Soc. Change*, 138, 261–278. https://doi.10.1016/j.techfore.2018.09.030

Hicks, D., Wouters, P., Waltman, L. et al. (2015) Bibliometrics: The leiden manifesto for research metrics. *Nature*, 520, 429–431. https://doi:10.1038/520429a

Hoffman, S. G. (2011) The new tools of the science trade: Contested knowledge production and the conceptual vocabularies of academic capitalism. *Social Anthropology*, 19 (4), 439–462. https://doi.org/10.1111/j.1469-8676.2011.00180.x

Kapetaniou, C., & Lee, S. H. (2017) A framework for assessing the performance of universities: The case of Cyprus. *Technol. Forecast. Soc. Change*, 123, 169–180. https://doi:10.1016/j.techfore.2016.03.015

Kirp, D. L. (2003) *Shakespeare, Einstein, and the bottom line: The marketing of higher education*. Cambridge, MA: Harvard University Press.

Kotosz, B., Lukovics, M., Molnár, G., & Zuti, B. (2016) How to measure the local economic impact of universities? Methodological overview. *Reg. Stat.*, 5. https://mpra.ub.uni-muenchen.de/73725/1/MPRA_paper_73725.pdf

Laursen, K., & Salter, A. (2004) Searching low and high: What types of firms use universities as a source of innovation? *Res. Policy,* 33, 1201–1215.

Leshner, A. (2003) Public engagement with science. *Science,* 299 (5609), 977. https://doi.org/10.1126/science.299.5609.977

Marcinkowski, F., Kohring, M., Furst, S., & Friedrichsmeier, A. (2014) Organizational influence on scientists' efforts to go public: An empirical investigation. *Science Communication,* 36 (1), 56–80.

Marginson, Simon. (2018) Global trends in higher education financing: The United Kingdom. *International Journal of Educational Development, Global Trends in Higher Education Financing*, 58 (gennaio), 26–36. https://doi.org/10.1016/j.ijedudev.2017.03.008.

Marradi, A. (2007) *Metodologia delle Scienze Sociali*. Bologna: Il Mulino.

Miller, K., McAdam, R., & McAdam, M. (2018) A systematic literature review of university technology transfer from a quadruple helix perspective: Toward a research agenda. *R&D Manag.*, 48 (1), 7–24. https://doi.org/10.1111/radm.12228

Mora, J.G., Ferreira, C., Vidal, J., & Vieira, M.J. (2015) Higher education in Albania: Developing third mission activities. *Tertiary Educ. Manag.*, 21 (1), 29–40. https://doi.org/10.1080/13583883.2014.994556

Slaughter, S., & Leslie, L. (1997) *Academic capitalism: Politics, policies, and the entrepreneurial university*. Baltimore, MD: Johns Hopkins Press.

Tukey, J. (1977) *Exploratory data analysis*. Reading, PA: Addison-Wesley.

Urdari, C., Farcas, T., & Tiron Tudor, A. (2017) Assessing the legitimacy of HEIs' contributions to society: The perspective of international rankings. *Sustain. Account. Manag. Policy J.*, 8 (2), 191–215.

Vagnoni, E., & Oppi, C. (2015) Investigating factors of intellectual capital to enhance achievement of strategic goals in a university hospital setting. *J. Intellect. Cap.*, 16, 331–363.

Vakkuri, J. (2004) Institutional change of universities as a problem of evolving boundaries. *High. Educ. Policy*, 17 (3), 287–309. https://eric.ed.gov/?id=EJ752117

Von Humboldt, W. (1854) *The sphere and duties of government*. Translated from the German of Baron Wilhelm von Humboldt by James Coulthard, London: John Chapman.

Part III

National Situation and Profiles

Chapter 8

The Communication of Research in Italy

The Efforts of Academia and Research Institutes

Giuseppe Pellegrini and Barbara Saracino

Introduction

For some years, now we have witnessed mobilisation on the part of scientists and research institutions, both in Italy and in other countries, and not only in Europe, aimed at intervention in the public debate by means of initiatives relating to the dissemination of information, communication and public engagement (Bauer and Jensen, 2011). At the same time, the public at large and organisations within civil society are becoming increasingly interested and willing to acquire knowledge and intervene with regard to issues related to the sciences and technology that directly concern them.

This double movement which is identified with respect to such issues as energy and the environment, health and infrastructures highlights the fact that, sometimes, a process of clarification and mutual understanding develops but, sometimes, there may also be short circuits in communication or even a dialogue between people who are hard of hearing. It is possible to distinguish between two principal motivations that are driving research institutions to develop processes relating to the dissemination of information, communication and public engagement.

First, one should note that many researchers have acquired an increasing awareness of the social value of science and technology in the transition from academic science to a post-academic science (Ziman, 1996). The pervasiveness of research activities and the (occasionally undesired) consequences of scientific discoveries have stimulated numerous debates and controversies and, to a large extent, have contributed to the evolution of communication methods, engaging institutions, researchers, communicators and the public in a process of change. In this respect, in recent years, initiatives have been implemented to bring the world of science and technology closer to civil society through activities involving the dissemination of information, communication and an active involvement of the various segments of society (Entradas and Bauer, 2017). Second, scientific institutions have understood that they cannot avoid investing in public communication, employing adequate resources and instruments. The different types of communication

DOI: 10.4324/9781003027133-11

involve specific strategies, with respect to which the role of the experts may be more or less decisive. In general terms, however, it has to be recognised that numerous initiatives have been proposed in recent years that have attracted the attention of the public through festivals, weeks for scientific culture, "science cafés", Researchers' Nights and many other events.

This movement, moreover, engenders the need to investigate the level of development of the various forms of communication and the encounter between science, technology and the public, in particular, with regard to the communication of research carried out by scientific institutions. At the same time, it is important to consider the structural context in which the theme of communication with the public and the dissemination of results has developed in recent years.

The Italian research system is composed of universities, public research bodies (many of which are supervised by the Ministry of Research – MIUR) and the private sector. To these are added other types of subjects, public or private, which may fall within the broader definition of "other research organisations". These subjects have different nature and purposes and carry out research activities of different types: universities and institutions deal with more "fundamental" or "basic" research, while companies engage in research with more productive purposes and applications. However, they collaborate and interact with each other, in a "system" logic that MIUR itself stimulates and favours (MIUR, 2020). The Italian university sector consists of 61 state universities to which 90% of students are enrolled, 30 non-state universities, of which 11 are online universities, and 6 special schools (ANVUR, 2018). There are 96,126 professors and researchers. From a gender perspective, it is noted that in higher academic positions there are fewer women. The National Research Council (CNR) operates alongside the university sector and is the biggest research institution in Italy, employing 5,418 researchers, 46.5 of whom are women. The CNR is divided into seven departments and 91 research institutes with a turnover of 900 million euros per year (CNR, 2020).

Considering the entire sector of research and development, research personnel in Italy are mainly employed in the private sector (43.6%) in lower percentages in universities (37.6%) and in the public sector (15.6%). They are mainly focused on research activities and career development which is determined by the ability to produce scientific results to be published in the most prestigious journals of the various disciplines. This type of orientation does not obviously encourage researchers to seek a relationship with the public even if, in recent years, some institutional conditions have changed and many initiatives to meet the public have been carried out such as various scientific festivals and many initiatives linked to the European Researchers' Night (Pellegrini and Rubin, 2020). In this perspective, we can consider these different initiatives and the institutional path of many states which over the last 20 years have launched projects to promote the development of the "third mission" of universities and research institutes, committing funding in order to facilitate the process of the transfer of knowledge and the social and economic impact of the results of research (Molas-Gallart et al., 2002).

This research and innovation policy has also involved the European Union, which, since 1990, has implemented a vast research programme called Science in Society to improve the relationship between the world of research and society by initiating paths of communication, participation and studying methods of research governance. The commitment has evolved over time and has resulted in the action taken by the programme Science with and for Society (SwafS) within the context of Horizon 2020, the EU Framework Programme for Research and Innovation (Kim and Yoo, 2019).

In Italy, the third mission has recently been a subject of interest and has led to the introduction of some legislative measures aimed at developing a system for evaluating the activities of universities and research institutions. This institutional process resulted in an initial method of assessment managed by the Italian National Agency for the Evaluation of the University and Research Systems (ANVUR) in the period 2004–2010. Subsequently, an assessment trial was initiated, concluded on 30 January 2015, and the collection of data started via the platform of the Cineca consortium on 2 March 2015 (ANVUR, 2015). This process will make it possible to test the indicators and direct the system towards stabilisation in view of a periodical evaluation that will occur on an annual or biennial basis. The ANVUR introduced assessment activities, focusing on two areas of interest within the sphere of the third mission: the valorisation of research and the production of public assets of a social nature.

The first area includes the management of intellectual property (patents), the start-up of spin-offs, external (third-party) contractual activities and intermediation structures (incubators, consortia, science parks). The second area encompasses all the activities of a social, educational and cultural nature that produce public assets not directly linked to the initiation of innovative business processes. It is a very heterogeneous set of activities managed by institutions, departments and individual researchers in very different ways and with resources that are highly variable and to a large extent conditioned by the various territorial socio-economic systems of our country.

Methods

In this chapter, we present the main results of the MORE-PE study conducted by Observa Science in Society, which involved more than 300 Italian research institutes and departments. The organisation of universities in Italy provides for their sub-division into departments corresponding to the various areas of study. The University Reform Act of 2010 introduced changes in the governance of Italian universities, whereby the structures responsible for research activities in the various departments were identified and their number was significantly reduced (Legge 240, 2010).

Sample

In addition to the 900 university departments, there are approximately 300 research institutions operating in Italy. The most important organisation is the Italian National Research Council (CNR) as we introduce in the introduction. Other important state research institutes are the National Institute for Nuclear Physics (INFN) with four national laboratories and 20 sections, the National Institute for Astrophysics (INAF) with 16 research structures, the Italian National Agency for New Technologies, Energy and Sustainable Economic Development (ENEA) with 14 research centres and the National Institute of Geophysics and Volcanology (INGV) with three departments and seven sections. Other private research institutes are also active within the national territory.

In Italy, the survey was conducted between October 2017 and February 2018; during this period, the Italian university departments and research institutes were contacted and invited to compile an online questionnaire. Observa oversaw the identification of the series of institutions involved and established contact with the various departments and research organisations, creating a data bank of 1,200 institutions. The final mapping led to the presentation of the survey questionnaire at 1,120 research units (Table 8.1), subdivided according to the various disciplines on the basis of the OECD classification: natural sciences, agricultural and veterinary sciences, engineering and technology, medicine and health sciences, social sciences, humanities and the arts. The study was carried out in Italian universities and the main Italian public research centres. About 366 institutes responded to the survey. In the analyses conducted for this chapter, the sample comprised 347 cases, all of whom completed the questionnaire (see Table 8.1).

Analysis

The chapter presents the descriptive analysis of the main topics covered by the survey and the results obtained through the Categorical Principal Components Analysis (CATPCA) classification method. The CATPCA represents a set of variables in a restricted number of dimensions, with most of the information available being reproduced using optimal scaling. For the CATPCAs, we considered ten-item batteries of the questionnaire: Public events (11 items); Traditional media channels (13 items); New media channels (8 items); Supporting communication activities in house or through outsourcing (7 items); Audiences (8 items); "Active" researchers in public communication activities (9 items); Rationales for communication (5 items);

Table 8.1 Sample characteristics by research area (OECD)

Research area (OECD)	Type of institution		Total	%
	University departments	Research institutes		
Social sciences	243	22	252	22.5
Humanities and the arts	129	18	154	13.8
Medical and health sciences	183	49	227	20.3
Natural sciences	166	69	247	22.0
Engineering and technology	126	61	188	16.8
Agricultural and veterinary sciences	39	13	52	4.6
Total RI	886	232	1120	100

Statements concerning media coverage of the research conducted (5 items); Opinions regarding the public (2 batteries, 13 items). In each item battery, a CATPCA was performed with Kaiser-Varimax rotation. For each dimension obtained by the CATPAs (on every single battery), we constructed an index, and the score of the index was normalised (between 0 and 1).

For each main topic covered by the survey, we will analyse the variance of the constructed synthetic indices, and we compare the Italian research institutions, using as independent variables the main research area of the institutes. In this chapter, we only show the relevant CATPCAs and the statistically significant differences between research institutions in terms of OECD research areas.

Public communication activities

Public events

Italian university departments and research institutes are engaged in various public communication activities with varying levels of frequency during the course of the year. Analysing the data, it emerges that at least once over the last year more than 90% organised an open-day, a workshop or a guided tour of their main premises and promoted a public conference; over 80% were involved in debates or workshops at a junior or a secondary school and wider-ranging initiatives that also take place annually at the international level; more than 70% participated in events organised by private institutions and science fairs or festivals. Over the last 12 months, about one in three institutions have indicated that they have never organised a science café or a similar event, an exhibition and a scientific project with the involvement of the public and more than half have never attended a participatory and/or deliberative policymaking event where the presence of experts is required.

The data, therefore, highlight that the communication initiatives in which the Italian research departments and institutes are engaged are above all strongly influenced and determined by the internal objectives of the institution, with a prevalent interest in the dissemination of research results. Using the Categorical Principal Components Analysis, it is possible to classify the responses provided by the Italian research institutions, identifying three dimensions of public events: *institutional activities, engagement activities and dissemination activities* (Table 8.2). The first group includes the UNESCO International Year, FameLab, the European Researchers' Night and similar types of international events; National Science Week and similar formats of national events; Science Festivals and Science Fairs. The second set of activities includes events organised by private institutions; deliberative and participatory events on policymaking; citizen science projects and exhibitions. Among the dissemination activities, we find Open Days, workshops, guided visits and similar types of events; talks and workshops at junior and secondary schools; public lectures; Science Cafés and other similar events.

Table 8.2 Public events (CATPCA; n=347)

	Rotated component loadings – CATPCA		
	Institutional activities	Engagement activities	Dissemination activities
Public lectures	0.068	0.342	0.672
Exhibitions	0.456	0.631	0.158
Open Days, workshops, guided visits and similar event formats	0.071	0.009	0.790
Science Festivals/Science Fairs	0.746	0.090	0.165
National Science Week and similar formats at national events	0.757	0.251	0.142
Science Cafés and similar formats at public discussion events	0.396	0.181	0.603
UNESCO International Year, FameLab, the European Researchers' Night and similar types of international events	0.853	−0.033	0.096
Deliberative and participatory events on policymaking	0.052	0.795	0.160
Events organised by private institutions (business organisations/industry/corporations)	−0.063	0.826	0.064
Talks and workshops at junior/secondary schools	0.135	0.114	0.702
Public science projects	0.327	0.644	0.230

Source: More-Pe Survey 2018 (most significant values in grey).

Considering the main research area with which a research unit is associated, it is noted that departments and institutes specialising in the social and humanistic sciences are more active than others in organising public conferences and science cafés or public discussions. The research units focusing on humanistic subjects are also those that organise the largest number of exhibitions. On the other hand, the departments and institutes dedicated to the natural sciences, engineering and technology participate more than the others in science festivals and fairs and in events organised at the national or international level that take place on an annual basis. The natural sciences research units are also more active in organising open days, workshops or guided visits in their laboratories, while the engineering and technology research units organise the largest number of events in collaboration with private organisations. Finally, it is noted that the institutions relating to the natural and the human sciences participate more frequently in debates and workshops held at junior or secondary schools (Table 8.3).

Traditional media channels

Italian universities and research institutes use various instruments and channels of communication to disseminate research results and news about their institutional activities. In particular, press releases published in daily newspapers constitute the communication channel most frequently adopted. With the Categorical Principal Components Analysis, it is possible to clearly distinguish the communication channels and methods used among traditional media-related instruments and specific target instruments (Table 8.4).

Table 8.3 Public events by OECD research areas (mean of the indices between 0 and 1; n=347)

	Institutional activities	Engagement activities	Dissemination activities
Natural sciences	0.353	0.243	0.478
Engineering and technology	0.277	0.343	0.421
Medical and health sciences	0.183	0.238	0.307
Agricultural and veterinary sciences	0.382	0.405	0.534
Social sciences	0.292	0.311	0.483
Humanities and the arts	0.319	0.347	0.561
Total (Mean)	0.295	0.296	0.454
F	4.874	3.793	10.228
Sig.	0.000	0.002	0.000

Source: More-Pe Survey (2018).

Table 8.4 Traditional media channels (CATPCA; n=347)

	Rotated component loadings – CATPCA	
	Traditional instruments (media)	Specific target instruments
Interviews for newspapers	0.866	0.201
Interviews for the radio	0.867	0.105
Interviews for the TV	0.877	0.183
Other TV (shows, programmes, etc.)	0.665	0.175
Press conferences	0.581	0.334
Press releases	0.735	0.344
Newsletters	0.059	0.770
Brochures/leaflets/publications for a non-specialist readership	0.181	0.722
Articles in magazines and newspapers for a non-specialist readership	0.731	0.290
Multimedia/videos/films/podcasts	0.487	0.546
Popular books	0.358	0.564
Policy papers/briefings on policy issues for industry, politicians, policymakers	0.255	0.659
Materials for schools (curriculum, textbooks, etc.)	0.166	0.659

Source: More-Pe Survey 2018 (most significant values in grey).

With respect to the traditional media channels, at least once a month over the last year, about 30% of research institutions produced a press release and participated in an interview or also in the composition of an article for a daily newspaper. On the other hand, interaction with TV or radio facilities occurs less frequently, reaching a level of approximately 15% per month. These results indicate a partial discrepancy between sources used by the public, in general, and those preferred by research institutions. Most information and news concerning scientific topics is, in fact, acquired by Italian citizens through television programs, as revealed by the results of a permanent monitoring scheme which Observa has been conducting since 2003, with an annual sample survey on the behaviour and the opinions of Italians regarding issues related to science and technology (Bucchi and Saracino, 2019).

Researchers working in the area of the human sciences are those most likely to be interviewed on the radio and offer their contribution in the drafting of articles for daily newspapers and magazines which become available to a non-expert readership; however, they are also the subjects that are the most active in the production of multimedia materials and products for schools. It would appear, moreover, that other communication instruments

Table 8.5 Traditional media channels by OECD research areas (mean of the indices between 0 and 1; n=347)

	Traditional instruments (media)	Specific target instruments
Natural sciences	0.369	0.262
Engineering and technology	0.364	0.338
Medical and health sciences	0.316	0.214
Agricultural and veterinary sciences	0.484	0.380
Social sciences	0.399	0.306
Humanities and the arts	0.421	0.412
Total (Mean)	0.375	0.303
F	1.819	4.038
Sig.	0.110	0.002

Source: More-Pe Survey 2018.

which are particularly useful for disseminating news regarding their institutional activities, such as brochures, leaflets and newsletters, and which, in fact, have been used by more than half of the research units in the last year, do not depend on the main scientific areas to which the units pertain (Table 8.5).

New media channels

If in the past ten years, according to data acquired by the Observa Science and Technology Observatory, the percentage of Italian users of scientific-technological content available via the web has grown by 20 points and such behaviour has become increasingly frequent (Bucchi and Saracino, 2019), Italian research institutions have also invested considerably in the dissemination of news through this medium. Websites, blogs and social media are now commonly used to convey information regarding the activities of researchers and the study and research programmes. At least once a week, more than half of the research units update their website, two out of five publish a post on Facebook and more than two out of ten post information on Twitter. Less widespread, however, is a frequent use of such instruments as YouTube and Instagram, which are currently preferred by the younger sections of the population.

For Italian research institutions, their website, Facebook page or Twitter profile represent high-interaction instruments, while YouTube and Instagram remain in the range of medium-level interaction instruments (Table 8.6). We may consider these two modalities as two levels of interaction with the public, the first of which is stronger and the second weaker. From an analysis aimed at determining which departments and research institutes use high-interaction or medium-interaction instruments more than others, it would appear that there are no significant differences between the main scientific areas to which the units pertain, except in the case of updating websites. As occurs with respect to all of the main communication activities that have been listed, the least active are researchers working in the field of medicine and health sciences (Table 8.7).

The question, in fact, arises as to who is actually in charge and what action is taken by these individuals with respect to the dissemination of information concerning the activities of the Italian research departments and institutes. The preparation of the information and data to be communicated is mainly managed within the research unit (of the department or

Table 8.6 New media channels (CATPCA; n=347)

	Rotated component loadings – CATPCA	
	Medium interaction instruments	High interaction instruments
Website updates (events, content...)	0.010	0.968
Blogs	0.710	0.264
Facebook	0.303	0.834
Twitter	0.499	0.638
Google+	0.743	0.392
Instagram	0.661	0.280
YouTube	0.614	0.473
Podcast	0.927	−0.130

Source: More-Pe Survey 2018 (most significant values in grey).

Table 8.7 New media channels by OECD research areas (mean of the indices between 0 and 1; n=347)

	Medium interaction instruments	High interaction instruments
Natural sciences	0.146	0.446
Engineering and technology	0.139	0.476
Medical and health sciences	0.092	0.286
Agricultural and veterinary sciences	0.094	0.438
Social sciences	0.101	0.469
Humanities and the arts	0.159	0.474
Total (Mean)	0.128	0.434
F	0.766	3.102
Sig.	0.575	0.010

Source: More-Pe Survey 2018.

institute). More than half of the research units manage their own website independently and in an autonomous manner prepare the materials which they communicate, curating also the presentation and display of data. Only a few units rely on resources that are external to the institute they form part of; in cases where this happens, it will occur on account of technical issues, such as the construction of a website or the organisation of an event. Finally, it is interesting to note that in 52% of cases, when training on public or media communication was arranged, this occurred within the host institution.

Audiences

Research institutions turn their attention to various public audiences: public administration, schools, businesses, the media, organisations within civil society and the general public. It is possible, through CATPCA, to classify the answers of the interviewees, identifying three audience groups: stakeholders and civil society (members of local municipalities, councils, associations; industrial delegates; governments, politicians, policymakers; non-governmental organisations); students inside or outside scholastic/academic settings; the general public and communicators (Table 8.8).

By analyzing the collected data concerning the frequency with which communication occurs with these subjects, it is possible to compile a ranking of the various audiences mainly targeted by the universities and research centres. In Italy, the first recipients of frequent or very frequent events of

Table 8.8 Audiences (CATPCA; n=347)

	Rotated component loadings – CATPCA		
	Stakeholders and civil society	Students	General public and communicators
General public (whoever may be interested)	0.077	0.216	0.871
Schools	0.051	0.983	0.167
Students outside the scholastic/academic context	0.171	0.967	0.202
Members of local municipalities/councils/ associations	0.654	0.150	0.471
Delegates from industry	0.882	0.052	−0.100
Governments/politicians/ policymakers	0.805	0.096	0.384
Non-governmental organisations (NGOs)	0.669	0.113	0.402
Media and journalists	0.380	0.175	0.758

Source: More-Pe Survey 2018 (most significant values in grey).

communication are schools and also students outside the setting of their courses of study. These subjects are followed by the general public, businesses and local administration bodies. It is, thus, evident that the scholastic/academic world and young people are among the primary targets addressed for the communication of the results of research.

Analyses conducted on the basis of areas of study allow for the identification of significant differences between the departments and the research institutes, depending on the scientific area to which they belong. The general public and members of local municipalities, committees and citizens' associations are the targets most frequently reached by researchers working in the human sciences. Schools and students are the main recipients of the communication activities of researchers dedicated to human sciences but also those active in the natural sciences, while the business world is involved to a greater extent by institutions that focus on engineering and technology and the representatives of the institutions have more frequent contacts with social scientists (Table 8.9).

The relationship with the media is an indispensable element for the communication of the content of research activities. Moreover, scientific institutions normally develop privileged relationships with communicators, journalists and media operators. How are the relations between researchers and communicators organised? Over the last year, almost three out of ten institutions were contacted three to five times and an equal number were contacted more than ten times; only 5% were never contacted by journalists in the past year. However, it must be noted that the Italian universities differ considerably with regard to their size. There are very large departments

Table 8.9 Audiences by OECD research areas (mean of the indices between 0 and 1; n=347)

	Stakeholders and civil society	Students	General public and communicators
Natural sciences	0.410	0.735	0.592
Engineering and technology	0.530	0.654	0.548
Medical and health sciences	0.413	0.539	0.548
Agricultural and veterinary sciences	0.549	0.688	0.600
Social sciences	0.528	0.604	0.576
Humanities and the arts	0.465	0.729	0.686
Total (Mean)	0.471	0.660	0.588
F	4.365	8.262	3.243
Sig.	0.001	0.000	0.007

Source: More-Pe Survey 2018.

that are well equipped in terms of communication resources and some other recently formed departments are very small. This consequently also determines significant imbalances in the methods employed to contact and communicate with the media.

Relations with the media are normally developed on the basis of trust and through constant interaction with journalists. Regardless of the scientific area to which the research unit mainly pertains, within the framework of the survey, it is noted that most departments and institutes develop personal relationships with journalists and have at their disposal a list for periodical contacts; only 19% do not have a list of journalists or media contacts. Contact occurs especially through direct relations established between communicators and researchers; the press offices of the institutions or personnel responsible for communication are consulted primarily in only 35% of the cases. The subjects engaged in communication activities who were interviewed expressed a favourable opinion in their consideration of the significant efforts made to communicate research activity to the public which would be useful to enhance the mission of their research unit, but most of the subjects referred to also stressed that there is a need to invest more in public communication.

Evaluation of activities

Finally, the capacity to monitor and assess communication activities and public involvement was investigated, and it was found that only 18% of the research units stated that such action was never or rarely undertaken. A general overview of the recipients of communication and the relations that the research units establish with the media allows us to reach various conclusions. It is evidently easier for Italian researchers to contact schools and students – also outside the scholastic or academic settings – or also the general public, with activities such as open days, workshops, guided tours or public conferences, especially within their own research unit premises, while contacts with the media occur frequently or very frequently only in 36% of cases. The press office operators and subjects responsible for communication activities, who are, in fact, present in almost all research institutions (see the following section), are referred to by journalists only on rare occasions and often contact occurs by means of direct relations between the communicators and the researchers.

Public communication resources

Normally, research institutions staff responsible for maintaining relations with the press and carrying out communication activities, conveying the results of the research undertaken. There is a considerable variety of resource distribution at Italian organisations, and the survey has revealed that only

27% of the research units count on their own staff members for communication activities. For 67% of these units, however, it is possible to count on the support of internal staff in their institution, while 5% have no access to personnel dedicated to communication.

Among the research units with members of staff responsible for communication activities, regardless of the scientific area in which they operate, in only 54% of the cases considered are these subjects dedicated exclusively to such duties. However, more than 80% of the staff have a permanent contract and work on a full-time basis at the research unit or the institution where they are employed, and for this reason, in about half of the cases, they focus on communication two to three days a week. The professional/academic background of staff members assigned to duties relating to communication is rather diverse: more than 60% have an educational background in the humanities and about 80% have acquired previous professional experience although just over one in four have assumed a professional role in the field of marketing, public relations and communication or have worked as a journalist and writer specialising in the sciences. Only slightly more than half of the staff forming part of the team responsible for communication in the research units whose members were interviewed has actually undergone formal training and has attended (even short) courses on communication.

Analyzing the time dedicated to communication activities, in the last five years, the work related to communications has increased. In fact, more than seven out of ten departments state that they have carried out activities for the non-specialist public in an increasing manner. The percentage of researchers who have participated in public involvement activities over the last year is estimated to be between 10% and 20% in three out of ten cases, between 20% and 40% in two out of five cases and is higher than 40% in more than a third of the cases considered. The economic resources used for communication activities (website maintenance, printing of materials, the organisation of conferences and public events, without considering costs for salaries), in about 40% of the institutes amounts to 5% of the annual research budget available to the research unit. About one department or institute out of ten does not allocate resources to these activities, while 13% of the units reserve between 5% and 10% of available resources for communication activities.

When asked to offer their observations on their commitment to public communications, the representatives of the Italian departments and research institutes declare that further resources should be made available. Analysing the data, it is evident that for all current levels of investment, there is a need to allocate more resources for public communication activities at both the economic level and in terms of personnel. The departments and research institutes that believe more than others that the number of communication activities for the public have actually increased are those involved in the natural sciences and the humanities. However, in research units that focus on the social and human sciences, we do find the highest percentages of researchers engaged in communication activities and the highest proportions of budget allocations spent on communication. In the latter groups, there is a stronger belief that economic resources invested in such activities are insufficient and should continue to grow.

Public communication rationales

Scientific institutions are induced by various forms of reasoning to communicate the results of their research. The survey carried out allows for a verification of the different perspectives and makes it possible to establish an order of priorities on the basis of which the Italian research units engage in public communication.

The first three rationales indicated are of the institutional and economic type. The respondents to the questionnaire declare that the researchers of their unit are dedicated to communication because, in this way, they comply with the mission of the university and the research centres they work for, publicly disseminate information concerning their research and are, thus, able to attract funding. Only for fewer than 5 out of 100 is public communication undertaken to encourage involvement on the part of the public. In comparison with other fields, for researchers occupied in the fields of medicine and social sciences, it is more important to respect the mission of their institution, while for researchers in the natural and human sciences, in engineering and technology, there is a greater need with respect to others to disseminate their research to the public. More specifically, in the survey, certain topics were investigated which concern the selection of information to be presented to the public, defining aspects deemed to be successful in public communication. The researchers believe that, above all, the innovative aspects – relevant with respect to daily life and current debates – are those which orientate their communicative action and also state that pleasant aspects deemed to be of immediate interest, also linked to social relations, are the least important. Those who believe that it is more important to communicate what is pleasant and interesting are above all the researchers who work in the fields of the natural sciences, engineering and technology, while researchers in the social and human sciences claim that it is more important to focus on what is relevant in ongoing debates.

Analyzing the impact of communication of information concerning research, the representatives of research institutions believe that visibility in the media is an important objective to be pursued; for this reason, they stress that the media should place greater importance on their activities and the results of their work. Only less than 15% believe that the research which is carried out is of little interest to journalists and that maintaining relations with them is not a task researchers should engage in.

Public perceptions

However, ultimately, one may ask how institutes perceive the public. Indicating their level of agreement with a series of statements, in their responses, the representatives of the research institutes, without any clear distinctions in terms of pertinent academic or scientific areas of study, highlighted some aspects that form an articulated framework. Adopting the Categorical Principal Components Analysis, it is possible to identify the emergence of three dimensions in the responses provided: three images of the public, which signal the unfolding and use of three communication models. An initial model, which we would refer to as an "exclusive model", presents a type of communication strongly centred on the researchers themselves with a low level of involvement of the public. In this case, in accordance with the underlying stance, communication is established as unidirectional and produced in an exclusive manner, starting from the results and convictions of the researchers. A second model, referred to as the "educational model", presents an approach of the didactic type in which there is a tendency to emphasise the educational role of the researchers, whereby they communicate the results of their work in an adequate manner. This orientation is supported by a basic assumption: the public must be educated in science in order to understand it. Finally, a third model defined as the "participative model" develops starting from the recognition of the contribution offered by the non-expert public through initiatives of involvement and public debate (Table 8.10). The three models cut across the units considered. Using as an independent variable the main scientific area of the institutes, the variance of the constructed synthetic indices between the groups is not statistically significant (Table 8.11).

Considering the individual items, we can point out that the most significant element is the agreement with the need to have a well-informed public to support research (82.5%). More than half of the respondents also believe that the public needs to be educated and would like the public to be more actively involved in discussions on the implications of research, but not necessarily in decisions regarding the actual direction of the research. However, only 34.5% believe that the public trusts science and scientists and 40% are not sure whether the public would like to make a contribution to science. More than two out of five respondents disagree with the idea that a vast body of citizens cannot be expected to be interested in research and that the public is not eager to learn about science. For more than three out of five respondents, in fact, the public is, indeed, interested in research that is carried on and needs to understand the general picture; we cannot communicate with the public selectively in order to avoid problems because it is not true that citizens are interested only in a part of the subjects of research activity. More than half would like the public to be more actively involved in decisions to be made regarding research but also believe that members of

Table 8.10 Opinions regarding the public (CATPCA; n=347)

	Rotated component loadings – CATPCA		
	Exclusive model	Participative model	Educational model
The public is not interested in the research carried out at our unit	0.771	−0.013	−0.217
The public is not eager to learn about science	0.657	−0.142	−0.092
The public is interested in a limited range of research topics, such as dinosaurs, dolphins and disasters	0.700	0.010	−0.095
The public wants to contribute to science	−0.120	0.551	0.286
We cannot expect a vast public to become interested in the research we carry on	0.792	0.085	0.046
If the public know more about our research, they will be more likely to support it	−0.365	0.187	0.614
The public do not need to understand the full picture; we explain what we think is appropriate	0.706	0.030	−0.026
The public need to be educated by those who are knowledgeable	0.092	−0.220	0.798
We recognise that the public is what it is and we adapt our communication activities accordingly	0.616	0.339	0.327
We would like the public to become more actively involved in decisions about the research conducted at our institute	0.141	0.633	−0.299
We would like the public to become more involved in discussing the implications of the research that we do, but not necessarily in decisions about the direction of our research	−0.043	0.700	−0.103
The public trusts science and scientists	−0.069	0.460	0.068
The public does not need to be scientifically literate to discuss the implications of our research	0.219	0.508	−0.034

Source: More-Pe Survey 2018 (most significant values in grey).

Table 8.11 Opinions regarding the public by OECD research areas (mean of the indices between 0 and 1; n=347)

	Exclusive model	Participative model	Educational model
Natural sciences	0.418	0.538	0.636
Engineering and technology	0.514	0.558	0.629
Medical and health sciences	0.434	0.554	0.633
Agricultural and veterinary sciences	0.472	0.520	0.667
Social sciences	0.449	0.536	0.597
Humanities and the arts	0.459	0.511	0.609
Total (Mean)	0.453	0.538	0.626
F	1.534	0.469	0.383
Sig.	0.179	0.799	0.860

Source: More-Pe Survey 2018.

the public need a "scientific culture" and knowledge to be able to discuss the implications of the research that is carried out.

Conclusions

The results collected through the MORE-PE survey in Italy provide robust evidence on basic trends observed in recent years such as processes and actions (Loroño-Leturiondo and Davies, 2018). Research institutions have included the public communication of science among their priorities, and for this reason, they are committed to transmitting information on research activities through initiatives that have an impact on various sections of the population. These proposals are organised largely on the basis of institutional conditions with a strong emphasis on the dissemination of the results of the research undertaken and to a marginal extent with the aim of actively involving the public.

The information channels and instruments for communicating with the public have changed over time and cover a broad spectrum of interest, making it possible to reach various types of recipients. In some areas, however, a path of strong involvement, as in the case of social networks, is still in its infancy. To disseminate the results of research, most research departments and units use internal resources and often those of the institutions of which they form part (universities and institutes), a fact that highlights the institutional intention to develop a structural and stable commitment to communication, avoiding improvisation and precariousness. The resources available, however, are still not considered sufficient by the respondents and the staff is not fully trained to effectively interpret the role of a public research communicator and use the most innovative instruments.

Considering the various disciplines and fields of research, we note that the departments dedicated to the medical sciences are among the least active on the communication front. Among the reasons for this weak commitment, it is possible to recognise the constant relationship that physicians and researchers have with people undergoing treatment and patient associations. As a result, research departments and institutes in general do not take action using all of the instruments of communication mentioned in the survey and reinforce purely institutional activities.

The recipients of communication concerning research are mainly schools and the general public. To a lesser extent, it is possible to contact the business world, public administrations and the organisations of civil society. The capacity to broaden the range of users of information that is communicated to the public, thus, represents a real challenge for the future, to be addressed with adequate resources from an economic and professional point of view, but also using methods and instruments that will take into account questions arising within the social and economic context in which the research institutions operate.

To conclude, it may be said that research institutions have accepted the invitation from civil society and the national and European programmes to activate new and effective channels of communication. In the coming years, the effort to keep up with these requests will have to be rendered more intense if we intend to improve the relationship between science and society, overcoming concerns regarding the dissemination of results solely with a view to ensuring consent and developing the processes of consultation and discussion on issues of common interest.

References

ANVUR (2015). *La valutazione della terza missione delle università italiane. Manuale per la valutazione*, Roma, ANVUR.

ANVUR. (2018). *Biennial Report 2018 – ANVUR – Agenzia Nazionale di Valutazione del Sistema Universitario e della Ricerca*. https://www.anvur.it/en/biennial-report/biennial-report-2018/

Bauer, M. W., and Jensen, P. (2011). The mobilization of scientists for public engagement. *Public Understanding of Science*, 20 (1), 3–11.

Bucchi, M., and Saracino, B. (2019). Scienza, tecnologia e opinione pubblica in Italia nel 2018, in Pellegrini, G., Saracino, B. (Eds.), *Annuario Scienza Tecnologia e Società 2019*, Bologna, Il Mulino.

CNR (2020). Consiglio Nazionale delle Ricerche, https://www.cnr.it

Entradas, M., and Bauer, M. W. (2017). Mobilisation for public engagement: Benchmarking the practices of research institutes. *Public Understanding of Science*, 26 (7), 771–788.

Kim, J., and Yoo, J. (2019). Science and technology policy research in the EU: From framework programme to HORIZON 2020. *Social Sciences*, 8(5), 153. https://doi.org/10.3390/socsci8050153.

Legge 30 dicembre 2010, n. 240, Norme in materia di organizzazione delle università, di personale accademico e reclutamento, nonché delega al Governo per incentivare la qualità e l'efficienza del sistema universitario, https://www.camera.it/parlam/leggi/10240l.htm

Loroño-Leturiondo, M., and Davies, S. R. (2018). Responsibility and science communication: scientists' experiences of and perspectives on public communication activities. *Journal of Responsible Innovation*, 5 (2), 170–185. https://doi.org/10.108 0/23299460.2018.1434739.

MIUR (2020). Il Sistema della ricerca in Italia, https:// www.miur.gov.it/web/guest/sistema-della-ricerca

Molas-Gallart, J., Salter, A., Patel, P., Scott, A., and Duran, X. (2002). Measuring Third Stream Activities, Final Report to the Russell Group of Universities, SPRU, University of Sussex.

Pellegrini, G., and Rubin, A. (2020). The Long and winding path of science communication, in Gascoigne, T., Schiele, B., Leach, J., Riedlinger, M., Lewenstein, B. V., and Broks, P. (Eds.), *Communicating science. A global perspective*, Canberra, Australian University Press.

Ziman, J. (1996). 'Postacademic science': Constructing knowledge with networks and norms. *Science Studies*, 9(1), 67–80.

Chapter 9

Public Engagement at Research Institutes in the Netherlands

Fertile Territory or Terra Nullius?

Pedro Russo, Robert Bergsvik, and Julia Cramer

With contributions from Anne Kerkhoven and Selina van den Oever

Public communication with science landscape in the Netherlands

'Science communication' describes the myriad of ways in which both the research process and its results can be shared with the public. It is an important dialogue between science and society regarding how research contributes to societal and economic growth. As that dialogue provides citizens with the knowledge to tell fact from fiction, they also get the opportunity to contribute ideas for future scientific developments. From the 1950s and onwards, the popularisation of science and technology has been the main strategy of the Dutch government for bridging gaps between science and society.

At the outset, science communication with the Dutch public sprang out of the public information campaigns occurring in the agricultural sector. The first focus on science communication in the Netherlands can be traced back to the Bender Commission of 1957, a collaborative effort with representatives of different Dutch universities (Dijkstra, van Dam, and van der Sanden, 2020). The aim of this commission was to establish closer ties between universities and different societal groups that universities depend upon to increase public trust. While the government started this as a pure public relations campaign, the commission quickly took on a democratic rationale: Dutch citizens are entitled to gain both knowledge and information about scientific endeavours, which enables them to make informed decisions about the impact of science and technology on society. Since the Bender Commission, the motivations for promoting one-way and two-way science communication have reflected democratic, economic and cultural considerations (Dijkstra, van Dam, and van der Sanden, 2020). This dialogue is as valuable for the research field as it is for society. Actively involving citizens in the research process may provide scientists with new inspiration for research questions and possibly help them solve ethical dilemmas (Engelshoven, 2019).

DOI: 10.4324/9781003027133-12

Even though Dutch public confidence in science is high, some controversies (e.g. the anti-vaccination programme) illustrate that providing scientific evidence does not automatically lead to acceptance. "Science alone is simply not enough" (Blankesteijn et al., 2014). From the 'Trust in Science' debates, organised in 2014 by the Rathenau Institute, the Royal Netherlands Academy of Arts and Sciences (KNAW) and the Netherlands Scientific Council for Government Policy, we have learned that the dialogue between the scientific field and society can, indeed, be improved (de Jonge, 2014).

Democratic rationale

The promotion of science communication during the 1970s was rooted in a democratic rationale. In the 1970s, Minister of Research Policy Fokele Trip championed the idea that scientific activity should not occur outside of societal context. This entailed a demand for close contact between researchers and other concerned actors. Trip's 1975 report, *Nota Wetenschapsbeleid*, was an elaborate account of how research policy is closely connected to the activity of science communication. The Office of Science Information was established because of Trip's report. The core message of the report was that citizens have the right to know and understand the wider effects of scientific activity and research (Dijkstra, 2008; Stappers et al., 1983). The main tool for creating closer ties between science and society during this period was the use of science shops. This concept originated from the Netherlands, rooted in the idea that universities could play a vital role in providing solutions to social problems. Additionally, these shops were focused on providing disenfranchised groups, such as minorities and financially weak groups (Dijkstra, van Dam, and van der Sanden, 2020). Initially, these shops were run by volunteers with support from university employees. They started to receive funding from the universities in 1978, but by the 2000s, support subsided, and many shops closed as a result of cuts to university funding in the Netherlands. While these shops have largely disappeared from the Netherlands, they have been exported to universities across the world (De Bok and Mulder, 2004; Lürsen et al., 2000; Mulder and Straver, 2015).

Economic rationale

The early 1980s were marked by several public debates in the Netherlands around issues such as nuclear energy and the environment. This led Minister for Education, Culture and Science, Wim Weetman to ramp up efforts to disseminate information to the public surrounding scientific and technological innovation, with the aim of further integrating the public's role in social decision-making around these complex issues.

However, a new rationale underpinned these efforts: It was increasingly recognised that scientific and technological progress were essential ingredients in achieving economic growth (Dijkstra, van Dam, and van der Sanden,

2020). Developments in this decade include the replacement of the Office of Science Information with the Foundation for Public Information on Science and Technology (later renamed as the Dutch Science and Technology Association). The Rathenhau Institute replaced the Netherlands Organisation for Technology Assessment, becoming the body in charge of assessing the social and ethical aspects of science and technology. Its other functions were to provide insights for policymakers about the outcomes of scientific research and stimulate public debate about emerging issues (Dijkstra, van Dam, and van der Sanden, 2020). The fostering of science literacy was now high on the agenda of Minister Deetman. He argued that this was key for enabling the public to keep up with rapidly moving developments within science and technology. As a result of these developments, the existing gap between science and society was in danger of becoming wider. To counter this, Deetman introduced Science and Technology Week. Between 1993 and 2001, six public debates were held on issues surrounding biotechnology, culminating with a debate on genetically modified (GM) food in 2001 (Dijkstra, 2008). According to Dalderup (2000), the growing focus on how science and technology are inherently connected to society, which occurred in the 1990s, is reflective of a new cultural rationale, in addition to the existing democratic and economic rationales underpinning the Dutch public engagement efforts.

Since the 2000s, more conservative forces have dominated Dutch politics. These governments have been less inclined to fund outreach and science communication. This was especially the case after the 2008 financial crisis. With the Dutch government scaling back its role as the main promoter of science communication, other actors have entered the stage, such as associations consisting of both private and voluntary actors.

Institutionalisation of Dutch science communication

Dijkstra, van Dam and van der Sanden (2020) summarised in detail the main stakeholders and milestones in the development of science communication in the Netherlands. In terms of stakeholders, three main groups are central: (1) science centres and museums, with national, regional and local focus, have been the initiators of several exhibitions but also educational programmes and public events; (2) universities, mainly through the formal training of science communicators (practitioners and researchers), some outreach programmes, press information offices and activities promoted by researchers and (3) media, newspapers, magazines (some of them popular science), radio, TV and Internet have been promoting science (specifically Dutch research) and working closely with research institutes (RIs) and researchers. Table 9.1 provides an overview of the main milestones of science communication in the Netherlands over the past two centuries.

Table 9.1 Science communication timeline in the Netherlands (adapted from Dijkstra et al., 2020).

Date	Related science communication milestones
1820	National Museum for History of Science established in Leiden
1930s	Science appears in media
1957	National report on science communication and start of national governmental program to support science communication
1966	Interactive science centre Evoluon opens
1969	National association for science journalists
1976	First PhD graduate in science communication and courses to train scientists in science journalism
1981–1983	Public debates on nuclear energy and environmental issues
Late 1980s	Science communicators organise themselves in various professional groups
1986	National Science Week organised for the first time in Utrecht
1997	Science Museum, NEMO, opens in Amsterdam
Since 2000	Masters and research degrees in science communication at research universities
2000s	Public celebration of international years of science (2005, World Year of Physics; 2009, International Year of Astronomy)
2010s	Societal Dialogue regarding Nanotechnology
2015	Launch of Dutch Research Agenda
2015 and 2019	Research policy documents highlighting the importance of science communication, Science Vision for 2025 (2015) and Curious and Committed—the Value of Science (2019)

The Dutch research landscape

In the Netherlands, about 25% of the research takes place at universities, about 15% at National Research Institutes and most (about 60%) within commercial enterprises. As in several other European countries, the New Public Management shift also impacted the university landscape in the Netherlands. While the post-World War II years were characterised by significant government involvement and regulation, the period from the mid-1980s and onwards is reflective of a strategy described as "steering from a distance" (de Boer, Enders, and Schimank, 2007:142). The university sector was now expected to exercise a practice of self-organisation, with a shift away from detailed ex ante measures set by the government to an ex post evaluation system. On the one hand, it introduced greater autonomy for how universities could decide to develop their organisational activities. Specifically, this meant greater administrative and financial control of property and buildings, appointment and management of staff and other internal matters. On the other hand, government regulation went from being concentrated around various directives to financial incentives.

In line with broader neoliberal approaches to administration and organisation, university funding was now more reliant on performance-based indicators (de Boer, Enders, and Schimank, 2007). Competition was now more central to how universities were to organise their activity. This included competing for research grants and for attracting students to the universities (Jongbloed, 2003).

For this study, we focused on the RIs at the national and university levels, comprising 40% of the research intensity of the Netherlands. The Ministry of Education, Culture and Science (Dutch: Ministerie van Onderwijs, Cultuur en Wetenschappen; OCW) is responsible for all the national policies regarding education (from kindergarten to higher education) and science and innovation. The Netherlands has 14 research universities that provide education, conduct (fundamental and applied) research and stimulate knowledge transfer. Two intermediary organisations, Netherlands Organisation for Scientific Research (NWO) and the Royal Netherlands Academy of Arts and Sciences (KNAW), have an important role in managing research funding and the quality of the research, respectively. NWO's mission is to support and enhance the quality and innovation of fundamental scientific research and promote the dissemination of research results. KNAW provides advice and policies to the Dutch government to ensure high research quality.

The Dutch national research agenda

In 2015, the Dutch National Research Agenda (NWA) was launched to strengthen the dialogue between science and society and increase the societal impact of research. The NWA is based on scientific questions that were submitted by the public. A total of almost 12,000 questions were summarised and processed to create 25 'NWA-routes'. These routes represent the societal themes and research areas in which the Netherlands wishes to excel (NWA, n.d.; OCW, 2014). Also, it is research within these specific areas that has the highest potential for large societal impact. Aiming to encourage research that fits into one or more clusters of the NWA, the Netherlands Organisation for Scientific Research (NWO) provides extra funding opportunities. Since 2018/19 the Ministry of Education, Culture and Science (OCW) invested more than 200 million EUR in research connected to the NWA (Engelshoven, 2019). Another 3 million EUR was allocated for the communication of NWA research results to the general public.

Open science as a policy-driver for public engagement

The Dutch government called for an 'Open Science' framework where science serves the interests of society. This new Open Science framework (Ministry of OCW, 2014) is meant to greatly benefit society as it enables everyone—including local and regional authorities, healthcare professionals, patients and citizens—to access and use scientific information. In turn, non-professional scientists can also submit questions or ideas or even help collect scientific data (Open Science, n.d.). Scientific research performed by amateurs or non-professional scientists is also referred to as 'Citizen Science'. As science communication is an important means to bridge the gap between science and society and restore public trust in science, it is increasingly subject to (inter)national science policy. A fascination for science should not be confined to scientists themselves. Appropriate communication about science and technology will keep the general public in touch with the field and abreast of developments. Everyone, young and old, will be well informed and enthusiastic about all aspects of science and technology. Science must be visible (2025—Vision for Science, choices for the future, Ministry of OCW, 2014).

Within the Netherlands, the priorities of Dutch science are set by the ministry of OCW. Other key players within Dutch science policymaking are intermediary organisations such as the NWO and KNAW, and higher education and research institutions. The science policy ecosystem also includes an array of advisory councils and representative organisations, such as the Rathenau Institute and the Association of Universities in the Netherlands (VSNU).

Research that is based in universities is funded by three different flows of funds; the first flow of funds is provided by the ministries, approximately two-thirds by the ministry of OCW. The second flow of funds is distributed by the NWO and KNAW, which also receives their budgets from the government. Whereas the NWO distributes its budget in the form of competitive grants, the KNAW funds researchers in the form of an Academy Professor Prize. The third and also the largest flow of funds is provided by the European Union and (inter)national public and private sources (Ministry of OCW, 2012). Next to their role in research funding, the KNAW has an important advisory role within the Dutch science system. Advice to the government on scientific endeavour can either be given on request or at its own initiative (Ministry of OCW, 2012). Additionally, together with the NWO and VSNU, the KNAW is responsible for setting out criteria that are used to assess the quality of scientific research. These criteria are published every six years in its standard evaluation protocol.

Research funding and public engagement policies

The merits of Dutch research are widely recognised abroad, and the NWO remains committed to maintaining and strengthening this position. As science is facing challenges in terms of reliability from some sectors of society, the NWO aims to adopt a more connecting role between science and society with the purpose of anticipating developments that shape both sectors. Following the policies set by the Dutch government, the NWO has developed a Strategy 2019–2022 for 'Connecting Science and Society'. In this strategy document, the organisation emphasises its commitment to strengthening the relationship between citizens and scientific actors for the purpose of solving new and complex societal challenges (NWO, 2019). However, this strategy document does not adequately address how the NWO will achieve a closer dialogue between scientists and citizens. There is no mention of concepts such as 'public engagement', 'science communication', 'public outreach' or 'knowledge dissemination'. While it stipulates that one of the ambitions of the NWO is to create a stronger connection between society and science (the nexus ambition), it does not include any concrete information about how to go about doing this, beyond ensuring open access to publications and research results. When comparing strategy documents from the Dutch Ministry of Education, Culture and Science and the NWO, we discovered that the number of terms and plans connected to science communication mentioned in NWO (domain) policy documents has been strongly reduced. Even though all included NWO policy documents contain a chapter called Nexus (connecting agendas, science and society), the vision document of the NWO Sciences domain (ENW) fails to mention society—or similar terms—in this chapter or the rest of the document. Moreover, all three vision documents of the NWO domains—ENW, Applied and Engineering Sciences (TTW) and Social Sciences and Humanities (SGW)—fail to adequately address, or emphasise the importance of, science communication. Although support for current science communication initiatives is evident in TTW and SGW vision documents, a discrepancy exists between the intentions of the government and the execution of their plans.

Resources for public communication at the institute level

While the study and practice of science communication have not yet been implemented at all levels of the scientific community in the Netherlands, it gained significant traction in later years. This is especially in connection with the ambition to further develop societal support for science and technology (Dalderup, 2000; Dijkstra, 2008). As this issue progressed, economic rationales became more prevalent in science information campaigns, at the expense of democratic and cultural rationales. The main drivers for science, technology and innovation are economic profit, which is again reflected in the efforts of science communication and public engagement. This has only intensified under the current conservative Dutch government. There have also been a few large-scale debates concerned with the social impact of technologies, such as nanotechnology and gene-modified food products. However, science communication remains largely as a peripheral focus in the Dutch scientific community (Dalderup, 2000; Dijkstra, Seydel and Gutteling, 2004; Wiedenhof, 1978).

Methods

Sample

In the Netherlands, the large facilities of RIs fall under faculties within universities. For this study, we used the NARCIS database of institutes provided by Dutch National Centre of Expertise and Repository for Research Data (DANS). DANS maintains a detailed list of Dutch RIs; based on this, 821 institutes were identified for this study. All research centres and departments of the universities and institutes in the list were mapped using information from universities' websites and then classified according to OECD areas of research. The OECD research areas were coded by one coder, based on the information on the institute's websites (Table 9.2).

For the study and because the total number of Dutch RIs was smaller than the 200 per OECD group (target sample), it was decided by the authors and editors to survey (in English) all the institutes in the Netherlands (n=821). To announce the survey and identify the correct contact person, an email was sent to the institutes in January 2017. In the Netherlands, the questionnaire was sent in February 2017. Two reminders were sent and, whenever possible, a reminder by phone and a last reminder, sent in June 2017. The data analysis presented in the next section was based on χ^2 statistical tests, to identify differences across different categories of the survey, for example, public communication activities, policies, and channels.

Results

As Entradas et al. (2020) show, the Dutch RIs are within the most active in public communication among the surveyed countries. In the sections below, we look in detail on the type and frequency of public communication

Table 9.2 Overview of the research institutes sampled for this study.

	Sampling frame (N)	Sampling frame (%)	Number of recorded responses	Response rate (%)
Natural Sciences	116	14.1	29	25.0
Engineering and Technology	105	12.8	13	12.4
Medical and Health Sciences	164	20.0	22	13.4
Agricultural Sciences	12	1.5	1	8.3
Social Sciences	302	36.8	30	9.9
Humanities	122	14.9	22	18.0
Total	821	100.0	117	14.3

activities that RIs have performed the previous year, how resources for these activities were allocated and also rationales for implementing these activities. For the analysis of activities, we compare activity between institutes with and without a communication policy.

Public events

The most commonly organised event by Dutch RIs is Open Days (Figure 9.1) and its workshops (39%). Closely followed are events for policymakers (36%) as well as representatives of industry and the corporate sector (36%). Indicators of a close connection between Dutch RIs and the corporate sector can also be found in the Rathenau Institute's TWIN report on the state of Research & Development (R&D) funding towards Dutch science. In 2017, the total expenditure on R&D in the Netherlands was 14.7 billion euro, of which the business community contributed 7.7 billion (52%) and the Dutch government contributed 4.6 billion (31%), while the remaining funds came from non-profit organisations. The Dutch corporate sector is also the largest performer of R&D, accounting for 59% of the research. Less successful is the organisation of Citizen Science by Dutch RIs, with 36% responding that they never organised such activity. "Such activity" is viewed as a cornerstone in forging stronger ties between science and society as well as a key component in Responsible Research & Innovation in society (Ministry of OCW, 2014).

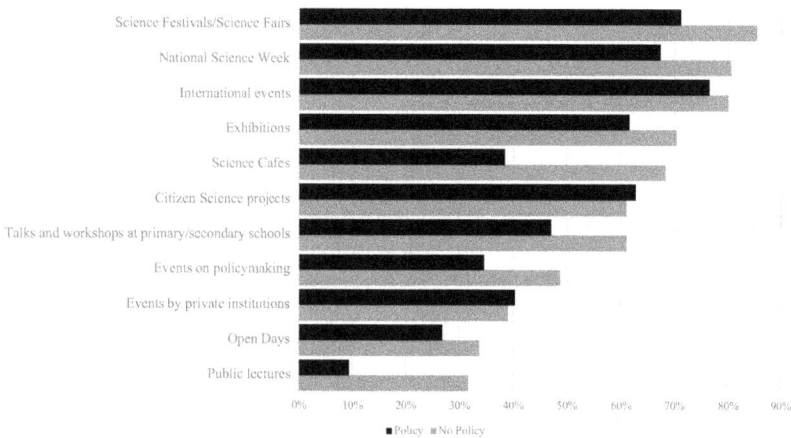

Figure 9.1 Public events by policy. The bars represent the sum of "Never" and "Annually" responses, expressed in percentages. Lower percentages indicate higher frequency of events (e.g. 32% of institutes with no policy never conduct or conduct annually, public events compared to 9% of institutes with policy that do not conduct public events).

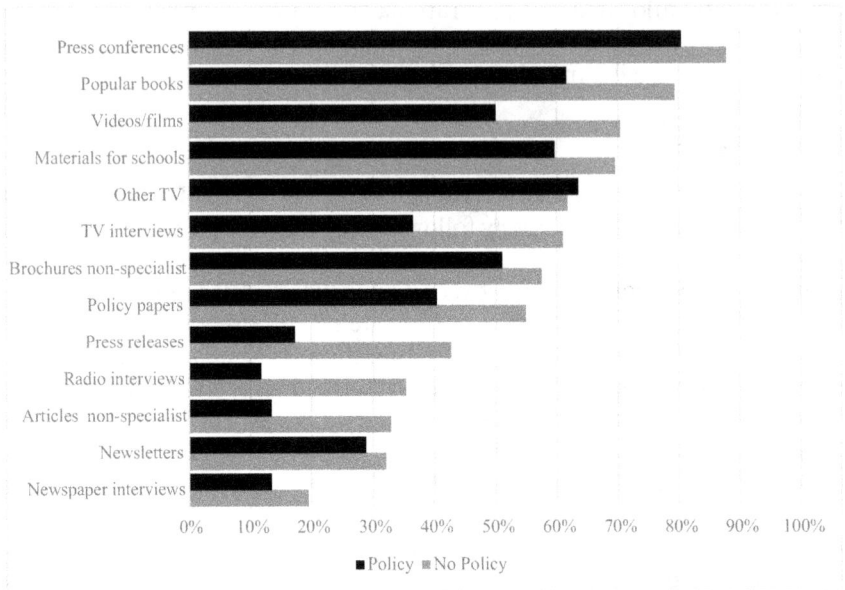

Figure 9.2 Traditional media channels by policy. The bars show the percentage of "Never" and "Annually" responses together. Lower percentages indicate higher frequency.

Traditional media channels

The most used channel for communication with non-specialist audiences among Dutch RIs is Interviews for the radio (45%, occasionally). This category is closely followed by interviews for newspapers (43%, occasionally) and articles in newspapers and magazines (42%, occasionally). The results in the final category further support the claim that the Dutch research community has a strong relationship with Dutch media and journalists.

There are clear differences between the subsamples of no public engagement policies and with communication policies (Figure 9.2); the χ^2 tests are significant for most of the channels. It is, however, important to highlight that the differences are especially significant for media-related communication channels. Overall, institutes with communication policies use more channels and more frequently than RI without public engagement policies.

New media channels

The most common form of new media activity is website management (31%, monthly), followed by the use of Twitter (27%, weekly) and Facebook (23%, weekly). While websites containing information about the scope of the project and its developments are important, large parts of the public are in danger of missing out because they have moved onto new media platforms as their prime

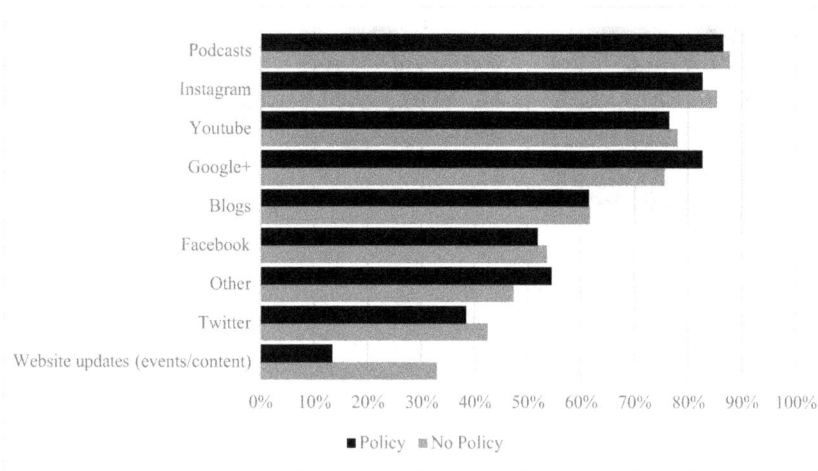

Figure 9.3 New media channels by policy. The bars represent the percentage of "Never" and "Quarterly" responses together. Lower scores indicate higher frequency.

source of information (López-Goñi and Sánchez-Angulo, 2018). Combining the two types of online activity could potentially increase the level of engagement. The low use of new media could have a connection to the relatively higher use of magazine and newspaper articles, where the format for conveying information is significantly different. Channels such as YouTube are used less frequently (36%, quarterly), but this also makes sense since it often requires significantly more time and effort to produce video/graphics content (see Figure 9.3).

Dedicated communication staff

As Figure 9.4 shows, 58% of the RIs with a communications policy responded that they have a dedicated communication staff within their unit. It is more common for those institutes without a communication policy to make use of their institution's communication staff when conducting communication activities (68%), and these differences are significant (p<0.01). As for budget (32%) report spending 1–5% of annual budget on public communications and (30%) report spending 1%; when asked about what they should spend, 38% of the institutes reported that they should spend between 1% and 5%.

As most of the institutes in this survey belong to one of the 13 university institutions in the Netherlands, it comes as no surprise that many of them rely on the use of the communication staff at university level. Staff is an important determinant of the level of communication of an institute (Entradas et al., 2020), and having access to such staff is vital for outreach with non-specialists. We found an association between having a communication policy and having access to communication staff.

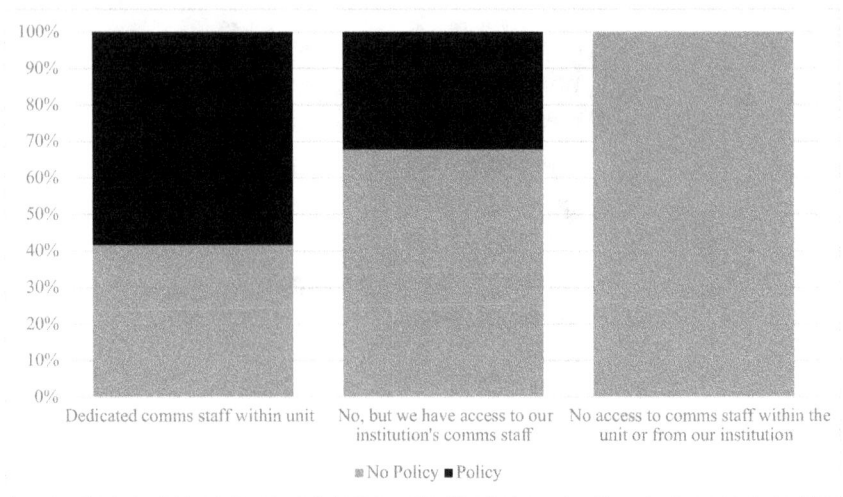

Figure 9.4 Communications staff by policy.

Public perceptions

In this section of the survey, we provided a series of agree/disagree statements about how the public is perceived by personnel working at the institute (Figure 9.5). Our results show that there is a significant desire to further involve the public in some of the activity of the institute; however, they also show that there is also some scepticism, especially with regards to involving the public in decisions about deciding which direction research should be conducted. For instance:

- We would like the public to become more actively involved in decisions about the research conducted at our research unit *(53%, Agree)*.
- We would like the public to become more involved in discussing the implications of the research we do, but not necessarily in decisions about our research directions *(53%, Agree)*.
- The public does not need to be scientifically literate to discuss the implications of our research *(53%, Agree)*.
- The public does not need to understand the full picture; we explain what we think is appropriate *(35%, Disagree)*.
- The public wants to contribute to science *(38%, Agree)*.

Barriers to researchers' engagement in the communication of RIs

We report on barriers to researchers to engage in the activities of the institute because as Entradas et al. (2020) show scientists are one of the most important contributors to institutes level of public communication.

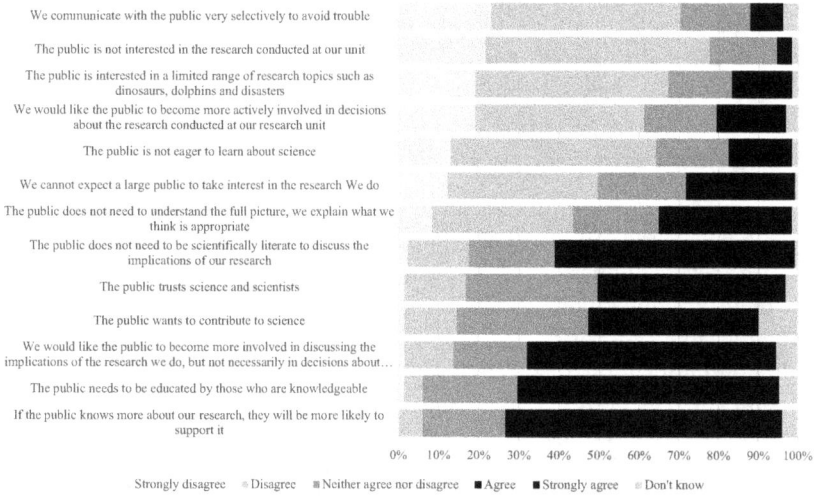

Figure 9.5 Public perceptions identified in Dutch institutes.

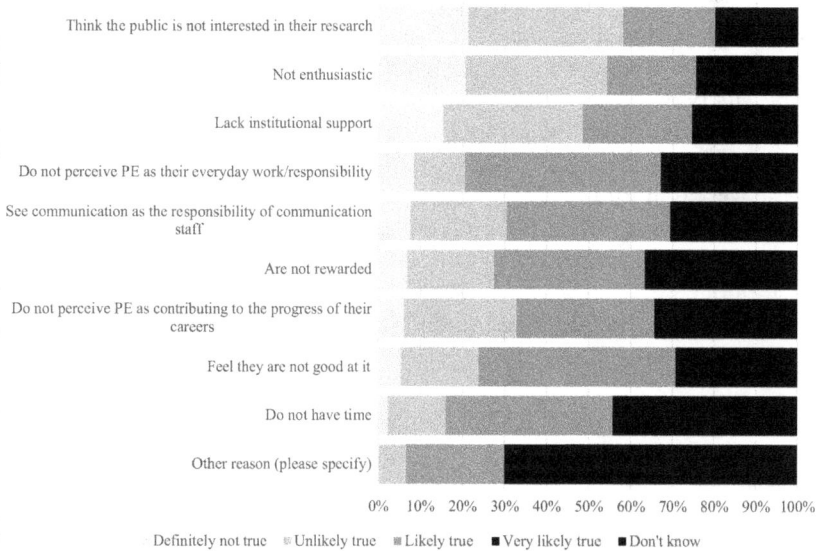

Figure 9.6 Barriers to researchers' engagement in public communication.

Our results show that time constraints and perceived lack of responsibility are clear barriers to communication at the institute level. This is also connected with gaining recognition for the work put into doing PE. On the question of citing time constraints, 36% answered "likely true" and 24%

"very likely true". In reference to responsibility, 47% answered that it is "likely true" that they do not feel it is their daily responsibility to conduct PE—rather, it is the responsibility of the communication staff (either of the unit or of the institution). Finally, 33% answered "likely true" that doing PE is not perceived to be contributing to career advancement (Figure 9.6).

Conclusions and future work

It is clear from our findings that the Dutch RIs host an active and fertile research community in terms of public engagement. We found evidence that RIs with specific public engagement plans perform better than RIs that do not have them. Having a policy provides a certain number of guidelines on how to conduct PE, and it may also create more awareness of the importance of public communication for the dissemination of research beyond the scientific environment. The policy could also provide clear guidelines for allocation of resources, which could contribute to solving issues such as time constraints and training and personnel not receiving a wage for doing outreach. Despite the support at the policymakers level, Dutch RIs are not required to have specific public engagement policies by their home organisation (e.g. universities) or by the policy institutes, such as the ministry or funding agency.

This leads to a mismatch between the expectations by policy setting in Dutch Research Agenda and the public engagement policy implementation at RI levels. There is a *Terra nullius* a "nobody's land" in terms of institutional public engagement implementation and coordination. This *Terra Nullius* sits between an "arms race" to public engagement resources at the level of RIs, "boots on the ground" of individual researchers and the "top marching orders" from the top policy stakeholders—namely, the Ministry and the Research Funding organisations.

But what are the factors keeping this empty land *Terra Nullius*? Why are the policies not implemented by funding agencies? Why do universities not enable public engagement with appropriate resources and policies? These questions deserve further investigation.

References

Blankesteijn, M., Munnichs, G. M., and van Drooge, L. (2014). *Science as a battle-ground: Public controversies around science and policy.* The Rathenau Institute.
Dalderup, L. (2000). Wetenschapsvoorlichting en wetenschapsbeleid in Nederland 1950–2000. *Gewina: Tijdschrift voor de Geschiedenis der Geneeskunde, Natuur-wetenschappen, Wiskunde en Techniek*, 23, 165–192.
De Boer, H., Enders, J., and Schimank, U. (2007). On the way towards new public management? The governance of university systems in England, the Netherlands, Austria, and Germany. In Dorothea Jansen (Ed.), *New forms of governance in research organizations* (pp. 137–152). Berlin: Springer.
De Bok, C., and Mulder, H. (2004). Wetenschapswinkels in de kennissamenleving. *Tijdschrift voor Hoger Onderwijs*, 22, 123–139.

de Jonge, J. (2014). *Report on the series of debates trust in science.* Rathenau Instituut. http://www.rathenau.nl/uploads/tx_tferathenau/Verslag_Debatreeks_Vertrouwen_in_de_Wetaliteit_2014.pdf (Accessed 17 December 2020).

Dijkstra, A. M. (2008). Of publics and science. How publics engage with biotechnology and genomics (PhD thesis). University of Twente.

Dijkstra, A. M., Seydel, E.R. and Gutteling, J. M. (2004). Effectieve wetenschapscommunicatie: een communicatievraagstuk. Kennisdagen Communicatie 2003. *Effectieve publiekscommunicatie: hints voor de Wetenschaps- en techniekcommunicatie. Papers, positiepapers en tips.* Amsterdam: Stichting Weten.

Dijkstra, A. M., van Dam, F., and Van der Sanden, M. (2020). The Netherlands: From the first science information officers to the Dutch Research Agenda. In *Communicating science: A global perspective.* Canberra: ANU Press.

Engelshoven, I. (2019). Nieuwsgierig en betrokken – de waarde van wetenschap. (Parliamentary Report) The Ministry of OCW, The Hague.

Entradas, M., Bauer, M. W., O'Muircheartaigh, C., Marcinkowski, F., Okamura, A., Pellegrini, G., ... & Li, Y. Y. (2020). Public communication by research institutes compared across countries and sciences: Building capacity for engagement or competing for visibility?. *PloS One, 15*(7), e0235191.

Jongbloed, B. (2003). Marketisation in higher education, Clark's triangle and the essential ingredients of markets. *Higher Education Quarterly, 57*(2), 110–135.

López-Goñi, I., and Sánchez-Angulo, M. (2018). Social networks as a tool for science communication and public engagement: Focus on Twitter. *FEMS Microbiology Letters, 365*(2). https://doi.org/10.1093/femsle/fnx246

Lürsen, M., Mulder, H., and Lieshout, M. (2000). *Kronkelpaden en afslagen; ontwikkelingen in en om wetenschapswinkels.* Gewina: Tijdschrift voor Geschiedenis der Geneeskunde, Natuurwetenschappen, Wiskunde en Techniek, 23, 207–213.

Ministry of OCW. (2012). The science system in the Netherlands – an organisational overview. The Ministry of OCW, The Hague.

Ministry of OCW. (2014). 2025 – Vision for science choices for the future. The Ministry of OCW, The Hague. https://www.government.nl/documents/reports/2014/12/08/2025-vision-for-science-choices-for-the-future

Mulder, H., and Straver, G. (2015). Strengthening community–university research partnerships: Science shops in the Netherlands. In B. Hall, R. Tandon and C. Tremblay (Eds.), *Strengthening community university research Partnerships: Global perspectives* (pp. 181–196). Victoria, British Columbia: University of Victoria and PRIA.

NWA. (n.d.) https://vragen.wetenschapsagenda.nl/

NWO. (2019). Connecting science and society. Strategy 2019–2022. https://www.nwo.nl/en/about-nwo/strategy

Stappers, J. G., Reijnders, A. D., Möller, W. A. J., and Hesp, L. A. T. M. (1983). Wetenschap als gemeengoed. Een studie van de wetenschapsvoorlichting in Nederland's Gravenhage, Staatsuitgeverij.

Trip, F. (1975). Nota wetenschapsbeleid. https://repository.overheid.nl/frbr/sgd/19741975/0000202941/1/pdf/SGD_19741975_0004061.pdf

Wiedenhof, N. (1978). *Wetenschapsvoorlichting: een bijdrage tot beeld-, oordeels- en besluitvorming.* Eindhoven: Technische Hogeschool Eindhoven.

Chapter 10

US American Scholars Are Finding Paths to Engagement through their Research Institutes and Centers

John C. Besley and Anthony Dudo

Introduction

Many scholars in the United States participate in public engagement activities (Bennett, Dudo, Yuan, & Besley, 2019) and there is substantial discussion across the academic and science community about the importance
of ensuring strong connections between science and society (e.g., Leshner,
2003; Scheufele, 2013). However, no over-arching national policy mandates
such activities (Besley, Dudo, Yuan, & Lawrence, 2018; Rainie, Funk, &
Anderson, 2015) and we know little about the structures in place that might
help scientists set and achieve their engagement goals (Besley, 2020). Individual organizations and funders, sometimes, have policies that may encourage or enable communication activities by scholars, but most of these
tend to vary greatly and give scholars a wide range of freedom to pursue
their own interests. And, while there appears to be a growing constellation
of people in the scientific community who want to find ways to increase both
the quality and quantity of communication activities, the focus has largely
been on helping individual scientists (Dudo, Besley, & Bennett, 2020; Dudo,
Besley, & Yuan, In press) rather than research-focused organizations that
sit at the meso level (Entradas & Bauer, 2016; Entradas et al., 2020). This
key level—operationalized in this chapter as research institutes and centers
focused around specific topics and challenges—sits between individuals and
larger-scale institutions such as universities in which scientists interact on a
daily basis and seems central to any effort to improve the effectiveness of
public engagement efforts. The current chapter, therefore, attempts to provide a novel look at communication from the perspective of research centers
and institutes in the United States with the argument that science communication scholars—as well as the broader scientific community—might benefit
from greater attention to this meso level. First, however, it is worth briefly
highlighting the dearth of federal (i.e., national) policy in this area.

The focus on individual scientist rather than organizations is especially
apparent at the federal level. This is the level around which most American research discussions take place because of the dominant role that the

DOI: 10.4324/9781003027133-13

federal government plays in nonbusiness research funding (National Science Board, 2020). A range of federal-level funding-related policies could serve as weak drivers of overall communication activity by individuals rather than organizations. Perhaps most prominent is the National Science Foundation's (NSF) requirement that individual grant applications include a description of both the "intellectual merit" of a research program and a discussion of how the research will have "broader impacts" beyond academia. This broader "research impact" requirement (by the country's primary funder of nonhealth, nondefense research) often sees research teams commit to various communication- and education-related activities aimed at ensuring that the research is useful and used by relevant communities. However, the quality of these activities varies widely and is the subject of substantial discussion (Kamenetzky, 2013; Watts, George, & Levey, 2015; Wiley, 2014). Further, while research organizations, sometimes, apply for "center" grants that fund multiple research teams, the more standard structure appears to be for individual scientists to apply for specific projects as principal investigators and meso-level research organizations to seek multiple grants from a range of sources.

Although it does not appear to have been the subject of focused research, this may make it difficult for organizations to fund full-time communication support teams as discussed below. One exception to this pattern is grants the NSF, sometimes, funds for specific communication- and engagement-focused activities and research through a range of different divisions, including those focused on education, social and behavioral sciences, and the natural sciences. The National Institutes of Health (NIH)—the largest funder of health-focused research—does not have a similar policy, but the health-promotion focus of much of the research may mean that there are practical reasons why researchers might wish to build relationships with geographic, policy, or patient communities (e.g., community-based participatory research) (O'Fallon & Dearry, 2002). Again, with some exceptions, the NIH model largely appears to be built around individual and team-based grant proposals rather than organizational proposals. One potential counterpoint to the individual focus of federal grant policy is the fact that, for historical reasons, each state has a "Land Grant" university that provides both agricultural and other "extension" services to residents. These universities were created through the ceding of federal land to endow universities across the country and have continued to receive federal agriculture funds to support their activities, including extension services that see faculty and other university experts helping people like farmers and community groups make use of evidence-based best practices (Mcdowell, 2003).

Beyond the federal government, individual organizations also have a range of policies that may make it more likely that scientists will take part in activities that see them communicate outside of their academic community. Again, however, these policies focus largely at the individual rather than meso level. Core among these is the fact that most tenure-track faculty in the United States are judged in terms of their research, teaching, and service. While the service component can include a range of things, it can often include external-facing activities (Green, 2008; Meara, Eatman, & Petersen, 2015). This could include attempts to engage a range of stakeholders in communities and government although it also includes inward-focused activities such as peer-reviewing journal articles and volunteering with scientific societies. Many universities also employ a range of professional communicators who could help research groups, departments, colleges, and the overall university interact with people outside of the department. For example, the 4,500-student College of Communication Arts and Sciences at Michigan State University where this chapter was partly written employs a team of about five communicators. These communicators help faculty and administration engage with alumni, potential employers of graduates, communities that faculty may wish to study, communities who may use research produced in the department, and listeners of college-owned public radio and television stations. In some cases, these audiences may be potential donors of money or other resources, whereas, in other cases, they allow the college to fulfill educational or research missions of the university. In practice, inasmuch as these communicators report to the college dean (the top academic administrator of the college), much of their work is focused on college-level priorities such as fundraising rather than the priorities of individual researchers. Further, the colleges typically assess faculty based on their individual service to their unit or field rather than their contribution to an organization's impact and the communication team largely works on behalf of the college dean rather than individual researchers.

Researchers could also obtain communication support through the scholarly or professional societies to which they belong. These groups may provide things such as training, speaker bureaus, or other ways of making it easier for scholars to find communication opportunities and succeed at communication (Yuan, Dudo, & Besley, 2018). For example, the American Geophysical Union (2019) hosts both in-person and online communication training as well as opportunities to write and speak with nonspecialist audiences, including policymakers. Scholars can also join advocacy-oriented groups such as the International Union for the Conservation of Science or March for Science. Unfortunately, these groups are often quite resource-limited and, thus, have limited capacity for providing communication support (Dudo et al., 2020; Yuan et al., 2018).

Overall, the core feature of public engagement in the United States is that it is largely up to individual scholars to decide how to best share their science and perspectives on public issues. Further, it is likely that such scholars face resource-scarcity in their efforts. There are, however, various efforts to better understand the overall landscape of public engagement-focused science communication in the United States, but this work is still in its exploratory phases (Besley, Newman, Dudo, & Tiffany, In press; Dudo et al., 2020; Dudo et al., In press; Gentleman, Weiner, Cavalier, & Bennett, 2018; Risien, Nilson, & Smith, 2018; Yuan et al., 2018; Yuan, Dudo, & Besley, 2019). Groups such as the National Academies of Sciences, Engineering, and Medicine—again, funded partly by foundations—have also recently taken an interest in finding ways to improve the quality and scope of the science community's engagement work (e.g., National Academies of Sciences, 2016; Scheufele, 2013). At the government level, there is also some funding provided to try and increase the effectiveness of broader impact activities (e.g., National Alliance for Broad Impacts, 2019). The current study, thus, occurred in a period of transition for the United States and it will be interesting to see if patterns change over time.

Methods

For the American survey, the focus was on research centers and institutes at top universities. The initial population of the survey was, thus, defined as institutes or centers at the 115 universities that the Carnegie Foundation classifies into doctoral universities with the "highest research activity" (sometimes known as "Research 1" or "R1" universities), a select group of the country's top doctorate-granting universities. To develop the sampling frame, a list of institutes and centers was made by visiting the institutional websites of each university. In almost all cases, these universities publish a list of these units and contact information was collected and the units were classified by the research team using the six-category OECD classification schema. Then, for each university, a random number generator was used to select two units from each university for each category (i.e., 12 units per university or 230 per OECD group). In cases where a university did not have a relevant center or institute, units were selected from other universities at random. Contact information was obtained from the Internet for selected units although it ultimately was not possible to identify enough agricultural science units such that the final contact list included 1,366 potential respondents. The survey asked that the person responsible for managing communication activities—whether the overall leader of the organization or a communication-focused individual—completes the survey, where possible.

To implement the initial survey, an initial introductory email was followed by six follow-up emails. This resulted in an initial sample of 210 responses. Next, a subset of 200 respondents was sent a traditional mail survey (shortened from the original, online survey) with a follow-up post-card, and a second copy of the instrument to boost the final sample and ensure that email nonrespondents were statistically similar to nonrespondents (few substantive differences were found). Respondents who wrote back to indicate that they were not appropriate contacts were updated in the database and a new letter was sent. Response rates were 30% for the natural sciences (n = 59), 12% for Engineering and Technology (n = 24), 19% for the Medical and Health Sciences (n = 38), 11% for the agricultural sciences (n = 23), 25% for the social sciences (n = 50), and 18% for the humanities (n = 36). Given both the small overall sample size and the size of the subsamples by category, the task here is not to make strong inferences about point estimates for the underlying populations. Instead, the hope is to provide an exploratory sense of the current public engagement infrastructure landscape as well as to make comparisons between groups.

In addition, one noteworthy set of choices underlying this research is that we did not seek to survey academic departments, colleges, and centralized communication groups. This was largely done in order to focus on a single type of organization and a sense that, whereas departments and colleges have faculty focused on a range of activities, institutes and centers are more likely to have identities that would lend themselves to targeted communication or public engagement activities related to their area of scholarship. For example, whereas a college-level communicator reporting to a dean might, sometimes, emphasize a specific researcher's work, they might also be responsible for the college's alumni and student communication and might, thus, spend a relatively limited amount of time focused specifically on communicating about scholarship. The approach used was also meant to be relatively consistent with the approach used in the United Kingdom. That being said, it should be recognized that a project focused on the broad range of communicators at American universities might result in somewhat different results.

Results

The initial picture that emerged in the data is that only a third of research centers and institutes in the United States have access to specialist communication staff. Further, the nature of this staff is limited in size. Specifically, of those who responded to the survey, 8% indicated that they had no access to communication staff, but 56% said they rely on communication staff outside of the unit. This left just 36% who had access to their own communication staff. Of this third, 16% said they had access to two or more communication staff members, 24% said they had access to one full-time communication staff member, and 40% said they only had access to part-time communication help. The amount of staff held by the remaining 20% is unclear. About half of the research centers or institutes were between 4 and 29 people (data not shown), bigger organizations had more support.

Beyond size, the background of the communicators available to research center and institutes also varied widely. Specifically, within the 36% of meso-level organizations that said they had their own communication staff, about half (54%) said they had a communication staff member with humanities training, 40% mentioned social science training, 26% mentioned natural science training, 14% mentioned medical or health training, and 13% mentioned engineering or technological training. Respondents further noted that about 40% of communicators had previously worked in advertising, public relations, or marketing, 29% had staff that had previously worked in design, 25% had staff that had previously worked in research, 19% had staff with previous work in journalism, and 19% said their communication staff had no previous professional experience. Overall, the picture is one of substantial diversity in terms of size and background. The fact that less than half of communicators had a background in communication strategy (i.e., they had worked in advertising, public relations, or marketing) may be noteworthy inasmuch as it might speak to a focus on tactical rather than strategic activities (Besley, O'Hara, & Dudo, 2019; Grunig, Grunig, & Dozier, 2006).

Below, we provide additional details aimed at telling the national story. Limited comparisons are presented for research area although most of these comparisons highlight similarities between categories rather than differences. Additional comparisons are not provided for parsimony but the biggest (and predictable) difference is that larger centers and institutes tend to do more engagement and have more engagement resources.

What types of activities are American scholars doing?

American scholars, through their institutes and centers, are participating in a range of public engagement activities despite a lack of dedicated staff. Traditional types of activities such as public talks (perhaps unsurprisingly) seem to be the most common but many other types of communication activities also occur. Below, public engagement activities are reported by mode.

For face-to-face engagement, the most common type appeared to be public lectures, with only about one in ten research institutes/centers saying that they have not done a public lecture in the previous year (Table 10.1). A further two in ten say they have participated in some sort of open day, such as lab tours. The remaining categories are less common but still mean that a substantial number of scholars are participating in a range of activities. The fact that it is possible to create a summative scale of activities suggests that research institutes/centers whose affiliates participate in one type of activity are also more likely to participate in other types of activities. ANOVA tests suggest there are often limited differences between fields (i.e., research areas) although there are some types of activities where it appears that there may be a difference between those in some aspects of the natural sciences and those in the social science or humanities. For example, those in the agricultural sciences appear more likely than others to host open days and exhibitions, which makes sense given the focus of their work. It is also

Table 10.1 Public events and traditional media channels engagement activities by research institutes/centers

	Public events engagement activities										Traditional media channels engagement activities						
	Scale (a = .76)	Publ. lect.	Open days	Sci. cafes	Exhibit.	Indus.	School talks	Delib. events	Cit. Sci.	Festiv.	Scale (a = .86)	Newspa.	Mags.	Press Rel.	Radio	TV	Ent. TV
Total (Never %)	2.07	12%	23%	39%	39%	41%	47%	47%	50%	63%	2.14	21%	25%	29%	34%	43%	72%
Nat. Sci.	2.21	14%	22%	45%	45%	35%	40%	53%	41%	50%	2.17	28%	29%	25%	36%	46%	70%
Eng. & Tech.		13%	13%	18%	18%	30%	38%	35%	38%	36%		22%	30%	35%	38%	43%	67%
Med./Health Sci.		14%	32%	39%	39%	26%	56%	49%	50%	50%		14%	19%	25%	36%	36%	71%
Agricultural Sci.		14%	0%	17%	17%	24%	42%	42%	53%	60%		10%	5%	10%	19%	29%	78%
Social science		11%	34%	53%	53%	58%	48%	33%	52%	88%		10%	21%	35%	21%	31%	74%
Humanities		8%	22%	32%	32%	58%	56%	67%	65%	78%		39%	40%	36%	54%	67%	75%
Total (Mean)	2.07	3.06	2.68	1.77	1.98	2.08	2.03	1.99	1.99	1.53	2.14	2.59	2.37	2.41	2.18	1.94	1.45
Nat. Sci.	2.21	3.02	2.79	1.91	1.91	2.16	2.18	1.91	2.25	1.74	2.17	2.54	2.24	2.57	2.12	2.02	1.55

Eng. & Tech.	2.33	3.30	3.04	2.00	2.55	2.48	2.29	2.17	1.95	1.91	2.14	2.65	2.35	2.39	2.08	1.96	1.42
Med./Health Sci.	2.06	2.86	2.50	1.81	1.85	2.37	1.75	2.06	2.06	1.69	2.09	2.56	2.51	2.33	2.14	.20	1.53
Agricultural Sci.	2.02	2.81	3.48	1.63	2.44	2.33	2.05	1.79	1.63	1.45	2.39	2.80	2.90	2.95	2.38	2.06	1.33
Social Sci.	1.99	3.17	2.34	1.60	1.66	1.79	2.00	2.37	2.04	1.21	2.03	2.96	2.53	2.25	2.52	2.04	1.42
Humanities	1.89	3.19	2.42	1.71	2.03	1.64	1.94	1.53	1.62	1.31	1.81	2.03	1.91	2.17	1.77	1.56	1.36
F	1.69	0.85	3.58	0.73	3.78	3.59	0.86	2.66	1.69	4.98	1.91	3.28	3.12	0.41	1.69	2.44	1.39
Significance	.14	.52	<0.00	0.60	<.00	<.00	.51	.02	.14	<.00	.09	.01	.01	.84	.14	.04	.23
DF1 = 5, DF2 =	172	215	214	205	203	212	214	213	204	210	194	214	212	209	213	216	212

Note: Response categories for means where: 0 (1), Once a year or less (2), 2–6 times/year (3), 7–20 times/year (4), 21+/year (5).

noteworthy that deliberative events, citizen science, and science festival participation are somewhat rare despite being a seemingly popular topic of the science communication literature. It may be that such activities will become more frequent over time.

Beyond face-to-face engagement, many organizations also appear to be interacting with various types of traditional media (Table 10.1). The most common is newspapers—who do much of the primary reporting in the United States (Golan, 2006)—but other types of media are also common. The data also suggest that many groups of scholars are seeking coverage. Specifically, about two of three research institutes/centers indicate that they have issued a press release in the previous year. Slightly more than half had been involved in television news or documentaries. There did not appear to be many substantial differences by research areas, but it did appear that humanities-focused research institutes and organizations were especially likely to be relatively more involved in all types of media activities, whereas agriculture-focused meso-level organizations were relatively less active.

Many research institutes and centers in the United States reported taking part in some type of online activity alongside other types of engagement activities (Table 10.2). Almost all maintained a website and about two in three maintained a Facebook page. Other channels were somewhat less common although it is notable that about half of research institutes centers had done something using YouTube and one in five research institutes/centers appeared to be taking part in podcast activity. In other words, they are getting involved in video and audio production, not just written or image-based communication. There were very few substantive differences by OECD category.

Largely consistent with the activities data, general audiences are the most common group that research centers and institutes say they reach. Journalists and students also seem quite common (Table 10.2). The responses seem to indicate that meso-level organizations are only sometimes trying to specifically reach industry, government, and NGOs as audiences. This might suggest a lack of strategic targeting of their efforts. Again, however, the differences—if any—between research areas seemed to be relatively modest.

Audience perceptions

How American research institutes and centers seek to communicate is only part of the story. It may also be important to understand how respondents at these organizations view their audiences as these may shape how they communicate. In this regard, those responding on behalf of the organizations indicated that they collectively held a range of potentially negative and positive perceptions about the audiences that they might reach through engagement activities. Importantly, however, negative perceptions appeared to be relatively less common than positive perceptions. Put differently, the

Table 10.2 New media channels engagement activities by research institutes/centers

	New media channels engagement activities*								Perceived audiences for engagement**							
	Scale (a = .81)	Website	Faceb.	Twitt.	YouTu.	Blogs	Podcast	Insta.	General public	Journal.	Studen. (Infor.)	Studen.	Local gov't	Non-local Gov't	Industry	NGOs
Total (Never %)		8%	31%	44%	56%	61%	79%	84%	3%	5%	6%	9%	16%	17%	19%	19%
Natural science		12%	45%	45%	56%	57%	84%	84%	7%	10%	3%	10%	32%	19%	14%	22%
Engineering and Tech.		4%	33%	38%	30%	64%	75%	82%	0%	8%	13%	8%	17%	8%	13%	14%
Med./Health Science		3%	43%	64%	73%	68%	77%	88%	5%	5%	11%	17%	19%	19%	17%	24%
Agricultural Science			33%	52%	45%	74%	85%	81%	4%	0%	20%	5%	10%	5%	5%	15%
Social science		13%	22%	43%	65%	56%	77%	90%	0%	2%	2%	9%	2%	10%	26%	10%
Humanities		6%	8%	23%	54%	58%	78%	74%	0%	3%	0%	3%	9%	31%	35%	23%
Total (Mean)	2.40	3.28	2.82	2.53	1.70	1.68	1.28	1.35	3.64	3.07	3.46	3.16	2.75	2.93	2.88	2.91
Natural science	2.32	3.21	2.50	2.45	1.61	1.72	1.20	1.38	3.54	3.07	3.71	3.25	2.47	2.76	2.81	2.81
Engineering and Tech.	2.66	3.70	2.79	2.79	2.39	1.50	1.38	1.50	3.71	3.00	3.39	3.17	2.96	3.33	3.83	3.00
Med./Health Science	2.14	3.14	2.46	2.11	1.45	1.56	1.29	1.26	3.49	3.08	3.09	2.94	2.49	2.83	3.11	2.81
Agricultural Science	2.29	3.29	2.76	2.19	1.85	1.37	1.25	1.57	3.39	3.05	2.90	2.95	2.67	3.00	3.36	3.00
Social science	2.45	3.19	3.08	2.59	1.65	1.77	1.35	1.20	3.81	3.27	3.57	3.23	3.15	3.38	2.49	3.25
Humanities	2.63	3.39	3.39	3.00	1.63	1.89	1.28	1.37	3.86	2.89	3.63	3.22	2.86	2.40	2.32	2.63
F	1.56	0.86	2.38	1.58	3.15	1.05	0.42	0.78	1.02	0.64	2.54	0.48	2.59	3.56	6.13	1.22
Significance	.17	.51	.04	.17	.01	.39	.83	.57	.41	.63	.03	.79	.03	<.00	<.00	.22
DF1 = 5, DF2 =	208	215	217	214	207	206	212	210	221	218	212	217	218	217	216	214

Notes: *Response categories for means are where: 0 (1), Once a year or less (2), 2–6 times/year (3), 7–20 times/year (4), 21+/year (5). **Responses categories for means were "Never" (1), "Rarely" (2), "Occasionally" (3), "Frequently" (4), "Very frequently" (5).

responses indicated relatively positive views about audiences. Only a small proportion of respondents from the surveyed centers and institutes, for example, indicated that they thought potential audiences were uninterested in the type of work done at their unit or lacked desire to learn about science (Table 10.3). Overall, the most common negative perception seemed to be that audiences might be interested in a relatively small range of topics, but even this view was seen as "likely" or "definitely true" by only about a fifth of centers and institutes that responded. In contrast, positive views about audiences appear to be widely held. Specifically, most organizations— nearly three-quarters—indicated that they thought that the public wants to contribute science and half (or more) said that potential audiences want to discuss their unit's work or that audiences were able to discuss this work (Table 10.3). That being said, only about a third of centers and institutes said they think the public trust science and scientists and about a fifth indicated that they thought the public wanted any active role in decision-making. There were only a few minor differences in terms of research areas.

Rationale for engagement

The survey included two types of questions aimed at understanding centers and institutes' explanations about both why their organization participated in public engagement activities and factors that might drive researchers within the organizations to avoid engagement. The responses suggest a general desire to share insights rather than to advance their own interests or advance their field. Consistent with studies of individual scientists, the responses also suggest that a primary barrier is resources and not a lack of inclination to engage (Besley et al., 2018).

On the first set of questions, the most common reason given for engagement had to do with a general desire to share research (Table 10.4). About four in ten respondents gave this response as the first or second most important reason for organizing engagement activities. The next most common response was to respond to policies of their home institutions and about three in ten respondents gave this answer. About two in ten respondents each indicated that their purpose was to listen to the public, get support, or raise their profile. At the low end, the least common responses were related to responding to funding policies, national policies, or to attract young people to sign; all of these were given by about one in ten of respondents, or less. For this set of questions, no analysis by research area was done because of the relatively small responses to any one category.

The second set of questions asked more specific questions about why the respondents thought that scholars at their organizations might avoid engagement (Table 10.5). The most common types of responses were associated with resources, with eight in ten respondents noting time concerns, and about seven in ten noting both a perceived lack of support and lack

Table 10.3 New media channels engagement activities by research institutes/centers

	Scale (a = .74)	Negative perceptions (strongly agree/agree) %					Positive perceptions (strongly agree/agree) %				
		Public not interested in unit's research	Public not eager to learn about science	Public does not need full picture	Cannot expect large audience interest	Interest is limited to small range of topics	Aud. wants to contribute to science	Want public to discuss unit work implicati.	Aud. does not need not lit. to discuss unit's work	Aud. trusts science and scientists	Would like public to be more active in unit decisions
Total		4%	12%	15%	18%	20%	74%	57%	50%	33%	21%
Natural science		0%	6%	17%	10%	22%	53%	56%	61%	39%	17%
Engineering and Tech.		5%	21%	21%	16%	11%	50%	58%	48%	24%	26%
Med./Health Science		6%	15%	21%	24%	20%	47%	61%	56%	28%	22%
Agricultural Science		11%	17%	22%	11%	17%	67%	55%	60%	45%	37%
Social science		0%	10%	5%	26%	15%	43%	61%	60%	28%	25%
Humanities		7%	16%	11%	21%	35%	59%	57%	55%	34%	25%
Total (Mean)	2.33	2.14	2.32	2.38	2.46	2.47	3.53	3.53	3.28	3.10	2.71
Natural science	2.19	2.04	2.04	2.37	2.22	2.37	3.80	3.59	3.06	3.13	2.57
Engineering and Tech.	2.40	2.11	2.53	2.63	2.32	2.42	3.21	3.28	3.33	3.28	2.56
Med./Health Science	2.33	2.09	2.36	2.52	2.61	2.50	3.34	3.48	3.24	2.97	2.77
Agricultural Science	2.49	2.33	2.56	2.61	2.44	2.50	3.35	3.61	3.17	2.83	2.78
Social science	2.24	1.93	2.20	2.17	2.55	2.34	3.67	3.48	3.50	3.21	2.93
Humanities	2.61	2.57	2.68	2.26	2.71	2.85	3.39	3.68	3.44	3.08	2.64
F	1.67	2.85	2.35	1.08	1.20	.83	2.39	.55	.93	.78	.66
Significance	.15	.02	.04	.37	.31	.53	.04	.74	.46	.57	.66
DF1 = 5, DF2 =	172	184	179	185	185	179	174	184	182	178	183

Notes: Response categories for mean were "Strongly disagree" (1), "Disagree" (2), "Neither agree nor disagree" (3), "Agree" (4), "Strongly agree" (5).

Table 10.4 Reasons for taking part in public engagement activities

	First reason	Second reason	Least important
We want to disseminate our research to the public	26%	18%	3%
We aim to respond to the policy/mission of our host institution/university	17%	11%	10%
We want to listen and involve the public in our research	10%	10%	10%
We want to get public support for the research we do	9%	13%	3%
We want to raise our research profile	7%	12%	9%
We want to attract funding	6%	9%	10%
We aim to respond to the policies of our funding bodies	5%	5%	7%
We aim to respond to national policies of public engagement	4%	4%	15%
We aim to recruit new generations of scientists	3%	4%	19%

Notes: N = 230. Respondents could select one option for each question.

of resources. A high proportion of respondents also noted more individual-level reasons for potential lack of engagement, including both sense of self-efficacy (i.e., would be good at engaging) and the sense that it might be someone else's responsivity and not contribute to their careers (all of which were seen as potentially true by about half of respondents). Less than a third of respondents seemed to think that low engagement was caused by a lack of scholars' enthusiasm or perceived public interest. There were, again, few substantive differences by field.

Overall experience

Two items in the survey might serve as summary indicators of current engagement efforts. First, it appears that slightly more than half of the people who responded felt that their engagement efforts had been successful or very successful (Table 10.6). However, this still left a third of respondents who were ambivalent (i.e., they felt their engagement efforts were neither successful nor unsuccessful) and a small proportion-about a tenth of respondents—who felt they were unsuccessful in their engagement. More emphatic is the finding that almost all respondents said they thought their organization was either putting the right amount of resources into engagement or that additional resources would be beneficial (Table 10.6), with slightly more respondents suggesting that more resources were needed. As

Table 10.5 Perceived reasons for lack of engagement

	Individual perceptions							Resource-related perceptions (Very likely true/likely true %)			
	Scale mean (a = .81)	[N]ot ... their ... responsibility	They feel they are not good at it	[S]ee ... as the responsibility of the comm. staff ...	[Does not] ... contribut[e] to ... careers	They are not enthusiastic about [it] ...	They think public not interested ...	Scale mean (a = .70)	They do not have time for it	They are not rewarded for [it]	They lack institutional support...
Total (very/likely %)		57%	55%	51%	44%	30%	25%		79%	69%	65%
Natural science		56%	68%	41%	46%	26%	25%		88%	64%	71%
Engineering and Tech.		77%	65%	60%	43%	32%	14%		87%	81%	64%
Med./Health Science		56%	56%	50%	60%	21%	22%		76%	68%	71%
Agricultural Science		47%	65%	67%	41%	44%	31%		68%	56%	67%
Social science		59%	42%	54%	40%	45%	26%		66%	74%	56%
Humanities		48%	38%	42%	27%	15%	30%		86%	74%	57%

(Continued)

Individual perceptions

Resource-related perceptions (Very likely true/likely true %)

	Scale mean (a = .81)	[N]ot ... their ... responsibility	They feel they are not good at it	[S]ee ... as the responsibility of the comm. staff ...	[Does not] ... contribut[e] to ... careers	They are not enthusiastic about [it] ...	They think public not interested ...	Scale mean (a = .70)	They do not have time for it	They are not rewarded for [it]	They lack institutional support...
Total (mean)	2.26	2.58	2.78	2.29	2.37	2.02	2.02	3.05	3.33	2.83	2.98
Natural science	2.36	2.91	2.65	2.60	2.38	1.91	1.86	3.10	3.35	3.1	2.86
Engineering and Tech.	2.29	2.71	2.44	2.41	2.63	1.85	2.03	2.98	3.00	3.00	3.00
Med./Health Science	2.53	2.47	2.88	2.89	2.35	2.13	2.13	2.85	2.89	2.72	2.83
Agricultural Science	2.27	2.56	2.28	2.51	2.30	2.16	1.92	2.89	2.98	3.05	2.74
Social science	1.98	2.19	2.21	2.17	2.05	1.62	2.11	2.91	3.07	2.85	2.71
Humanities	2.27	2.57	2.52	2.44	2.36	1.96	2.01	2.97	3.12	2.93	2.87
F	1.09	1.41	2.35	.155	.85	1.32	.36	.39	1.41	.57	.45
Significance	.37	.22	.04	.17	.52	.26	.88	.85	.22	.73	.82
DF1 = 5, DF2 =	136	183	158	175	175	175	173	171	189	180	185

Notes: Percentages are for respondents indicating very likely true and Likely true. Response categories for means were "Definitely not true" (1), "Unlikely true" (2), "Likely true" (3), and "Very likely true" (4).

Table 10.6 Perceived successfulness of public engagement efforts and need for public engagement resources, both by research area

	Perceived successfulness					Resource adequacy			
	Mean	Very unsuc. (1)	Unsuc.	Neither succ. nor unsucc.	Succ.	Very succ. (5)	Should devote fewer resources	Devotes the right amount ...	Should devote more ...
Total	3.53	3%	7%	34%	48%	8%	1%	45%	54%
Natural science	3.63	2%	3%	36%	49%	10%	0%	47%	53%
Engineering and Tech.	3.42	4%	17%	29%	33%	17%	4%	42%	54%
Med./ Health Science	3.33	3%	11%	44%	33%	8%	3%	47%	50%
Agricultural Science	3.22	9%	4%	43%	43%	0%	0%	36%	64%
Social science	3.63	2%	8%	27%	51%	12%	0%	41%	59%
Humanities	3.69	0%	0%	31%	69%	0%	3%	50%	47%

Notes: N = 227. ANOVA for mean comparison by research area: $F(5, 221) = 1.71$, $p = .13$; N = 226, Chi-Square (10) = 6.32, $p = .79$.

with other questions in the survey, there was little substantive difference between research areas.

Conclusion

Overall, the public engagement picture that emerges in the United State when looking at research universities' centers and institutes is one in which a substantial amount of activity—much of it traditional in nature—is ongoing. Most organizations surveyed, in this regard, were doing most of the activities asked about in the survey at least once per year and had positive views about participants. It also seems clear that there is a great deal of similarity across research areas with only small, understandable variation. However, it is also a picture in which organizations are making do with limited resources and a sense that more resources would help.

If there is a societal desire for scholars to engage more deeply with their fellow citizens, this might suggest the need for additional attention on questions about whether the scientific community is doing enough to ensure that centers and institutes have access to the resources—including human resources in the form of professional communication support and dedicated time—that may be needed to ensure scholars can engage effectively (Besley,

2020; Davies & Horst, 2016). This focus on the meso-level organization—the level at which many researchers operate day to day—thus likely deserves far more attention than it has typically received. As noted in the introduction, science communication researchers and practitioners have tended to focus on the individual level when trying to understand science communication activity (Bennett et al., 2019). And, as noted, this is also the primary level for much of the funding in the United States.

References

American Geophysical Union. (2019). Sharing Science, from https://sharingscience. agu.org/

Bennett, N., Dudo, A., Yuan, S., & Besley, J. C. (2019). Chapter 1: Scientists, trainers, and the strategic communication of science. *Theory and Best Practices in Science Communication Training* (pp. 9–31). New York: Routledge.

Besley, J. C. (2020). Five thoughts about improving science communication as an organizational activity. *Journal of Communication Management*, 24, 155–161. doi: 10.1108/jcom-03-2020-0022

Besley, J. C., Dudo, A., Yuan, S., & Lawrence, F. (2018). Understanding scientists' willingness to engage. *Science Communication*, 40, 559–590. doi: 10.1177/1075547018786561

Besley, J. C., Newman, T., Dudo, A., & Tiffany, L.-A. (In press). Exploring Scholars' Public Engagement Goals in Canada and the United States. Public Understanding of Science.

Besley, J. C., O'Hara, K., & Dudo, A. (2019). Strategic science communication as planned behavior: Understanding scientists' willingness to choose specific tactics. *PLoS ONE*, 14, e0224039. doi: 10.1371/journal.pone.0224039

Davies, S. R., & Horst, M. (2016). *Science Communication: Culture, Identity and Citizenship*. London: Palgrave MacMillan.

Dudo, A., Besley, J. C., & Bennett, N. (2020). *Landscape of Science Communication Fellowship Programs in North America*. Princeton: Rita Allen Foundation.

Dudo, A., Besley, J. C., & Yuan, S. (In press). Science communication training in North America: Preparing whom to do what with what effect? *Science Communication*.

Entradas, M., & Bauer, M. M. (2016). Mobilisation for public engagement: Benchmarking the practices of research institutes. *Public Understanding of Science*, 26, 771–788. doi: 10.1177/0963662516633834

Entradas, M., Bauer, M. W., O'Muircheartaigh, C., Marcinkowski, F., Okamura, A., Pellegrini, G., … Li, Y. Y. (2020). Public communication by research institutes compared across countries and sciences: Building capacity for engagement or competing for visibility? *PLOS ONE*, 15, e0235191. doi: 10.1371/journal. pone.0235191

Gentleman, D., Weiner, S., Cavalier, D., & Bennett, I. (2018). Landscaping overview of U.S. facilitators of scientists' engagement communities. *Paper Presented at the Support Systems for Scientists' Communication and Engagement Workshop IV: Science Engagement Facilitators*, Monterey Bay, CA. https://www.informalscience. org/support-systems-scientists%E2%80%99-communication-and-engagement-exploration-people-and-institutions

Golan, G. (2006). Inter-media agenda setting and global news coverage. *Journalism Studies*, 7, 323–333. doi: 10.1080/14616700500533643

Green, R. G. (2008). Tenure and promotion decisons: The relative importance of teaching, scholarship, and service. *Journal of Social Work Education*, 44, 117–128. doi: 10.5175/jswe.2008.200700003

Grunig, J. E., Grunig, L. A., & Dozier, D. M. (2006). The excellence theory. In C. H. Botan & V. Hazleton (Eds.), *Public relations theory II* (pp. 21–62). Mahwah, NJ: Lawrence Erlbaum Associates.

Kamenetzky, J. R. (2013). Opportunities for impact: Statistical analysis of the National Science Foundation's broader impacts criterion. *Science and Public Policy*, 40, 72–84.

Leshner, A. I. (2003). Editorial: Public engagement with science. *Science*, 299, 977. doi: 10.1126/science.299.5609.977

Mcdowell, G. R. (2003). Engaged universities: Lessons from the land-grant universities and extension. *The ANNALS of the American Academy of Political and Social Science*, 585, 31–50. doi: 10.1177/0002716202238565

Meara, K., Eatman, T., & Petersen, S. (2015 Summer). Advancing engaged scholarship in promotion and tenure: a roadmap and call for reform. *Liberal Education*, 101, 52+.

National Academies of Sciences, E. A. M. (2016). *Communicating Science Effectively: A Research Agenda*. Washington, DC: The National Academies Press.

National Alliance for Broad Impacts. (2019). *Center for Advancing Research Impact in Society*. Retrieved from https://broaderimpacts.net/aris/

National Science Board. (2020). *U.S. R&D Performance and Funding. Science and Engineering Indicators*. Retrieved from https://ncses.nsf.gov/indicators

O'Fallon, L. R., & Dearry, A. (2002). Community-based participatory research as a tool to advance environmental health sciences. *Environmental Health Perspectives*, 110, 155.

Rainie, L., Funk, C., & Anderson, M. (2015). *How Scientists Engage the Public*. Washington, DC: T. P. R. Center.

Risien, J., Nilson, R., & Smith, B. (2018). Landscape overview of university systems and people supporting scientists in their public engagement efforts. *Paper presented at the Support Systems for Scientists' Communication and Engagement Workshop III*, Academic Institutions, San Diego, CA.

Scheufele, D. A. (2013). Communicating science in social settings. *Proceedings of the National Academy of Sciences*, 110, 14040–14047. doi: 10.1073/pnas.1213275110

Watts, S. M., George, M. D., & Levey, D. J. (2015). Achieving broader impacts in the National Science Foundation, Division of Environmental Biology. *BioScience*, 65, 397–407. doi: 10.1093/biosci/biv006

Wiley, S. L. (2014). Doing broader impacts? The National Science Foundation (NSF) broader impacts criterion and communication-based activities.

Yuan, S., Dudo, A., & Besley, J. C. (2018). Landscaping Overview of Organizational Support for Public Engagement from Scientific Societies: Study funded by the Kavli, Moore, Packard, and Rita Allen Foundations for the Support Systems for Scientists' Communication and Engagement Project.

Yuan, S., Dudo, A., & Besley, J. C. (2019). Scientific societies' support for public engagement: An interview study. *International Journal of Science Education*, Part B, 1–14. doi: 10.1080/21548455.2019.1576240

Public Communication in Japanese Research Institutes

Still Dark or Sunrise?

Asako Okamura

How science communities have related to social issues in Japan

In 1999, the Budapest declaration – that is, the Declaration on Science and the Use of Scientific Knowledge and the Science Agenda – was adopted by the World Conference on Science, calling for new scientific missions, including 'Science in Society and Science for Society' (World Conference on Science, 1999). This declaration seemed to have a certain impact on science policy in Japan. It has been cited many times by Japanese science policy documents. However, it is not clear whether its principles and practices have spread to science communities in Japan.

As in other countries, rising concerns about social issues such as pollution, climate change, mad cow disease, genetically modified organs (GMOs), etc., have led scientists in Japan to tailor their professional interests to such issues to some degree. However, there seems to be a certain distance between most scientists and the public, possibly due to recognition gaps such as the 'safety myth',[1] even on issues that face intense public concern such as GMOs. There may also be differences between the cultures of different scientific fields (e.g., Jensen, 2011; Besley, and Dudo, 2017). For example, the field of astronomy has historically nurtured a culture of active outreach (Entradas and Bauer, 2019), aiming to convey a pure enjoyment of science, and the field has naturally gained popularity. Other scientific fields may need to explain their usefulness to society (Entradas et al., 2019) or rely on the principle of voluntarism. The motivations for science communications, thus, differ across scientific fields.

The nuclear accident caused by the Great East Japan Earthquake on March 11, 2011, had a significant impact not only on Japanese society, in general, but also on the behaviours and perceptions of individual scientists. This event seemed to end the 'safety myth' about nuclear power plants and forced them to reach out to society and communities beyond their peers. After the accident, several new practices arose, such as the deliberative polling surveys on energy and environmental options in 2012 that drew citizens and researchers into a joint policy decision-making process (Sone, 2014).

DOI: 10.4324/9781003027133-14

However, since the last change of the ruling party, these efforts have been sluggish (Kudo et al., 2018).

The declining popularity of science and math among Japan's youth might have been another trigger for science communities to reach out to society, leading to the launch of the Super Science High School (SSH) programme[2] by the Ministry of Education, Culture, Sports, Science and Technology (MEXT) in 2002. This programme has motivated top scientists to visit SSH schools and talk to students.

However, these observations are anecdotal and not based on hard evidence. Much is unknown about how scientists interact with the public, especially regarding the degree to which such activities are institutionalised (or not). Therefore, the objectives of this chapter are as follows: (1) to explain the structure of public communication activities from a bird's eye view, as seen by research institutions (RIs); (2) to detangle the complex interactions among the different elements that may affect the level of public communication activities at research institutions; and to (3) understand how policies, institutional factors (context), and perceptions play a role in determining the level of public communication activities at research institutions. We use 'public communication' and 'public engagement', which infer science communications and engagement with non-academic, both in one way and interactive means. Where necessary, we use the term 'participatory' to indicate more interactive communications and engagement.

Our study and its research questions

Due to the inadequacy of existing data in Japan, we participated in MObilisation of REsources for Public Engagement (MORE-PE) survey to answer some specific research questions.

Because of rising social concerns about emerging technologies and governmental efforts to promote public communication, it may be natural to think that scientists believe they (or science communicators) need to communicate their research and its social implications to the public. However, little is known about how such communication is conducted at research institution levels. Especially, although existing policy measures in Japan emphasise the importance of the 'co-creation' of Science, Technology and Innovation (STI), it seems that participatory activities are probably not widespread and is practised only in limited cases. With this in mind, we created the following research questions:

> RQ1: What are the details of public communications in Japanese research institutions, and does this vary across different research areas?

With this question, we aimed to find out about the types of activities performed by RIs, the audiences addressed, the motivations, the resources allocated; and what institutional factors raise and lower the level of public communication activity and how it differs across research areas.

The second research question was:

RQ2: How is public communication activity affected by research settings policies and perceptions?

We aimed to determine what factors, including policies, motivate or demotivate the practices of public communications and what perceptions (perceived images towards researchers, media and the public) are behind these.

The chapter proceeds as follows. The next section explains Japan's national research system to provide a background for its public communication activities and details how science policy in Japan has shifted its focus to science's role in society. We then describe the sample and analysis. Results section describes the main results to our RQs, and the final section discusses the implications and limitations of our study for science communication in Japan.

National background

The Japanese research system

In 2018, the higher education[3] and government sectors accounted for 26% of research and development (R&D) expenditures in Japan, while the other 72.8% were generated by the private sector (NISTEP, 2020a). The higher education sector accounted for 18.8% of R&D performance in 2018. For the year, within the higher education sector in Japan, there were 783 universities, of which 86 were national universities, 85 were local public universities, and 612 were private universities (NISTEP, 2016a).

Overview of policies focused on science and society

Every five years since 1996, the government formulates a new Science and Technology (S&T) Basic Plan as a policy framework that uses a long-term perspective to implement systematic and integrated S&T policies. The relationship between science and society is one of the policy areas of these Basic Plans. The first S&T Basic Plan (1996–2000) stated that it was necessary to improve public literacy, understanding, and interest in science and technology was necessary. The second phase (2001–2005) added a focus on the communication necessary to build a new relationship between science and society as well as the creation of more interactive channels. This second plan also addressed the ethics and social responsibility of researchers. The third plan (2006–2010) introduced the concept of Ethical, Legal, and Social Issues (ELSI) and clearly stated that these issues must be dealt with. The fourth plan (2011–2015), enacted after the nuclear accident of 2011, further recommended the involvement of the public in the planning and promotion of policies.

Finally, the fifth plan (2016–2020) stated the necessity of stakeholder involvement in a process of 'co-creation' of STI through dialogue and collaboration.

Thus, from the initial S&T Basic Plan to the most recent, the policy focus has shifted from a unidirectional approach of promoting understanding to a more interactive one. The fifth plan relates to the similar concept of 'responsible research and innovation', which gained widespread attention through EU Horizon 2020, the EU Framework Programme for Research and Innovation enacted from 2014 to 2020. However, rather than merely observing that these concepts and philosophy are stated in the relevant policy frameworks, it is necessary to verify that they are actually put in place through concrete policy measures.

Policy measures

Japan has enacted a variety of concrete policy measures about science and society. These range from those on formal STEM education (e.g. SSH) to those on informal STEM education activities, including science fairs/museums, science communication, and the promotion of dialogue with and accountability to the rest of society. Some of these measures started as early as the 1950s, and others were only recently introduced in the new millennium (Watanabe, 2017).

Policy measures targeted at researchers. One of the policy measures directed at researchers was the 'Science and Technology Dialogue with the Public' measure, introduced by the Cabinet Office in 2010 (Cabinet Office of Japan, 2010). This mandates researchers to conduct science communication activities if they receive more than 30 million JPY in public funding per year.[4] Their activities are assessed before and after project completion. This measure implies that researchers receiving public funding are highly aware of their obligation to carry out science communication. As for individual research projects, from 2011 to 2017, the Japan Society for the Promotion of Science (JSPS) required all applicants to The Grants-in-Aid for Scientific Research (KAKEN) to include their plan for science communication with the public in their research proposals.

Policy measures targeted at research bodies and institutions. Graduate-level training courses in science communication were established to several universities in the middle of the 2000s, with government subsidies for limited years (five years). Also, many universities have recently incorporated relevant strategies or policies, including social contribution, social/local engagement, industry-university collaborations, which may partially include public communication of their research activities. However, little is known about the relevant policies of the subordinate institutions of research bodies.

Rationale of the MORE-PE survey in Japan

As explained before, several policies have been formulated to deepen the relationship between science and society in various ways, such as increasing students' science literacy, promoting science communication. However, there have been few empirical studies to verify the effect of such policies. Little is known about the degree to which science communities, whether individual or institutional, conduct science communication activities. Furthermore, we know far less about the ways in which science communities, whether individual or institutional, engage with the public and collaborate with other stakeholders to practise the 'co-creation' of STI, as encouraged by the 5th S&T basic plan.

The data necessary for policy evaluation are scarce; however, some measurements of the links between science and society have been conducted in Japan. The Japan Science and Technology Agency conducts surveys of science communication activities among scientists (CSC-JST, 2013, 2017). Also, the Surveys on Full-time Equivalency Data at Universities and Colleges (FTE survey) are conducted by MEXT and the National Institute of Science and Technology Policy (NISTEP), which could be a means to investigate the percentage of time (that is, effort rate) that researchers spend on activities for social contributions, including engagement with the public (Kanda and Tomizawa, 2015).

Another source that could be useful in determining the contribution to public communication activities at the research body level is the National University Corporation Education and Research Evaluation, conducted by MEXT and the National Institution for Academic Degrees and Quality Enhancement of Higher Education (NIAD-QE). This evaluation is mandatory for national universities every six years. It evaluates national universities' medium-term objectives and plans as well as their annual plans for education, research, and management. This evaluation scheme includes a section on collaboration with the public and contribution to society. Individual evaluation reports of the universities, including data such as the number of events held for public outreach about research results, are freely available online (NIAD-QE).

The above sources are derived from surveys at individual and research body levels. MORE-PE was the first nationwide empirical investigation of public communication activities at Japan's research institution level.

Methods

Sample

The statistical unit of analysis was set to the meso level of universities: that is, the subdivisions of the university such as departments, research institutions, and centres. The research areas covered were (1) Natural Sciences; (2) Engineering and Technology; (3) Medical and Health sciences; (4) Agricultural and Veterinary Sciences; (5) Social Sciences; and (6) Humanities and the Arts.

In Japan, no official surveys have created a comprehensive list of all universities' subordinate institutions, and this is especially true for non-departments. Thus, it is challenging to conduct a census of all university research institutions because roster information is not available. As universities' sizes and levels of research differ significantly, we believe that a simple random sampling at the university level is not necessarily desirable. Ideally, stratified sampling of universities should be conducted by establishing criteria such as region, university size, and the level of research.

However, as our primary objective was to observe the level of public communication associated with research activities, we decided to target only universities in which research was currently being conducted above a certain level. As there is no standard indicator for the level of research activity per university, we used information provided by the Grants-in-Aid for Scientific Research (KAKEN) about the amount of research funding allocated per university, as published on the JSPS website (JSPS). The KAKEN is a major, highly competitive public research fund in Japan and can be considered a proxy index to gauge the universities' research levels.

We first identified 66 universities from the JSPS list, which covered around 80% of universities that were granted KAKEN funding in FY2017 (including both continuing funding from previous years and new funding granted in 2017). Next, we identified all subordinate institutions of the targeted universities. For 32 of the 66 universities, we used the information about the subordinate organisations or institutions published by NISTEP (2016b).[5] For the other targeted universities, we identified the subordinate organisations from the organisation chart on the universities' websites. We chose the subordinate institutions as the second-level organisation for the universities. For example, in the case of the University of Tokyo, the University of Tokyo was the representative body (RB), and the Institute of Medical Science of the University of Tokyo were the subordinate institutions (RI).

In total, we identified 1470 RIs and collected the following information about them: name of the institution; URL; postal codes; postal address; telephone number; and name and e-mail address of the director (or representative) when available. We then identified the research area of each institution based on the descriptions from the website.

Survey distribution

Before distributing the online survey, we conducted a pre-survey. For those RIs of universities whose websites did not include an e-mail address (338 RIs), we sent a postal mail to collect this information. For those RIs of universities whose websites did include an e-mail address (1149 RIs), we sent e-mails to confirm the contact e-mail address and the person in charge. Some RIs declined to participate in this survey. After the pre-survey, the sample size decreased to 1134.

The online survey was conducted from February to May 2018, and two reminders were sent. One questionnaire was collected per institute and completed by the Unit's Director/Coordinator/Head of institute (39%); Researcher (28%); Management/Administrative staff (24%); or Other (6%); and Communications staff (1%); 6% did not answer. The number of years in these positions was an average of 8.7 years with SD = 9.6. We received N=321 responses for a response rate of 28%, as shown in Table 11.1.

Characteristics of the sample

The average size of RIs as given by the number of researchers working in the institute was 163.1 people, with a significant variance (SD = 394.8). Regarding the research budget per year, 40% were under 50 million yen; 12% were between 50 and 100 million; 25% were between 100 and 500 million yen; 10% were over 1 billion yen; and 6% did not know. Regarding the split between research and teaching, 8% devoted exclusively to research; 31% engaged more in research than teaching; 12% more in teaching than research; and 42% engaged equally in research and education (7% did not answer).

The median length of RIs establishment was 21 years; the average 38 years with a slightly large variance (SD = 36.7). 2% responded that they had started the communication activity less than one year ago; 12% for between one to five years; 17% for between five to ten years ago; 52% for more than ten years ago; 16% responded 'don't know'.

Moreover, about 49% of RIs increased communication activities over five years: 33% stayed the same level; 4% decreased the level, with 12% for 'don't know'. Approximately 40% of Engineering and Agricultural RIs, 50% of Social Sciences RIs, and 60% of Natural Science and Humanities RIs have increased public communication activities over the past five years.

Table 11.1 Sampling frame and response rates (for universities)

Research area	Sampling frame (N)	Sampling frame (%)	Responses (N)	Responses (%)	Response rate (%)
Natural Sciences	245	22	64	20	26
Engineering and Technology	225	20	68	21	30
Medical and Health Sciences	194	17	67	21	35
Agricultural Sciences	42	4	16	5	38
Social Sciences	318	28	65	20	20
Humanities	110	10	27	8	25
NA			14	4	
Total	1134	100	321	100	28

Variables

We created composite indicators as 'variables' for the three main types of activities and audiences investigated in the study: public events, media engagement and audiences. For example, for public events, 11 types of events were classified into three: 'diffusion', 'institutional', or 'participatory'. The classification of variables was based on the results of principal component analysis and its interpretations. Then we calculated the activity level for each event type by converting categorical responses to numeric values and summation over type. The same manipulations were performed for media engagement (nine items)', classified into: 'traditional media', 'press relations', or 'promotion activities'; and audiences (eight items), classified as 'general public', 'industry, local government, and national government', or 'Non-governmental Organisations (NGOs) and media' (see Chapter 16 for description of variables). Table 11.2 shows the variables created.

To investigate RQ2, i.e. how policies, institutional factors, and perceptions played a role in determining the level of public communication activities, linear regression models were used. In these models, the dependent variable was the level of activity by type (diffusion, institutional, participatory); the explanatory variables were:

Institutional-context variables. (1) Research area (*fos*), (2) When RI started the public communication activities (*startscpe*), (3) Researchers' engagement in public communication activities (*res.engagement*), (4) Research budget (*research.budget*), and (5) Size (*researcher.no*)

Policy-related variables. Three variables related to policy in communications were included: (1) Having a public communication policy (*policy1*); (2) Having a public communication action plan (*policy2*); and (3) Expecting researchers to be involved in communication with the public (*policy3*).

Perception-related variables. Nine items related to perceived barriers to researchers engaging with the public (three items), media perceptions, and public perceptions (three items) were also included in the analysis (see Chapter 16 for full description of perceptions). These were (1) Researchers don't have motivations for communication activities (e.g. they do not think public communication is a part of their job) (*rnomotivation:*); (2) Researchers have no confidence in communications (e.g. the researchers consider themselves unskilled at public communication) (*rnoconfidence*); (3) Researchers do not communicate with public due to constraints (e.g. lack of financial resources, time, or institutional support) (*rconstrain*); (4) Having negative views on media (*mnegative*); (5) Having indifferent views on media (*mindifferent*); (6) Having positive views on media (*mpostitive*); (7) Having negative views on public (*pnegative*); (8) Having positive views on public (*ppositive*); (9) Having a deficit-model view on public (*pdeficit*).

In the regression models, nonresponse data were replaced with single imputation method where necessary.

Table 11.2 Variables by type of activity

Types of activities	Subordinate variables	
Public events	Diffusion type	1. Public lectures 3. Open days, workshops, guided tours 6. Science cafes and seminar formats 10. Talks/workshops at schools
	Institutional type	2. Exhibitions 4. Science Festivals/Science Fairs
	Participatory type	5. National event (National science week) 7. International events 8. Deliberative and participatory events 9. Events organized by private institutions 11. Citizen science projects
Media engagement	Traditional media	1. Newspaper 2. Radio 3. TV 9. Magazines, newspapers 11. Popular books
	Press relations	5. Press conferences 6. Press releases
	Promotions	7. Newsletters 8. Brochures/leaflets/publications
Audiences	General public	1. General public 2. Schools 3. Students outside teaching
	Industry, local government, and national government	4. Members of local municipalities/councils/associations 5. Delegates from industry 6. Governments/politicians/policy-makers
	Non-governmental organisations (NGOs) and media	7. NGO 8. Media and journalists

Note: The questions asked were for public events and traditional media, 'Roughly, how many times in the past 12 months has your research unit engaged in the following events/traditional media channels, either as organiser or contributor?'; for audiences 'How often has your research unit/researchers engaged with each of them in the past 12 months?'

Public communication activities

To understand the overall structure of public communication activities, let us start with the macro view: *what activities – public events and traditional media,* are used and how frequently, with whom (audiences) are activities being conducted, as well as the communication resources (staff, funding, and policies), and why. We focus on disciplinary differences, as shown by the field of science, indicated in the responses.

Public events

Generally speaking, RIs hold *'diffusion'* events most frequently, including public lectures, open days/guided visits, and talks at schools. Other types of events, such as *'Institutional'* events (e.g. exhibitions and science cafes) or *'Participatory'* events (e.g. citizen science projects), are less frequent (see Figure 11.1).

Figure 11.1 shows a variation between disciplines in terms of the types of events they carry. In the Social Sciences and Humanities, the frequency of public events is generally lower than in other fields. These results corroborate previous ones using data from institutes in Portugal (Entradas and Bauer, 2017), suggesting similar patterns of public events between these two countries in regard to scientific disciplines. On the other hand, participatory-type events, such as 'Deliberative and participatory events' and 'Citizen science projects', are infrequent across all disciplines, with little variation between disciplines.

Media engagement

'Traditional media' was most frequently used, which includes articles in magazines/newspapers, interviews for newspapers, and interviews for television (Figure 11.2). Press relations including press releases and press conferences were also used. In terms of the use of traditional media, the differences between the disciplines seem to be small, but when it comes to press relations, activity is higher in the Natural Sciences, Medical and

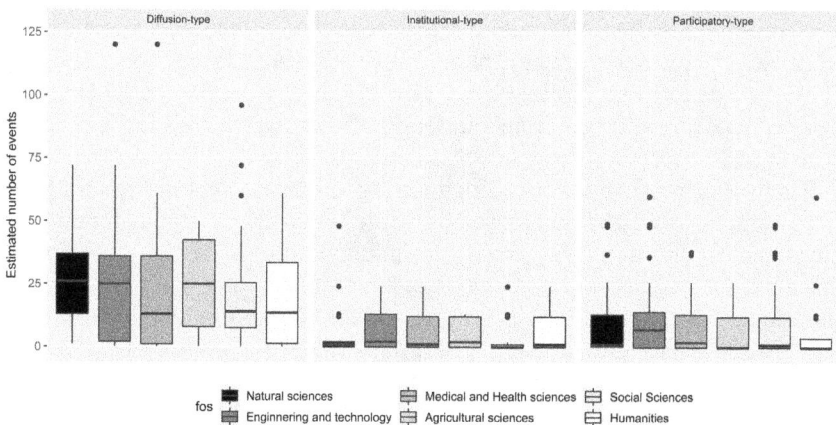

Figure 11.1 Public event activities by research area.

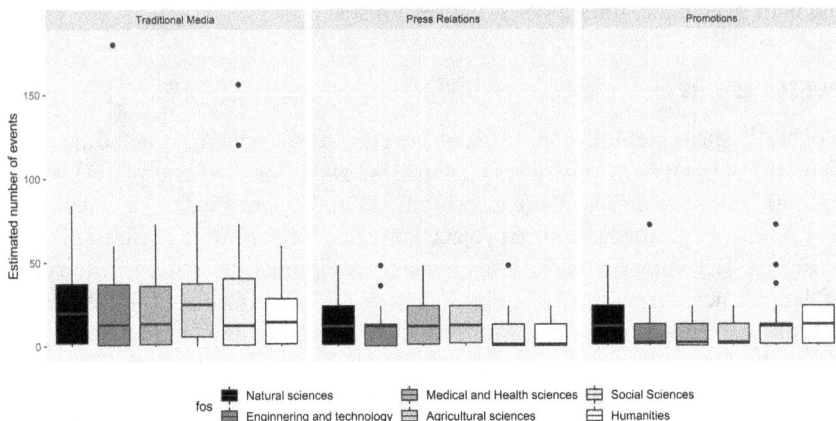

Figure 11.2 Media activities by research area.

Health, and Agricultural Sciences, and lower in the Social Sciences and Humanities (Figure 11.2).

In addition, we also asked about online engagement. Most of the institutions frequently updated their websites. However, social media – including YouTube, Instagram, Twitter, Facebook, blogs, and podcasts – were not frequently used. This is in line with findings in other countries as described in the national chapters. Due to the low engagement in these channels, we do not use these variables in our analysis.

Audiences

Regarding the audience of public communication, RIs most frequently engage with the category under general public, also including schools and students outside teaching, followed by industry and government (Figure 11.3). The delegates from industry accounted for the highest audience group in all type of audiences.

RIs in Engineering & Technology and Medicine & Health Sciences have less interaction with the general public, while Social Sciences and Humanities have more interaction. With regard to industry and government, the Social Sciences and Engineering & Technology have high levels of interaction, while the Humanities have low levels. Social Sciences have a slightly higher level of engagement with NGOs and the media than other disciplines (Figure 11.3).

Finally, it may be natural to find that correlations among key activities tend to be high. RIs with high intensity of diffusion-type public events were more likely to conduct other type of events, had more contacts with general public audiences and had more media engagement (especially for traditional

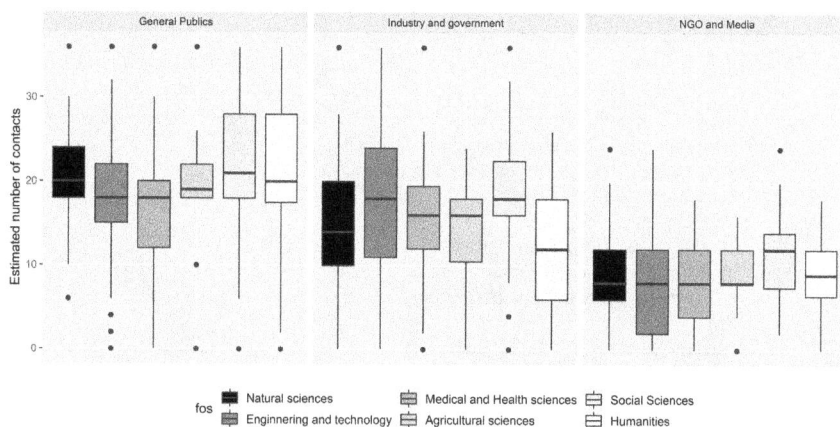

Figure 11.3 Interaction with audiences by research area.

media channels and promotion). Relatively high correlations were observed between (i) diffusion-type activities and institutional-type activities, (ii) traditional media channel and promotion activities, and (iii) audience engagement between industry & government and NGOs and media. RIs which actively engage with the website also use other online activities, including Twitter.

Communications staff and policies

Communication staff. In terms of staffing, 30% of RIs have communication staff in the RI, 49% do not have staff in the RI but have access to host institution staff, and 21% do not have staff in the RI nor access to host institution staff. Of the RIs that had communication staff, the average number of staff exclusively dedicated to communication activity was 0.98 (SD = 2.29), and the average number of staff who were partly engaged was 4.12 (SD = 5.06).

43% of communication staff had an academic background in the Humanities, 37% in the Natural Sciences, 35% in the Social Sciences, 33% in Engineering & Technology, 11% in Medicine & Life Science, and 10% in Agriculture. As for their previous work experience, 49% were researchers, 30% were in administration, 27% were in human resources, and 22% were in public relations and marketing. A significant number, 23% had no prior work experience. And, in terms of training in communications, 44% said they had attended workshops or short-term training, while 85% said they had not received any formal communications education.

Public communication policies/guidelines. Regarding the policy or action plans on public communication: 32% of RIs have a policy regarding public

Table 11.3 Public communication policies/action plans

%	False (%)	True (%)	Don't know (%)	NA (%)
Q22_1_We have a public comm policy	44	32	19	4
Q22_2_We have public comm action plans	40	39	17	4
Q22_3_We expect researchers to be involved in comm with public	12	66	18	4
Q22_4_Our comm efforts respond to national policies	13	39	44	4
Q22_5_We have no plan/policy but engage with public	28	51	16	5

communication, and 44% do not. The number of RIs with an action plan and those without were almost equal, around 40%. As for whether their institution's public communication activities are responsive to the national policy, 44% were not sure, 39% were responsive, and 13% were not responsive. Sixty-six per cent expected researchers to be involved in public communication activities, and 51% said their institutions were engaged in public communication activities even though they did not have a policy or action plan (Table 11.3).

Rationales for communication

As for the motives for communicating with non-experts, many respondents indicated that the dissemination of their research was the most important, followed by it being required by university policy and desiring public support (Figure 11.4). 'Raising the research profile' was the least frequently indicated, followed by 'involving citizens in the research', 'responding to the policies of funding bodies', and 'attracting funding'. These results imply that RIs did not consider that public communication impacts their research. Most of them considered the relationship with the public as unidirectional: they expected to diffuse their research, but did not expect to involve the public in their research. Also, these results may imply that funding bodies have not successfully incorporated the values of public communication into their funding policies.

Policies, institutional factors and perceptions

To address our RQ2, i.e. how do policies and institutional contexts affect the level of public communication activity and how the perceptions on researchers, the media, and publics influenced their activities on public communication, we run regression analysis for the three main types of public events (diffusionist, institutional, and participatory events), which we describe below (see Table 11.4 for results).

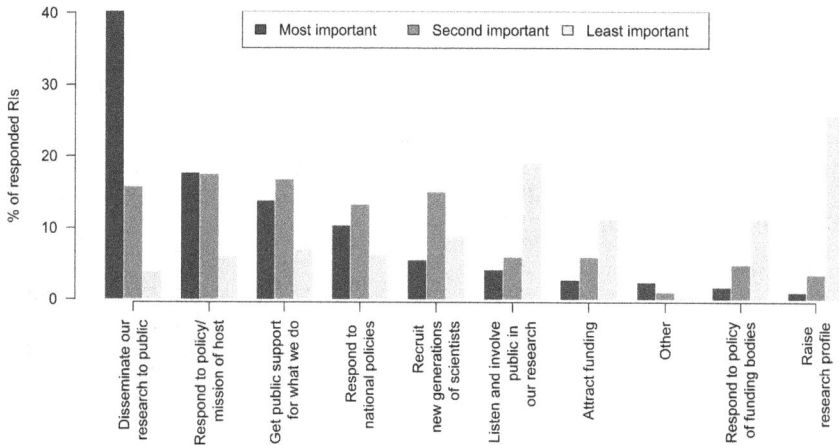

Figure 11.4 Reasons to undertake communication with non-specialist audiences.

It is important to note, however, that we rely on correlation rather than causal analysis.

Institutional context

RIs in the Engineering field have higher activity with statistical significance than Natural Science in all types of public communication activities: diffusion-, institutional-, and participatory-type events. RIs that have engaged in public communication for more than ten years have higher levels of activities, particularly in diffusion-type activity than those who have started less than ten years. The coefficient is negative (but statistically insignificant) regarding the participatory-type activity, implying that the length of public communication activity has a random or indifferent effect to such type of activity. The number of researchers has a positive coefficient with statistical significance for diffusion- and participatory-type activity.

Policies

RIs with policies and action plans have a higher level of activity than those without, but the coefficient is not statistically significant. RIs that expect researchers to participate in public communication activities have higher levels of activity in all types of public communication activities than RIs that do not. However, another regression exercise implied that when it comes to new social media channels, the results indicate that organisational policy may be a factor that discourages bottom-up activity: RIs without a public communication plan or policy are more active on social networking sites (including blogs, Twitter, Facebook, and YouTube) and RIs with a public

Table 11.4 Regression results

	D_event + I_event + P_event	D_event	I_event	P_event
Medical and Health sciences	0.067	−0.381*	0.237	0.210
	(0.515)	(0.222)	(0.214)	(0.238)
Agricultural sciences	−0.071	−0.117	0.203	−0.157
	(0.789)	(0.339)	(0.328)	(0.365)
Social Sciences	0.026	0.039	−0.138	0.124
	(0.516)	(0.222)	(0.215)	(0.239)
Humanities	−0.119	−0.076	0.132	−0.175
	(0.673)	(0.290)	(0.280)	(0.311)
Engineering and technology	0.984**	0.010	0.520**	0.453**
	(0.498)	(0.214)	(0.207)	(0.230)
Startscpe_10years more	0.417	0.452***	0.006	−0.041
	(0.370)	(0.159)	(0.154)	(0.171)
res.engagement_10–40%	0.376	0.165	0.218	−0.007
	(0.461)	(0.199)	(0.192)	(0.213)
res.engagement_40–100%	0.647	0.470**	0.201	−0.024
	(0.494)	(0.213)	(0.206)	(0.229)
research.budget	0.220	0.110*	0.051	0.059
	(0.148)	(0.064)	(0.0612)	(0.068)
researcher.no	0.250**	0.120**	0.043	0.088*
	(0.113)	(0.049)	(0.047)	(0.052)
policy1_com.policy2	0.517	0.282	0.120	0.115
	(0.549)	(0.239)	(0.229)	(0.254)
policy2_com.act.plan	0.690	0.153	0.360	0.177
	(0.529)	(0.227)	(0.220)	(0.244)
policy3_expect.researcher. engage	1.791***	0.547**	0.524**	0.720***
	(0.544)	(0.234)	(0.226)	(0.251)
Rnomotivation	0.350	0.153	0.028	0.169
	(0.405)	(0.174)	(0.169)	(0.187)
Rnoconfidence	0.761*	0.274	0.086	0.401*
	(0.448)	(0.193)	(0.187)	(0.207)
Rconstrain	−0.377	−0.055	−0.078	−0.244
	(0.349)	(0.150)	(0.145)	(0.161)
Mnegative	−0.102	0.042	−0.098	−0.046
	(0.315)	(0.136)	(0.131)	(0.146)
Mindifferent	−0.207	−0.087	−0.061	−0.058
	(0.235)	(0.101)	(0.098)	(0.109)
Mpositive	0.116	0.077	−0.006	0.045
	(0.289)	(0.124)	(0.120)	(0.133)
Pnegative	0.271	0.184	0.201	−0.114
	(0.406)	(0.175)	(0.169)	(0.188)
Pdeficit	−0.170	−0.161	−0.064	0.055
	(0.428)	(0.184)	(0.178)	(0.198)
Ppositive	0.282	0.152	0.096	0.034
	(0.395)	(0.170)	(0.164)	(0.182)
Constant	−3.421	−1.486	−0.940	−0.995
	(2.843)	(1.224)	(1.184)	(1.315)
Observations	321	321	321	321
R^2	0.261	0.315	0.144	0.159
Adjusted R^2	0.190	0.249	0.062	0.078
Residual Std. Error (df = 292)	2.746	1.182	1.143	1.270
F Statistic (df = 28; 292)	3.678***	4.785***	1.749**	1.967***

Notes: *** $p < 0.001$; ** $p < 0.01$; * $p < 0.05$

communication plan (but not a policy) are more active on social networking sites (including Twitter, Facebook, and YouTube).

Perceptions

Although there was no result with statistical significance for most of variables on perception, we obtained some interesting results. RIs that agree more that researchers are not motivated to engage in public communication had higher activity levels in all public communication activities. This was also the case in RIs that agrees more that researchers lacked confidence in their communication skills. These results may reflect some frustration feelings among respondents regarding researchers' low engagement in public communication. On the other hand, RIs that agree more on constraining reasons, such as lack of time and resources, actually had lower activity levels. RIs with a more negative view of the media or less willing to build relationships with the media had lower activity levels. RIs with a more positive view of the media had a higher level of activity. The more RIs agree to the deficit-model view of citizens, the lower levels of diffusion type, and institutional type activity. Both negative and positive views of citizens had positive coefficients on activity levels.

Discussion and conclusion

MORE-PE was the first nationwide empirical investigation of public communication activities at the research institution level in Japan. To build on the insights of MORE-PE, we addressed two main research questions concerning the public communication practices at Japanese research institutes, and the factors that affect that practice. We discuss the implications of the results presented here to policy.

Overall, the results show that science communication at Japanese research institutes seems to be at early stages. Despite policy frameworks in place such as the S&T Basic Plan that support public communication, these activities have not gained much momentum in Japanese research institutions. However, there are certain cases where institution-wide efforts seem to be expanding. Regardless of discipline, all RIs in Japan conduct public communication activities to some extent. Of the RIs that responded to our survey, the majority (approximately 50% of Engineering, Medicine and Humanities RIs, 70% of Natural Science and Social Sciences RIs, and 80% of Agricultural RIs) commenced public communication activities more than ten years ago. Approximately 40% of Engineering and Agricultural RIs, 50% of Social Sciences RIs, and 60% of Natural Science and Humanities RIs have increased public communication activities over the past five years, and RIs with longer experiences in public communications had more public communication activities, especially diffusion-type events. However, it seems that Japanese RIs are not so enthusiastic about such public communication activities, especially in comparison to institutions of other countries

(Entradas et al., 2020). Participatory activities are even less common in Japan, as these activities require more interactive ways of communication. Deliberative events, and citizen science projects are especially uncommon in most scientific fields in Japan. Thus, though diffusion activities are fairly common, other participatory activities with wider reach are not. This fact coincides with the rationale for conducting public communications: 'diffusing their research results' (unidirectional) was the most important reason, while 'listening and involving the public in research' was one of the unpopular ones.

Institutional policies seem to have mixed effects on actual level of activities. RIs with institutional policies had more diffusion type activities than those without. However, another regression exercises indicate that institutional policies may be a de-motivating factor for bottom-up activities, especially for SNS activities. RIs with no public communication plan or policy had more SNS (including blog, Twitter, Facebook, and YouTube) activities, and RIs with public communication action plans (rather than policies) had more contact with NGOs and the media. Rather than having policies, expectations matter more in determining activity levels: RIs expressing higher expectations on researchers' contributions to public communication had more activities in all type. The results also show that some frustration exists among respondents regarding researchers' engagement in public communication. Respondents that believed their 'researchers do not think public communication is part of their jobs' carried out more diffusion activities and had more contact with the public and the media. Additionally, RIs that indicated that constraints (e.g. a lack of financial resources, time, or institutional support) were the reason their researchers did not carry out public communication activities were less likely to conduct these activities.

Therefore, soft measures such as institutional reforms may be necessary to provide more financial resources, time, and institutional support, rather than merely enacting constraining policies. Changing researchers' perceptions (such as the perception that public communication is not a part of a researcher's job) may also be needed, even for RIs that already carry out many diffusion activities. Other types of efforts, including communication training and motivation training, may also help change researchers' views on public communication activities. Another critical factor is staffing. There was a lack of professionally trained communication personnel at RI levels. Thus, the institutionalization of public communication activities at RI levels seems underway but has not been successful yet. Indeed, 60% report neither success nor failure; only 32% report success for the self-evaluation of their public communication activities.

Moreover, national policies, including those of funding agencies, do not seem to motivate RIs to conduct more public communication activities. Also, RIs do not recognize that public communication activities enhance their research profile.

Thus fostering an understanding of the values of public communication to their research activities and societies would be necessary.

Our research specifically targeted university RIs. To understand the whole picture of public communication activities in Japan, this sample would need to be expanded to cover other actors, including public research organizations, science museums, funding agencies as well. Also, more research is needed on the relationship between public communication and industry, considering the role played by the private sector in Japanese research systems.

References

Besley, J., & Dudo, A. (2017). Scientists' views about public engagement and science communication in the context of climate change. *Oxford Research Encyclopedia of Climate Science.*

Cabinet Office of Japan (2010). 内閣府（H22）「国民との科学‧技術対話」の推進について（基本的取組方針）（平成22年6月19日決定）https://www8.cao.go.jp/cstp/stsonota/taiwa/index.html

CSC-JST (Center for Science Communication, Japan Science and Technology Agency) (2013). An Investigation into Scientists Involvement in Science Communication Activities. July 2013. https://www.jst.go.jp/sis/scienceinsociety/investigation/items/csc_fy2013_04.pdf

CSC-JST (Center for Science Communication, Japan Science and Technology Agency) (2017). An Investigation into Scientists' Involvement in Science Communication Activities. September 2017. https://www.jst.go.jp/sis/scienceinsociety/investigation/items/csc-report_2017researchers.pdf (in Japanese)

Entradas, M., & Bauer, M. (2017). Mobilisation for public engagement: Benchmarking the practices of research institutes. *Public Understanding of Science, 26*(7), 771–788.

Entradas, M., & Bauer, M. W. (2019). Bustling public communication by astronomers around the world driven by personal and contextual factors. *Nature Astronomy, 3*(2), 183–187.

Entradas, M., Marcelino, J., Bauer, M.W. et al. (2019). Public communication by climate scientists: what, with whom and why?. *Climatic Change,* 154, 69–85. https://doi.org/10.1007/s10584-019-02414-9

Entradas M, Bauer MW, O'Muircheartaigh C, Marcinkowski F, Okamura A, Pellegrini G, et al. (2020) Public communication by research institutes compared across countries and sciences: Building capacity for engagement or competing for visibility? PLoS ONE 15(7): e0235191. https://doi.org/10.1371/journal.pone.0235191

Jensen, E. (2011). Evaluate impact of communication. *Nature, 469*(7329), 162–162.

Kanda, Y., & Tomizawa, H. (2015). Changes in the ratio of time spent on work activities by university & college faculty members – a comparison of results of the "Survey of full-time equivalency data at universities and colleges of 2002, 2008 and 2013". *NISTEP Research Material, 236.*

Kudo, M., Yoshizawa, G., & Kano, K. (2018). Engaging with policy practitioners to promote institutionalisation of public participation in science, technology and innovation policy. *JCOM, 17*(04), N01. https://doi.org/10.22323/2.17040801.

NIAD-QE (National Institution for Academic Degrees and Quality Enhancement of Higher Education). National University Corporation Education and Research Evaluation. https://www.niad.ac.jp/english/unive/nuce.html

NISTEP (2016a). Japan Science and Technology Indicators 2016. NISTEP RE-SEARCH MATERIAL No. 251, National Institute of Science and Technology Policy, Tokyo. DOI: http://doi.org/10.15108/rm251e

NISTEP (2016b). The NISTEP Dictionary of Names of Universities and Public Organizations' (ver. 2016.1). https://www.nistep.go.jp/archives/30132

NISTEP (2020a). Japan Science and Technology Indicators 2020. NISTEP RE-SEARCH MATERIAL No. 295, National Institute of Science and Technology Policy, Tokyo. DOI: https://doi.org/10.15108/rm295e

Watanabe, M. (2017). From top-down to bottom-up: A short history of science communication policy in Japan. *JCOM*, *16*(03), Y01. https://doi.org/10.22323/2.16030401.

World Conference on Science (1999). Declaration on Science and the Use of Scientific Knowledge, July 1999. http://www.unesco.org/science/wcs/eng/declaration_e.htm

Sone (2014). 曽根 泰教, 原子力政策と討論型世論調査, 公共政策研究 (2014). 14 巻, pp. 37–50, https://doi.org/10.32202/publicpolicystudies.14.0_37

Acknowledgement

I want to thank all MORE-PE members, especially Marta Entradas and Martin Bauer, for their encouragement and patience. I am also grateful to Tatsuo Oyama and Kei Kano, who were members of the Japanese team and reviewed an earlier draft and provided valuable feedback and many insightful ideas. Last but not least, I also thank Mayuko Nakamura and Azusa Miyagi for their excellent assistance. Without their assistance, this work has not been completed.

Notes

1 The myth of nuclear safety is well known, as it is believed by many people to be "absolutely safe", even though the evidence is not clear. This is said to have been behind the Fukushima accident.

2 The SSH programme focuses on science and mathematics education with the aim of developing future international science and technology personnel. In addition to implementing advanced science and mathematics education in high schools and other institutions, the program promotes joint research with universities and initiatives to foster internationalism.

3 The higher education sector in Japan consists of universities (comprising undergraduate and graduate education), junior colleges, professional training colleges, etc.

4 The Cabinet Office provided the following three examples of 'public dialogue': (1) special lessons in science classes at elementary, middle and high schools; (2) presentation of research results to the public at universities and research institutions; and (3) symposia and lectures at museums and science museums.

5 NISTEP compiled a list of organisations which conduct research and development in Japan and also identified their subordinate institutions for the top 32 universities that publish many papers (NISTEP, 2016b).

Chapter 12

Communicative Dispositions of British Research Institutes

Martin W. Bauer

The University in Britain and elsewhere

It will be useful to start with briefly rehearsing some visions of what a research university that hosts research institutes might be, in order to see more clearly its present shape and needs to reach out into wider society.

Doring (2020) reminds readers of an ancient concept of research in higher education which nowadays seems very difficult to grasp because of its arcane elitism: the university is a hermetic Platonic idea. Any current embodiment of this 'idea' in time and place is 'mimesis' that falls short of its 'true original', and this insight is only available to the initiated few, i.e. the body of academics. A powerful implication of this vision is that academics are not working for a university but working for the 'idea of the university'. This 'truth' empowered Ernst Kantorowicz, a medievalist scholar, to utterly reject state interference into the affairs of the 'idea'. In the 1950s Cold War, Kantorowicz refused to take an oath against un-American activities and was promptly dismissed from Berkeley University. He was later offered a position at the new Princeton Institute of Advanced Studies, where he continued to contemplate his 'Two-Bodies of the King' problem, the ideal and the real one.

For Max Weber (1919), in addressing a student audience, the university is **a place to live the vocation for science**, i.e. a lifestyle for which the entry and the career path can be a wild gamble, but the internal politics of which seeks to defend the autonomy of the place vis-à-vis Church or State interference. To stay confident and to progress in this universe demands **relentless specialisation**. The University is also an institution with the **vocation of science to** serve society. Value-free enquiry examines inconvenient facts and elucidates presuppositions without advocating any of these. Scholarly researchers face decisions: *if you take position A, you gain facts Q, but you buy into presupposition x, y, and z.* More recently, we talk of paradigms or frameworks that hold the facts in place. And this contribution to society is mainly intellectual, i.e. radically **disenchanting the world**. Four contributions are, thus, highlighted: technical knowledge and calculations to increase the social control

DOI: 10.4324/9781003027133-15

of life; developing methods and making tools of inquiry; analysing 'world view' presumptions and their implications; clarifying the language and the choices between presuppositions. The academic works with a pantheon of gods (no monotheism) whose unending struggle he seeks to understand without fear or favour. This leads to a passionate plea to separate the role of the teacher from that of the charismatic leader; the teacher speaks in the lecture room and avoids politics; the leader appeals to the street in doing politics. Weber approves of the plain attitude of trading training for parental money, like trading groceries for cash, and thereby not accepting any moral indoctrination on top of it. Notably, Weber spoke of lecturing, and not seminar discussions, where the latter might well occur. We might speculate what Weber would have said about the modern academic researchers as a business entrepreneur.

Butterfield (1962) reflects on the post-WWII expansion of the British Redbrick universities. His concern is safeguarding **the tutorial system of teacher–student relations** which competes with the lecture syllabus and examination. He seeks to foster the non-utilitarian curiosity that cultivates imagination and intellectual flexibility that drives research. He worries about the '*hardening of teaching rituals into examination passing*' (i.e. skill training) that does not cultivate the intellectual elasticity which is needed to cope with circumstances where today's knowledge is already outdated. Students need to be able to **pursue knowledge for its own sake**, think outside the box, challenge the established framework and peek at the 'backside of the carpet'. He utterly opposes a dual career path of teacher or researcher which devalues the profession and cheats the socially mobile twice over: exploited by their oppressors and hoodwinked by those who try to offer opportunities (p. 25). Universities exceed knowledge transfer and convey a sense of belonging that leads to leadership functions that lay in wait for the unexpected outside any syllabus. Universities curate ideas. The Renaissance, Enlightenment and Romanticism (one is tempted to add nowadays Behavioural Science) were hedged outside the university, which became the location for their second stage revival, sifting and combining ideas into scholarly canons, thus, preserving identities through mutations.

Halsey and Trow (1971) track British academia after the Robbins reforms of the early 1960s and examine the academic in empirical detail, noting the expansion of the natural and applied sciences, and the social sciences. In 1957, there were 27 universities in the United Kingdom; by 1968, there were 46 (and by 2021 there are 132 listed). Oxford and Cambridge had long lost their monopoly but still dominate the hierarchy (p. 134). Overall academic staffing increased from below 10,000 in the 1950s to over 30,000 by the end of the 1960s (>430,000 by 2018). The career and lifestyle of university lecturers reflect tension between research and teaching, elitism and expansion. Their conclusion is a four-fold typology: **elitist or expansionist researchers, elitists or expansionist teachers**. While the elitists want to keep to small numbers of 'excellence' both for teaching and research, the expansionists accept an expanding role for the university. The expansionist teacher aspires towards a 'university without walls', taking up the extra-mural mission to reach an ever larger number of people and responding to increasing demand. They carry the arguments for universal education into tertiary education, are the likely drivers of future expansion and are the precursors of the later PE (public engagement) movement, which is our present concern. Halsey and Trow (ibidem, 451ff) point to five functions of the university and growing demand for all of them: transmission of facts (knowledge transfer), forming attitudes and values (culture transfer), development of new knowledge and ideas (R&D), selection and recognition of a social elite (providing status) and training of higher professions (lawyers, medicine, clergy, academics). Communication only enters (p. 363) as peer publications and conversations. Communication as PE is implicit in the political commitment of academics, mostly with leftist bias. Their observations prospect the expansion of the system without loss of quality that would confirm the conservative moral panic over 'more means worse'.

Similarly, Finkelstein (1984) traces the emergence of the modern academic in the United States, its long past and short history until the 1970s. This social role encompasses four tasks: teaching and tutoring, research, administration, institutional-professional-public services. The demographics of faculty is examined closely in terms of age, sex and religion: Presbyterian, Episcopalian and Jewish were over-represented; Catholics were underrepresented, so were women who made up a bare 25% of faculty. Throughout the post-war expansion, the triad of **teaching, research and service** crystallised more clearly. Academics split their week into 55% teaching (33% is PG, 22% UG), 24% research (writing 4.3 papers over four years), 17% administration and 5% community services (p. 88). The latter split into **extramural activities** and consultancy jobs (paid or unpaid, an integral part of 2/3 of careers; up to one day per week of salary complements is tolerated, if not interfering with normal duties). This 'external career' also includes **civic participation** which is much higher among university faculty than among the comparably educated population. They are three to four times more likely to sign petitions or write to governors. Social scientists are the most politically 'liberal' segment (p. 173). Academic work became a 'way of life', pervading leisure activities and eroding work-life separation (social scientists tend to bring their work to the dinner table, p. 174) and accentuating religious and ethnic ties, friendship patterns and politics. However, these characteristics are selected into rather than socialised at the university.

Noam (1996) predicted a major **shake-up of the university by the Internet revolution.** The arrival of computer networks changes the role of Universities, from a centre of information in a geographical location to becoming a hub in a global network. *In the past, people came to the information stored at the university, in the future the information will go to the people wherever* *they are* (p. 9). Universities morph from a local community of scholars to a hub where networks of scientific disciplines keep a foot; equally local libraries become access points in vast inter-library networks. Knowledge transfer can be privatised, while universities take on a different role: they shift from knowledge transfer to mentoring, internalisation–identification and role modelling, guidance, socialisation, interaction and group activities based on physical proximity and community building beyond the campus time, from lectures back to tutorial (to Butterfield's delight).

Shapin (2008) traces the historical shift of the academic stereotype from the aristocratic **amateur**, the nerd with independent means (Dr. Frankenstein), to the **professional scholar** who prices his or her intellectual independence from State or Church interference (Weber), to the modern-day **entrepreneur** who has one foot in the lab (publications) and the other in business (patents) without experiencing much conflict between these. Clearly, business operations demand and normalise a very different expectation of communication, namely, that of lobbying, branding and marketing.

Willetts (2017) sees in the university **a major socialisation agency**, i.e. the main way Western societies now manage the transition from adolescence into productive adulthood. The university, thus, replaces older pathways via military service or industrial apprenticeships. He defines as follows:

'A university is an independent corporation devoted to higher education. It is a community of scholars and students. Its autonomy, evidenced above all in the right to award its own degrees, set a university apart from other forms of higher education. For education to be higher it must be at the frontiers of knowledge: that does not mean it must include research but at least its teaching needs to be informed by new discoveries and current arguments' (p. 5).

Willets recognises behind all differences in delivery a shared **European heritage of higher education.** The system of European universities, as defined by the Bologna process, comprises three cycles: **Bachelor, Master and Doctoral**. However, his account wants to go beyond clarifying the character of the university and grasp its demanding environment. This opens our research question: how does the university project its activities in society to secure public support?

Finally, Mandler (2020) reconstructs the British university in the context of a global transition to mass education, comprising universal secondary education and the massive expansion of tertiary levels. For the historian, the university is a 'battlefield' where the meritocratic vision of educating the 'able and capable' and technocratic planning for the 'needed' labour force clash with the democratic aspiration for a better life for all those 'able and willing' to educate. Economists struggle over whether studying at university means acquiring a consumer good or making an investment.

This brief review shows that a university is, thus, many things:

- An arcane Platonic idea for the initiated few.
- A lifestyle and a vocation for science in the service of society making the world disenchanted place; any presuppositions are elucidated without fear or favour in relentless specialisation.
- A place where non-utilitarian curiosity is cultivated in order to prepare for changing circumstances where useful knowledge is quickly outdated.
- A place of massive expansion with tensions between research and education, elitism and widening access come to the fore.
- The workplace where time is distributed between research, teaching and community service; the question is raised: how does this academic body match society's diversity in terms of religion, class, ethnicity, gender and its intersections?
- The Internet revolution alters a geographical centre into a network hub: people no longer go to university but access the training provided from anywhere; the functions of the locality need to refocus on quality social interaction.
- The (self)-stereotypes of academics shift from that of a nerdy amateur, via the independent professional scholar, to that of an entrepreneur prospecting profits in a knowledge market.
- A university education supersedes the military career or an industrial apprenticeship as the main socialisation agency and pathway of social mobility in modern society.
- The university is a European tradition with three cycles: Bachelor, Master and Doctorate.
- If the university is a public–private partnership, it must be audited, and widely audible and visible whether it is a quality consumer good or a viable investment.

With this list of functions, we are reminded of Habermas (1988) who argues that a functionalist analysis, i.e. reducing the university to a societal subsystem with one key function, does not exhaust its societal role. The action field 'research university' is not functionally differentiated from other spheres of society but deeply rooted into and closely coupled with the life-world through a bundle of contributions. This cannot be reduced to a main function, e.g. R&D for economic growth. Even Big Science as it happens at CERN near Geneva or ITER in Southern France cannot operate without activities that go beyond research; research cannot guarantee its own conditions of possibility. The historical university has been ratcheting up several social functions:

1 A core task of doing research and recruitment of researchers (graduate schools)

2 Training of professionals (professional certification of law, medicine, management, etc.)
3 General education (knowledge transfer in disciplines, canons, colleges, textbooks)
4 Enlightenment of public discourse, elaborating a common understanding of the world and society (Intellectuals)

An expanding education sector in need of a communication function

After having examined some visions and projections about the modern university in Britain and elsewhere, let us now examine some empirical trends that describe the state of affairs of the university in the 21st-century Britain. For his purpose, we briefly examine three trends (Table 12.1).

Trend 1: Massive expansion of the UK sector

Figure 12.1 shows the expansion of the UK university sector in several phases since the 19th century. A first expansion takes place in the early 19th century when UCL and King's College London break the English monopoly of Oxford and Cambridge; Ireland and Scotland were always a world apart. The second expansion happened in the late 19th century and around WWI with the forming of new universities such as LSE or Imperial College. The third wave of expansion takes place in the 1960s in response to the 'Sputnik Shock' of 1957, which then stalled into the 1980s. The fourth and fifth waves of expansion follow into the new millennium. This massive expansion in higher education capacity led to an increased participation rate of the young age cohorts, from 1% to 2% by the 1870s, to 3% by the 1920s. Post WWII, this increased to 15% by the late 1960s. In the 1990s, the university participation rate increased to 35% and even to 50% in the new millennium. This expansion also changes what it means to 'do a university degree' in Britain, morphing from an elite education of below 5% of mostly young men, to nearly ½ of the 18–25 age cohort, majority women, as part of an international context of development that is far from

Table 12.1 Participation rates by period of expansion

Period of expansion	Participation rate of 18–25 years
1820s	1–2%
1890–1920	2–3%
1960–1970	3–15%
1990–2000	15–30%
2000–2020	30–50%

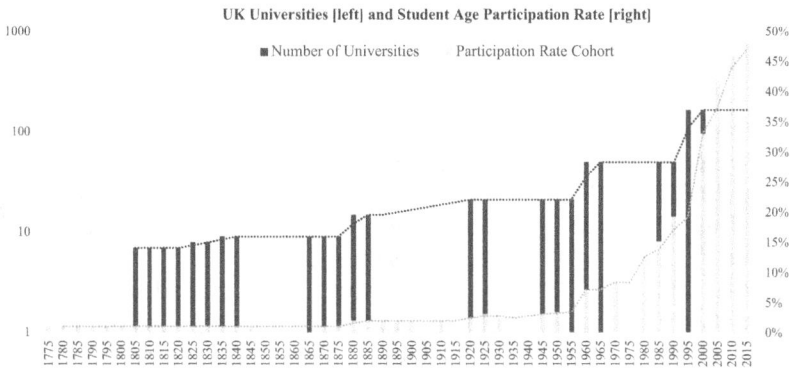

UK Universities [left] and Student Age Participation Rate [right]

■ Number of Universities   Participation Rate Cohort

Figure 12.1 The expansion of British higher education since the 18th century from age participation rates of 1–2%, to 48% by 2015 (source: THE, UK historical statistics).

being an island at the edge of a continent (Windolf, 1992; Schneider, 1982; Mandler, 2020).

Trend 2: Historically unprecedented spending on education, science and research

This expansion means a massive increase in staffing and budget volumes of the university sector from 1,500 academics in 1910, to about 30,000 by 1968, to over 430,000 by 2017/2018 (Halsey and Trow, 1971, p. 150ff; University UK, 2019). It is likely that the attention students get in terms of teacher interaction and money spent per student is declining over this expansion. The staff student ratio has nearly doubled from about 10:1 in the 1960s to 17:1 in 2010. This compares unfavourably with other OECD countries (UCU, 2012) though the trend to reduced spending per student is widespread in advanced industrial countries.

Willetts (2017), a former UK university minister, asks the question: how do we pay for all this? This question undoubtedly leads to increased pressures on university bodies for public accountability and to be audible in public. This means a massive shifting of the acceptable legitimation of science from curiosity (freedom of research) to social relevance and value for money (utility). According to the Law of Declining Marginal Utility, the cost of creating additional knowledge is increasing. This also means that international schemes of big and prestigious science come under increased public scrutiny. CERN, ITER and ESA, the European Space Agency, are under pressure to explain themselves to the public, and for this purpose, they all are building and expanding their communication teams.

Box 1. Basic figures on the UK university sector

- The UK public university sector counts 154 universities entering in REF 2014; there are a few other higher education providers which variously are or are not included in these counts. Recent newspaper published university rankings feature 132 units (2021).
- In 2017/2018, there were 2,341,425 students in higher education; 5.9% from EU, 13.6% non-EU countries; in the UK, the University entry rate of the 18-year cohort is thus 33% in 2018.
- 2017/2018, the UK universities employ 429,560 academic staff; 12.1% from EU, 8.4% from outside the EU and 54.4% are female.
- 2017/2018, income of the sector was £38.2 Billion mainly from teaching (55.6%), paid by Gov (7.8%) and via student fees (47.3%); expenditure reached £37.2 Billion, of which 53.7% are counted teaching related
- A comparator: in 2015, UK pharma sector generate £28 billion income with 63,000 employees
- The communication expenditure of universities is difficult to gage from official figures. It is hidden in a) 7.3% outgoings that fall on Administration and Central Services; 6.3% other expenditure and 5% accommodation and conferences, all of which might include 'communication and event making'.
- Since 1994, the UK sector splits into two sections: 24 'Russell Group' members (R-units) of 'world-class and research intensive' universities and 100+ other Universities (non-R units), many more in local character, which group variously into the University Alliance, Million+, GuildHE, 1994 Group, or the Cathedral Group.

Trend 3: Expanding science communication and the changing role of science writing in the United Kingdom

This pressure to be audited and to be audible and visible to wider society leads also to an expansion of science communication as a civic and commercially gainful activity (Bauer, 2017; Bauer and Jensen, 2011). While this mobilisation, historically called 'extra-mural education', has roots in 19th-century demands for technical education (Cunningham et al., 2008), the more recent professionalisation of British science writing can be traced to the late 1940s and the formation of the BSWA (British Science Writers Association). A renewal of science communication was catalysed by an influential report by the Royal Society (1985). The Public Understanding of Science or 'PUS movement' has since evolved with three controversies over 'mad cows' (early 1990s), 'GM food' (later 1990s) and 'MMR vaccines (early 2000s), each of which highlighted the relevance of science communication (Smallman et al.,

2020, Gregory and Lock, 2008). This is a success story of putting science into public attention by marking a routine presence in the mass media and public conversations, and this before the Covid-19 pandemic of 2020/2021.

However, this roll-out of science communication is likely to have reached its nadir in the mid-2000. Several indicators highlight also the risks of intensified public communication for science (Bauer, 2018) which dampens the enthusiasm. Three trends might be mentioned here: First, the boom of free content circulating in online media brings a crisis of science journalism whose economic base as 'gate-keepers' and 'reporters' on newspapers, magazines, radio and TV is fast eroded. This weakness of science journalism is compensated by a strengthening of the professional Public Relations mission for science so that one science journalist is served content by six or more PR professionals by 2010 (Bauer, 2013). Second, some science journalists respond with 'epistemic vigilantism', carving out a new role as knowledge consumer protectors. Ben Goldacre (2012) is doing exactly that by critically examining scientific practices and exposing dubious results and players. This can lead to journalistic pack hunting: for example, in 2015, Professor Tim Hunt (Nobel Prize of Medicine, 2001) was forced to resign from UCL after a light-hearted, but unfortunate joke in public exposed 'sexism in the lab', which created a media storm. Clearly, the notion is dawning that increased publicity is itself risky, not only about the risks of climate change or risky nuclear energy, and this awareness dampens the enthusiasm for public communication, as it has before (Bauer, 2012; Bauer, Pansegrau and Shukla, 2019).

University resilience in this changing context – REF, TEF, KEF

The shifting of the functions of universities from knowledge transfer and status machine, to an innovation breeder seems to be a US-led post-WWII development; though the research university was invented on a 19th-century Germany model of producing applied knowledge for industry. It is the latter which also necessitates a professional communication function: i.e. 'show and tell' to the world what we can do, what we have developed and that it works better. Thus, it is also important for the modern university, to attract attention to retain its staff and student body, but also to invite industrial collaborators and justify the tax-payers' money.

Increased national and global competition for research results, students and careers requires highlighting 'excellence' among a pool of elite researchers. Profit and research interests seem no longer to be in contradiction. The neo-liberalisation of academia encourages spin-offs, and commercial activities to define a niche in national and international markets. Though universities might keep a dominance in pure research, not least in the humanities, the sector tends to be bifurcated into types such as 'universities' and 'universities of applied science' (explicitly so in Germany or Switzerland) or different leagues, as in Britain, of the Russell versus non-Russell groups.

This logic, according to which science is too valuable to leave to scientists in autonomy, seems to require that the sciences and its institutions are made accountable, so they can be rewarded on measured performance. In 1992, the UK set up the HEFCE (Higher Education Funding Council for England) to administer the allocation of Government funding to Universities according to a system of performance audit and create a visible hierarchy of higher education based on 'explicit criteria'. HEFCE combined earlier agencies that had started this process; HEFCE itself was superseded in 2018 by the duality of the Office for Students (OSI) and UK Research and Innovation (UKRI). This audit culture of higher education moved in three steps since the early 1990s. First, RAE (research assessment exercise) went through six rounds of assessment from 1986 to 2008. Second, this was superseded by REF (research excellence framework) in 2014 which included the 'impact agenda' (Watermeyer, 2016); units could prove their social value beyond research excellence with impact case studies. Third, this finally consolidated in the triple audit of REF, TEF and KEF. The TEF (teaching excellence framework) compares and shames any faltering teaching provision. KEF (knowledge exchange framework) counts commercial spin-offs, business models and patents, arising from research.

In addition to national hierarchy building, we witness the emergence of an international ordering of universities as in the **QS World University Ranking**, a metric system developed since 2004 in collaboration with the UK magazine Times Higher Education (THE). Another one is offered by the Shanghai Jiao Tong **Academic Ranking of World University** (ARWU) ranking compiled since 2003 and sponsored by the Chinese Government (Pavel, 2015). These league tables offer visible cues of prestige and reputation that are noted by academics, alumni, trustees, sponsors and students.

On the other hand, this load of assessment could amount to the **crucifixion of British academia** on the triad of REF, TEF and KEF. Other names for this system are a 'Frankenstein's monster' (Martin, 2011), the 'tyranny of metrics' (Muller, 2018, 67ff) or the 'strangulation of universities in the noose of competitive accountability' (Watermeyer, 2019, 1ff). These metric systems create a spurious ordering of units, nationally and internationally, and variously define and visualise a global competition in what appears to be a zero-sum game over the limited resource of reputation. Only, that reputation is not a limited good that is only gained when somebody else loses out. The historical world of universities is not the game that is suggested by these rankings; to play this game is a trap with many unintended negative consequences (Brankovic, 2021).

However, playing this game or not, even trying to redefine this game, necessitates and stimulates a more astute and more professional communication function at research institutions and universities, which is capable to manage this visibility, to regain control over the narrative that is created by this audit system and to position the institutions favourably in the eyes of various national and international stakeholders. However, it might be difficult not to get caught in the system.

The disposition of research institutes to communicate more widely

Considering the above trends and contexts, we have been systematically studying the professionalisation of communication in 186 British university-based research institutes. Our sample is randomly stratified by disciplines and, thus, offers a representative picture across 24 Russell and 130 non-Russell institutions, from a sampling frame of >1,000 institutes (for details, see Section 3 of the book). For the research institute, we contacted a 'designated speaker'. In the absence of any previous data on this matter, we provide an initial scoping to benchmark for future inquiries (Entradas et al., 2020). Figure 12.2 shows the distribution of research units according to OECD disciplines and university type.

The public communication of research institutes is our present focus, often referred to as PE, or activities known to academics as 'doing scientific citizenship'. It comprises activities that are not focused on the immediate interests of the unit itself, but aim to serve the wider community, either symbolic-culturally (e.g. with events for a wider audience) or by providing civic utility through help to solve local problems (e.g. via science shops or ambulant advisory). For purposes of economic accounting, this third PE mission of universities is increasingly conflated with business relations, known as KEF. Knowledge engagement in this accountable sense is both a source of income and

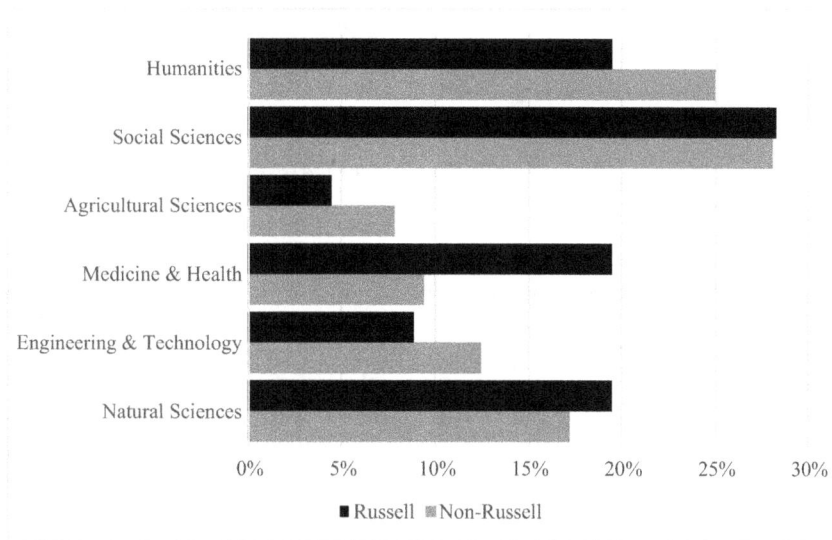

Figure 12.2 The stratified random sample of non-Russell research institutes (n=73) and Russell-based research institutes (n = 113) of our study; data collected between February 2017 and May 2018 with probability weighted response rate of 31% from a sample >1000 (see Entradas et al., 2020).

knowledge-transfer within a cycle of innovation and economic growth. The more this is audited, the more this is what research institutes will seek to do.

It remains ultimately an empirical question to examine how PE in terms of community services and civic engagement and the commercial activities of KEF type differentiate, conflate or compete in any one institution; this will be part of a follow-up study to our present scoping exercise. The results presented here concern the communication functions by research institutes, which is the meso level, in contrast to the level of individual researchers or the central level of university. We were asking the following questions: *who is staffing these activities, what is being done, to what rationale and for which audiences in your research institute?*

We consider three hypotheses for this UK National profile: Russell Group universities (R units) compete both nationally and internationally, we expect enhanced public communication in these research units compared to the ones in non-Russell universities (non-R units), hence:

- H1: R units show higher intensity of communication activities than non-R units;
- H2: R units have a different profile of rationales from that of non-R units;
- H3: R units show more professionalisation of communication than non-R units.

In the following, we will be using 'public communication' (PC) and 'PE' interchangeably, mainly for lack of more information about how people make a distinction that really makes a difference.

What is going on: public events, traditional media channels and new media channels

We asked speakers of research institutes about activities performed over the past three years to get a sense of the portfolio of communication that makes up their practice (Figure 12.3). Communication activities are expanding, a little bit faster among R-units than among non-R units: only 5% of R institutes report that activities decreased over the past five years (10% for non-R units); for 32%, there is no change (21% non-R); while for 67%, PC activities have increased (56% among non-R). Curiously, we also observe that communication activities increased particularly in the Humanities (68%) and in Medicine (77%). For the Social Sciences and Agriculture, there is little reported change. We differentiate the activities on the ground into event making, legacy media and social media as shown in Figure 12.3.

Public events refer to organising outreach events of which public lectures, open days or exhibitions are the most common. Letting people into the Labs at night, citizen science projects and participation in National Science Week is less common in this set. Most units report such event making. About 48% do at least quarterly public lectures, 17% even monthly. And 46% report taking

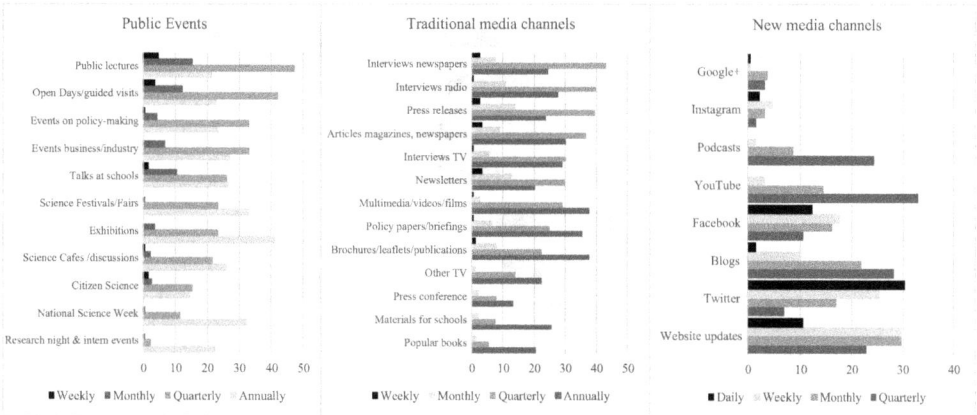

Figure 12.3 % of units performing public events, traditional media channels and new media channels; the stacked bars report the % of units that report activities over the 'past three years'.

part in National Science Week, and 38% engaging in citizen science projects. Public lectures, open days and events for business and policy are, for those who do, more frequent quarterly events. Overall, there is little difference between R and non-R institutes. There is a broad and diverse repertoire of event making in evidence, including read-a-thons, concerts, school workshops, debates, literacy festivals, public performances, round tables, public seminars, stands at local museums, industry conferences, history weekend, etc.

Traditional media channels activity refers to mass media contacts, of which press releases, articles and interviews are most typical. Less common are organising press conferences, writing popular science books or materials for schools, or being involved in TV productions. Most institutes report at least annual media contacts of some sort, 43% report quarterly and 10% report monthly or weekly media contacts in the form of interviews. For those who are media active, issuing press releases, writing articles for newspapers, giving radio and TV interviews are more frequent quarterly activities.

On the other hand, 75% never did a press conference, 80% of R units and 65% of non-R units say 'never'. Non-R units seem to be more media savvy. For many media activities, R-units report more media abstinence than non-R units. For example, 20% of R-units report never doing press releases, while only 8% of non-R units report the same. While media contacts seem to be more common among non-R units, for those who are active among R units, the reported media contacts are more frequent than among active non-R units. This is consistent with a 'mobilisation effect' in Russell universities, where the media efforts are stronger and concentrated on a few active units (see Entradas and Bauer, 2017).

New media channels activity, of which websites and Twitter are most typical, is reported by most units. One can say: everybody is on social media at least annually; 69% at least monthly and 35% weekly or daily. Not

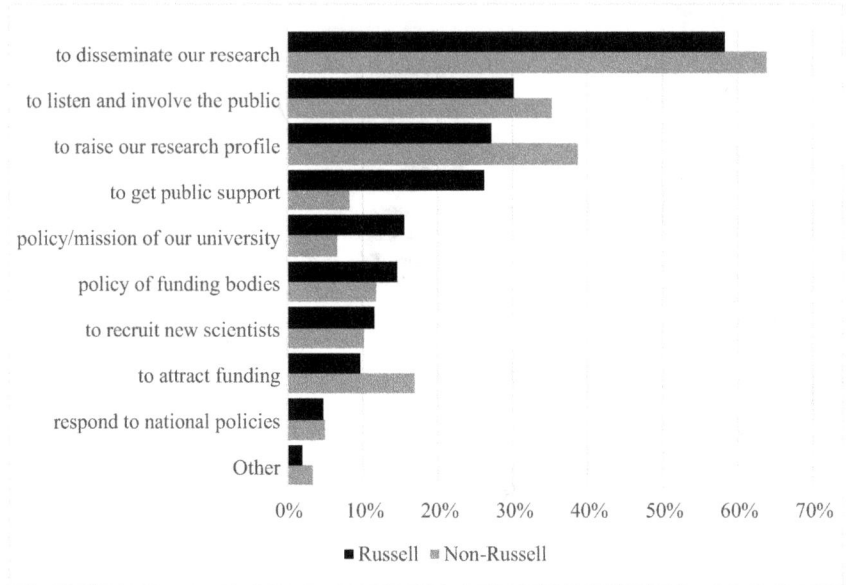

Figure 12.4 Different espoused rationales for communication activities for Russell and non-Russell units (n=187).

surprisingly, social media activities are more frequent than legacy media and event making. Less common are Google+, Instagram or podcasts. Tweeting is for many research institutes a daily activity; website update is more of a weekly thing; while blogging, YouTube or podcasts come less frequently on a monthly or quarterly cycle. Other social media activities include InMediaRes, LinkedIn groups, management of discussion groups, Storify, Vimeo, Vlogs and Webinars. Social media practice among research institutes is pretty much common ground; there is very little difference between research units. Maybe Instagram is more used in non-R units than in R units.

Why do research institutes go public

We asked the why-question: *what is your main rationale for doing PE?* The speakers for the institutes mark the most and second most important purposes. Most units aim to disseminate their research and that is what PE should do; this might go well together with raising the research profile and to listen to and involve the public as shown in Figure 12.4. To a lesser degree, PE also means seeking public support and attracting funding, or to respond to national policies, and recruit new researchers. This step function between prior and lesser reasons for public communication is more accentuated for non-R units than in R units, where all these potential aims are important on a gradient.

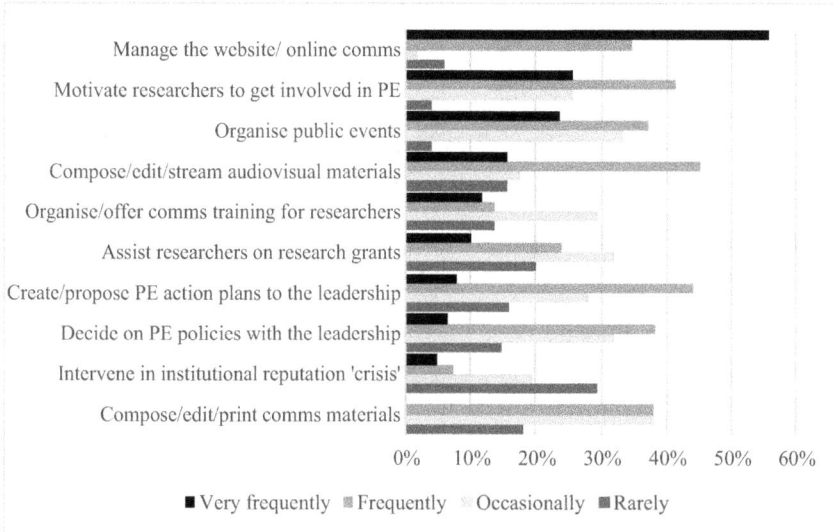

Figure 12.5 Main communication tasks and their intensity reported in % of units (n=187).

Our second question asked: *what does it take to achieve these aims?* We seek to understand the tasks of PC professionals employed in research institutes. This involves frequent website maintenance, organising events, to compose and edit contents but also to plan events and motivate research staff to get involved in PC as shown in Figure 12.5. To a lesser extent, communications staff assists in grant writing, offers training to staff, or gets involved in reputation management. Other reasons that people mention include doing politics, serving what our desires of the audience, addressing misinformation, supporting human health or relating and, thus, enlightening current issues through classical and ancient myths. Others give excuses for not doing much because 'we are not a Science Centre', or 'we are not a science-based unit'. These excuses reveal a stereotype: communication is best left to specialist outfits, mostly related to the natural sciences.

Communication professionals seem to tread much common ground here; there is little difference in how communication is done. Overall, these tasks cluster neatly, based on who does what and what else, into three streams: (a) developing and planning, (b) composing and editing content and (c) providing support and guidance for research staff.

The audiences: targets and their stereotype

All communication needs to consider the audience, more or less specific, for purposes of targeted activities. We asked about target audiences for the communication activities of the research institutes and to what extent these are in focus or not. Again, there is very little difference between R and non-R units with respect to these target audiences. The range and priority of targets is very much common ground among communication professionals as it seems. Whether this reflects real practice, or only a 'professional stereotype' that is easily gagged by questionnaire responses, is something additional qualitative research will reveal.

Our eight audiences are the focus for more than 80% of all research units either occasionally or frequently as shown in Figure 12.6. Overall, the general public and mass media journalists are top focus. If we consider 'frequent engagement', the general public, local industry and students outside teaching are top of the list. Local authorities and NGOs, or even schools, are more of an occasional contact target of research institutes.

Communication does not only seek an audience, but it also, willy-nilly, does so with a more or less stereotypical image of that audience. We investigate how research units 'rate' their audiences on a number of statements such as 'the public is only a source of trouble', 'the public trusts scientists', 'the public is interested in our research' and 'the public want to contribute to science'; 13 such presumptive claims in total. In total, these statements define an operational image of the public in which our informants are vested to various degrees.

With regards to this stereotype of the public, we can discern differences between the universities. Non-R units are more trusting of the public (the public trusts us!) but also more aware that communication needs to be strategically selective in order to avoid troubles down the line than R units. R

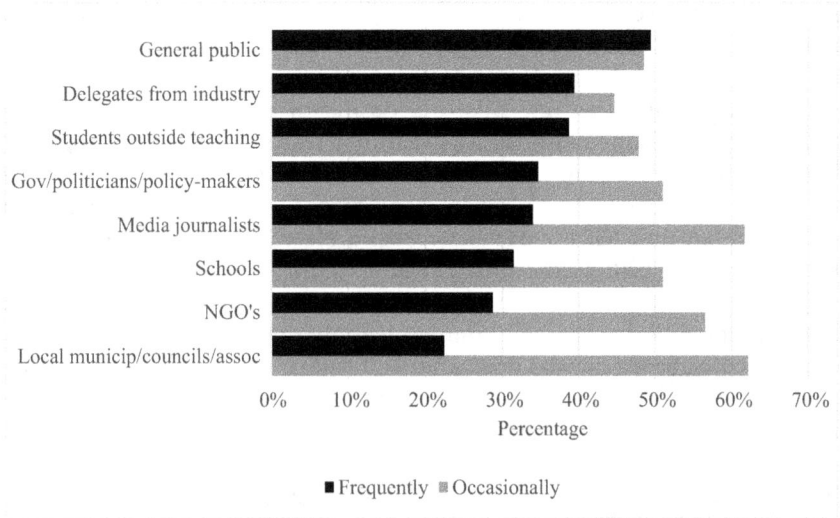

Figure 12.6 Target audiences and the general intensity of contacts in percent of units (n=157).

units are equally trusting and cautious in relating to the public but, at the same time, reject the notion of strategic selectivity. Non-R units, thus, seem to be more apprehensive of the public than R units.

Overall, these presumptions group into three dimensions which might be identified as the public (a) is interested in science, (b) is seeking participation and (c) is a trouble risk. Different research environments do hardly differ. Communication offered by research units seems to operate under the presumption that the public is interested and wants to get involved. However, on whether the public constitutes a trouble risk, R units see less risk than non-R units. This confirms observations above that non-R research institutes are more apprehensive of the public than R units.

Staffing levels: who is involved and with what background?

About 26% of all research institutes report having specialist staff for communication; slightly more, with 30% for R than the 20% of Non-R units. Proportionately more R units employ specialist staff for communication (though this difference might be within random fluctuations considering our sample size). Half of these, or 14% of all units, have specialist staff who are fully dedicated working on full-time contracts. In addition, another 24% units have staff that is partially dedicated to communication. In total, 48 of 187 units report employing a total of 136 PE staff, or 2.8 persons on average, full or part time.

For curiosity, we can also observe how institutes in different disciplines employ specialist communication staff. Indeed, different disciplines are investing differently here: 50% of medical units report having specialist staff; 27% of engineering and agriculture units report the same; 26% for Natural Science and 23% of Social Science units. With only 3% of humanity units reporting specialist staff, this seems not yet part of their aspiration.

Box 2. Estimates of the numbers of communicators in British research institutes

Trying to estimate more numbers from all this, we can say: 48 units (26%) report 84 full-time PC staff. Among these units, 23 (12%) report additional 52 part-time staff, of which 40% are employed ½ day, 40% one day and 20% two days per week. This suggests that a total of 136 people occupy 104 FTE (full-time equivalent) positions conducting communication across these 48 units, or 26% of all research units, with 2.8 persons per unit on 2.1 FTE on average.

Considering that our effective sampling ratio is around 30%, we estimate that something in the area of **250–350 [3 × 104 +/−50] FTE**

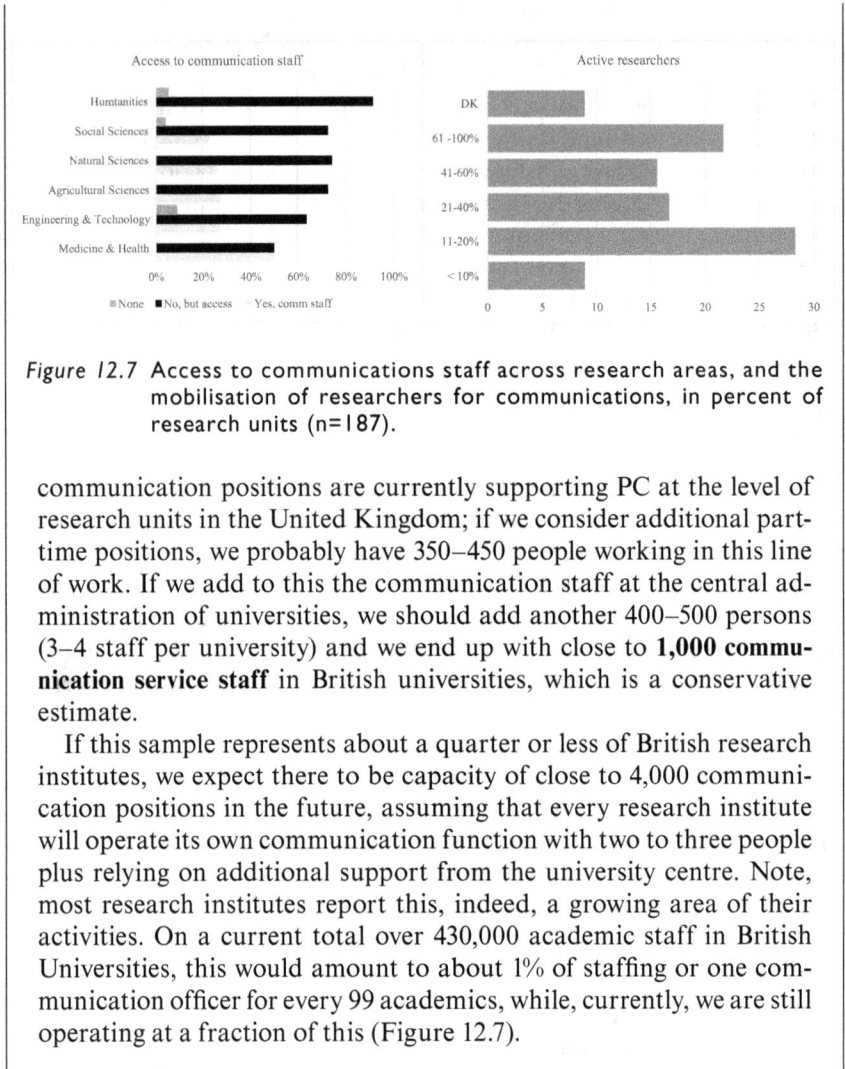

Figure 12.7 Access to communications staff across research areas, and the mobilisation of researchers for communications, in percent of research units (n=187).

communication positions are currently supporting PC at the level of research units in the United Kingdom; if we consider additional part-time positions, we probably have 350–450 people working in this line of work. If we add to this the communication staff at the central administration of universities, we should add another 400–500 persons (3–4 staff per university) and we end up with close to **1,000 communication service staff** in British universities, which is a conservative estimate.

If this sample represents about a quarter or less of British research institutes, we expect there to be capacity of close to 4,000 communication positions in the future, assuming that every research institute will operate its own communication function with two to three people plus relying on additional support from the university centre. Note, most research institutes report this, indeed, a growing area of their activities. On a current total over 430,000 academic staff in British Universities, this would amount to about 1% of staffing or one communication officer for every 99 academics, while, currently, we are still operating at a fraction of this (Figure 12.7).

Specialist communication staff is not the only and most important dispositive to mount PC activities. Many units do not have their own communication specialists and access such competence centrally as shown in Figure 12.7. Indeed, 71% report that they access communication staff only centrally. However, 20% also report that they are maintaining a media contact list, keeping their options open for a direct contact with the media. Furthermore, PC activities do not only come from specialist staff who are employed to do the communications as shown in Figure 12.7. Many PE initiatives are called for by funding agencies and seek to mobilise funded researchers into outreach activities (Bauer and Jensen, 2011). When asked, whether regular research staff gets involved in PE activities, about 50% of units report ½ their staff taking part in PE activities; another 35% of units report a majority of staff mobilised. But many do not know the real situation. Generally, non-R units report higher levels of staff mobilisation than R units, which probably reflects closer coupling with local communities.

A final question regarding communication staffing is: *who is working in these PE teams?* Of the total of professional PE staff in our survey, 48% [66] come with a background in Social Science or humanities; 40% [55] come from science and engineering and 13% [18] report no prior degree (n=139). In terms of previous occupations, 34% [45] were one-time researchers; 20% [27] transition from marketing; 17% [22] previously were administrators; 7% [9] journalists and another 7% [9] project managers; 5% [7] are multi-media designers or another 5% have no experience and finally, 3% [4] report coming from teaching or are librarians, are Chartered PR professional or artists.

There is no difference in the job age and experience of PE staffing in different types of research units, for those 26% of units who have a PE team: 62% of teams are five years or older; 31% one to five years and 2% of units acquired a team only in the past year. Most teams, thus, appear to be well established while being concentrated in one-quarter of all research units. Again, the pool of communication professionals seems the same for all research units, pretty much common ground for all those involved.

British research units declare spending on average 3.9% of their operating budget on communication activities; considering the margin of error, we might talk of about 3–5%; there is no difference between R and non-R units in this level of expenditure. This is expected to rise to 5–7%; all units expect to spend more on communication in the coming years.

We observe, however, that the richer and larger the units, in terms of annual communications funding, the smaller is their 'perceived' level of PE expenditure. 'Perceived' means asking 'how much of your budget do you spend on PE'. However, if we try to estimate real expenditure, based on staffing levels and project activities, we find that actual PE expenditure increases with larger units as a percentage of the overall budgets. Small units seem to overestimate their PC resourcing and larger units seem to underestimate this effort. Note these estimates do not yet include the large numbers of volunteering research staff who engage in PE communication activities.

Clearly, communication at the level of research institutes is a growth area, both in terms of staffing and in terms of budget that is allocated to their activities. This is clearly a space to watch in the coming years.

Conclusion

The university is an institution that has collected over its history a good number of different functions in wider society. This multi-functionalism is at risk to be reduced by the more recent audit culture to one single focus: providing economic utility. This tendency comes on the back of a massive expansion of UK universities since the 19th century, in five waves from 1% to 50% participation rate of the young age cohort. The bulk of expansion, from 15% to 50%, occurred since the 1990s. This leads to massive pressures to audit this system with REF, TEF and KEF in order to allocate resources based on measured performance, which by unintended consequence creates a monster that threatens to 'crucify' or 'strangulate' or 'tyrannize' British higher education, whatever term one might care to use here.

In this environment, the expansion and professionalisation of university communication, also known as 'Uni Comms', is a response to these challenges. This response engulfs both university central administration but also the level of research institutes and departments, what we called here 'meso level units', who must respond to this audit culture and mark presence, to be audible and visible in an increasingly competitive concert. About 26% of British research institutes report having professional staff to provide this communication function locally, employing in our estimates some 1,000 specialist staff by 2020, with a budget of 3–5% of the total unit expenditure. This in addition to the growing supply of communication at the level of central administration of universities.

We compared how 'Russell units' (research universities) and 'non-Russell units' (mixed universities) fare in this 'brave new world' in terms of activities, rationale for communicating, targeted audiences and presumptions, and staffing level and quality. There is overall much common ground between how R units and non-R units operate their communication function if they have one. Communication is generally a growth area. Common ground is in evidence with regard to the profile of activities, the rationale for PE and the professional tasks, in the presumptions about the audience and, finally, the level of staffing and backgrounds. However, there are discernible differences between types of units in the way they engage communication with the public:

- PC activities are generally expanding, a little bit faster among R-units than among non-R units.
- Non-R units seem to be more media savvy. For many media activities, R-units report more abstinence than non-R units.
- Social media being much in use, Instagram is more used in non-R units than in R-units.
- A step function between prior and lesser reasons for public communication is more accentuated for non-R units than in R units where all aims are on a gradient.
- Non-R units seem to be more apprehensive of the public than R units. On whether the public means trouble, R units see less risk than non-R units.

- Proportionately more R units employ specialist staff for communication than non-R units.
- Non-R units report higher levels of academic staff mobilisation for PE than R units.

These differences likely reflect an increasingly competitive environment which we explore elsewhere (see chapter 'arms race'). We ultimately wonder how the boundaries between different types of science communication might fall in this professional concert. The historical focus of the PUS movement was to increase civic engagement, to create a common understanding about society's futures; while the KEF audit fosters the PR logic of corporate positioning, reputation management and product marketing. Can a subtle boundary between multiple functions of the university and between different communicative aspirations persist? Only the future will tell, let us keep a vigilant eye.

References

Bauer MW (2013) The knowledge society favours science communication, but puts science journalism into a clinch, in Baranger P and B Schiele (eds) *Science communication today – international perspectives, issues and strategies [Journees Hubert-Curien]*, Paris, CNRS Edition, pp145–165.

Bauer MW (2017) Kritische Beobachtungen zur Geschichte der Wissenschafts kommunikation, in Bonfadelli H, B Fähnrich, C Lüthje, J Milde, M Rhomberg and MS Schäfer (eds) *Forschungsfeld Wissenschaftkommunikation*, Berlin, Springer Verlag, pp17–40 [German original]

Bauer MW (2018) Trust in Science after the BREXIT, in deMarec J and B Schiele (eds) *Culture of Science*, Montreal, Acfas, pp95–102. [English 2018]

Bauer MW and J Gregory (2007) From journalism to corporate communication in post-war Britain, in Bauer MW and M Bucchi (eds) *Science, Journalism and Society: Science Communication Between News and Public Relations*, London, Routledge, pp33–52.

Bauer MW and P Jensen (2011) The mobilisation of scientists for public engagement, *Public Understanding of Science*, 20, 1, 3–11.

Bauer MW, P Pansegrau, and R Shukla (2019) *The Cultural Authority of Science – Comparing across Europe, Asia, Africa and the Americas*, London, Routledge; Vol 40 Routledge Studies of Science, Technology & Society, pp2–21.

Brankovic J (2021) The absurdity of University Rankings, LSE Blog, March 22nd.

Butterfield H (1962) The university and education today – *The Lindsay Memorial Lectures*, University College of North Staffordshire, London, Routledge & Kegan.

Clay RA (2008) The corporatization of higher education, *Monitor of Psychology*, Dec, pp48–54.

Cunningham P et al. (2009) *Beyond the lecture hall – universities and community engagement from the middle ages to the present day*, Cambridge, University Faculty of Continuing Education & Victoria Press.

Doring (2020) On Ernst Kantorowicz, academic freedom and the secret university, *The Chronicle Review*, 20 November.

Entradas M and MW Bauer (2017) Mobilisation for public engagement: Benchmarking the practices of research institutes, *Public Understanding of Science*, 26, 7, 771–788.

Entradas M and MW Bauer (2018) Die Kommunikationsfunktion im Mehrebenensystem Hochschule, in Schaefer M et al. (eds) *Forschungfeld Hochschulkommunikation*, Wiesbaden, Springer Verlag, pp97–122 [original in German]

Entradas M, MW Bauer, C O'Muirchearteigh et al. (2020) Public communication by research institutes compared across countries and sciences: Building capacity for engagement of competing for visibility, *PlosONE*, 15(7). https://doi.org/10.1371/journal.pone.0235191

Finkelstein MJ (1984) *The American academic profession – a synthesis of social scientific inquiry since WWII*, Ohio, Ohio State University Press.

Gregory J and SJ Lock (2008) The evolution of 'public understanding of science': Public engagement as a tool of science policy in the UK, *Sociology Compass*, 2/4, 1252–1265.

Goldacre B (2012) *Bad Pharma – how drug companies mislead doctors and harm patients*, London, Fourth Estate.

Habermas J (1988) Die Idee der Universitaet – Lernprozesse, in Eigen M et al. (eds) *Die Idee der Universitaet – Versuch einer Standortbestimmung*, Berlin, Springer-Verlag, pp139–173.

Halsey AH and MA Trow (1971) *The British Academics*, Cambridge, Harvard University Press.

Mandler P (2020) *The crisis of meritocracy – Britain's transition to mass education since the second world war*, Oxford, OUP.

Martin BR (2011) The REF and the impact agenda: Are we creating a Frankenstein's Monster? *Research Evaluation*, 20, 3, 247–254.

Muller JZ (2018) *The tyranny of metrics*, Princeton, PUP.

Noam EM (1996) Electronics and the Dim Future of the University, *Bulletin of the American Society for Information Science*, Jun/Jul, 6–9 [reprint from SCIENCE, 1995, 270, 247–249].

Pace I (2018) The RAE and REF – resources and critique, Blog page https://ianpace.wordpress.com/2018/04/03/the-rae-and-ref-resources-and-critiques/

Pavel AP (2015) Global university rankings – a comparative analysis, *Procedia Economics and Finance*, 26, 54–63.

Royal Society (1985) *The Public Understanding of Science*, London, Royal Society

Schneider R (1982) Die Bildungsentwicklung in den westeuropaeischen Staaten 1870–1975, *Zeitschrift fuer Soziologie*, 11, 3, 207–226.

Smallman M, SJ Lock and S Miller (2020) UK chapter, in Gascoigne et al. (eds) *Communicating Science – a global perspective*, Acton, AUS, ANU Press, pp931–948.

Stern N et al. (2016) *Building on success and learning from experience – an independent review of the research excellence framework*, London, Department of Business, Energy and Industrial Strategy OGL.

Watermeyer R (2016) Impact of REF: Issues and obstacles, *Studies in Higher Education*, 41, 2, 199–214.

Watermeyer R (2019) *Competitive accountability in academic life – the struggle for social impact and public legitimacy*, Cheltenham, Eduard Elgar Publishing.

Willetts D (2017) *A university education*, Oxford, OUP.

Windolf P (1992) Cycles of expansion of higher education 1870–1985: An international comparison, *Higher Education*, 23, 3–19.

Weber M (1989 [1919]) *Science as a vocation* [ed. P Lassmann, I Velody, H Martins], London, Unwin Hyman.

Universities UK (2019) *Higher education in facts and figures*, London, Woburn House.

Chapter 13

Public Engagement Activities of German Research Institutes

A Tale of Two Worlds

Tim Belke and Frank Marcinkowski

Introduction

Just 30 years ago, German universities had two central tasks: research and teaching. Each institution was charged with balancing the search for new knowledge, on the one hand, and the dissemination of this knowledge and scientific methods to students, on the other hand. Over the course of the last few years, both the demands on universities and their self-understanding have changed. Further tasks have been added, which are summarised under the term 'third mission' (Roessler, Duong, & Hachmeister, 2015, p. 4). These include knowledge transfer as well as 'partnerships between industry and universities and research institutions' or, in short, 'linking universities and their members with civil society' (Roessler et al., 2015, p. 4). These new requirements were accompanied by, among other things, a deregulation of higher education laws (Hüther, 2010; Marcinkowski, Kohring, Friedrichsmeier, & Fürst, 2013, p. 257) and led to increased competition among universities (Friedrichsmeier, Laukötter, & Marcinkowski, 2015, p. 128; Marcinkowski et al., 2013, p. 257).

In order to hold their own in the intensified competition, many universities have begun to engage in 'public relations' or intensified their efforts in this area, which have so far been negligible. Demands for more and better science communication have always been heard from various quarters. However, this chapter deals with the extent to which these demands are actually being met where research and teaching are carried out – in the individual research institutes (RIs) of the German university landscape – and how the situation compares to that at non-university RIs.

National background

May 27, 1999 can be seen as a key moment for a broad awareness of the supposed need for additional efforts in science communication. On this day, major German science organisers, such as the Stifterverband für die Deutsche Wissenschaft, the German Research Foundation, the German Rectors'

DOI: 10.4324/9781003027133-16

Conference and other actors, published a memorandum. The memorandum states, in a prominent place, a demand for scientists 'to present their work in a form that is also understandable for non-specialists' (Stifterverband für die Deutsche Wissenschaft et al., 1999). According to the memorandum, science communication with laypersons is 'less developed in Germany than in other countries', and dialogue with the public should be given 'the high priority it deserves'. The need for professionalisation is mentioned: 'Universities and research institutions are called upon to provide the necessary infrastructure' (Stifterverband für die Deutsche Wissenschaft et al., 1999).

The PUSH memorandum can be described as 'the starting signal for a breathtaking upgrade of universities and scientific institutions with communication staffs and marketing departments' (Rehländer, 2018). The signatories founded the non-profit association Wissenschaft im Dialog, intended to improve the communication of science to society. However, 'many of the goals formulated in the memorandum as a voluntary commitment […] are still unachieved today' (Rehländer, 2018).

According to the Siggen Circle (2013), an influential group of science communicators, scientists and science journalists, almost 15 years later, 'science communication […] has established itself in Germany. Its role is recognised as increasingly important for science itself and for the public' (Siggener Kreis, 2013, p. 1). Internationally, however, science communication in Germany continues to appear 'little noticed and little networked' (Siggener Kreis, 2013, p. 2; see also Siggener Kreis, 2018, p. 6). Moreover, scientists themselves often shy away from the public: 'Only a few scientists are currently seeking the support of their communication departments for the development, expansion and professionalisation of their communication' (Siggener Kreis, 2013, p. 2).

The nature and orientation of communication with the general public is also viewed critically by the scientific community and even by those who are firmly committed to science communication. In an opinion piece for the journal Research & Teaching, Marcinkowski and Kohring explicitly point out the risks of an exaggerated orientation towards media publicity, including the mainstreaming of research, misallocation of resources and loss of university autonomy (Kohring & Marcinkowski, 2015). The authors draw on their extensive empirical research on the consequences of the so-called New Governance of science for the relationship between universities and the public, which can be summarised by two overarching trends: first, a disguised politicisation of universities which, as a result of their newly won freedoms, have to take charge of their own approval management, and second, a comprehensive medialisation of scientific organisations, since decision-makers within universities tend to copy the role model of public policy for this purpose (Marcinkowski et al., 2013). Recently, Zurich communication scientist Mike Schäfer (2018) has repeated this assessment. He states that communication is often 'no longer primarily about disseminating relevant research results, but rather about strengthening the institutional reputation of one's own university, maintaining its legitimacy in the eyes of stakeholders, and thus [...] securing the flow of resources'. The question of whether science communication actually serves to educate lay people or is more of a marketing exercise has been the subject of repeated critical debate in the German community 'for at least 20 years' (Hoffmann, 2018). The Siggener Kreis also deals with questions of this kind in its annual conferences. In their very first position paper, the members stated that actors in science communication are 'increasingly competing for reputation, funding and influence' (Siggener Kreis, 2013, p. 1).

However, the debate on science communication as a source of legitimacy not only focuses on the legitimacy of certain research fields or projects vis-à-vis certain stakeholders but also aims at the legitimacy of the science system as a whole. Thus, a loss of trust in science can be observed. Science communication can and must courageously counter these 'anti-Enlightenment developments: through science and science communication with integrity and a focus on the common good' (Siggener Kreis, 2017, p. 2). The PUSH memorandum of 1999 also spoke of 'informing the public to legitimise scientific activities' and 'actively seeking trust' (Stifterverband für die Deutsche Wissenschaft et al., 1999). Other authors see science not as being threatened by a largely distrustful population but rather by the increasing dependence of scientific institutions – especially universities – on politics and private industry. Fending off these encroachments and defending the freedom of science is the primary task of science communication, but it is notoriously neglected in the inter-organisational competition for the reputation of universities (Kohring, 2019).

Discussions on the pros and cons of science communication are still going on and largely revolve around the same issues that were already foreseeable 20 years ago. The calls for more and better communication with the general public are as loud as ever. The latest contribution to this discussion was made by the German federal minister of research and technology, Anja Karliczek, in November 2019 with a policy paper from her ministry on science communication.[1] In this paper, science slams, public dialogues and other measures of science popularisation are described as obligatory activities in which all research institutions and scientists should regularly participate in the future. In pointing out that active science communication should, in the future, become an integral part of the ministry's financial support for research, the minister implicitly threatens those institutes that refuse to comply with her request to withdraw funding. This obligation to popularise basic research, as critics argue, harbours the danger that a form of science will be favoured that is easy to communicate, as the news value of oddity promises a contribution to the solution of current problems (social impact). This development is at the expense of basic research that is difficult to access for the layperson. The ministerial directive also fails to recognise the reality of the communication departments of many German universities. Although the communication departments in some places have grown very large, they are often unable to perform this popularisation task because they are largely occupied by the university management with brand management tasks. This is also a consequence of the switch to competitive procedures for university financing (Marcinkowski & Kohring, 2014). After all, the question of whether the supply of science communication does not already far exceed the demand plays hardly any role in this debate.

Against this background, the communication activities of German RIs are described below. The extent to which German universities communicate publicly and which resources are made available for this at the management level are the subjects of various studies from the sociology of science and communication studies (cf. inter alia Friedrichsmeier, Geils, Kohring, Laukötter, & Marcinkowski, 2013; Friedrichsmeier et al., 2015; Kohring et al., 2013). How individual scientists in Germany communicate with the public has also already been the subject of research (Peters et al., 2008a; Peters et al., 2008b; Peters, 2013). This study explicitly deals with communication activities at the level between individual scientists and the top of the organisation. It is based on the assumption that members of research institutions do not normally communicate with the public on their own initiative but rather require the encouragement and support of the 'middle' organisational unit to which they directly belong (Marcinkowski et al., 2014). How and why individual RIs and centres activate their members in this way, what resources they make available for this purpose and how professionalised communication is organised at this level is the subject of this study. It builds on the pioneering work done in Portugal (Entradas & Bauer, 2017) that led to further replication in several other countries, including Germany. For the German case, not only RIs at universities are considered but also the publicly funded non-university research sector (see Section Research Landscape in Germany). Before these assumptions are empirically tested, the research landscape in Germany will first be briefly outlined in order to give a clearer definition of the object of the study.

Research landscape in Germany

Research in Germany is largely conducted at public universities and so-called universities of applied sciences. There are also four publicly funded research associations with a range of institutes, as well as research institutions from government ministries, and private companies and institutes that conduct research. This study focuses on both the public universities and the four publicly funded research associations. Research at private and ecclesial universities does not take an important role: the focus is clearly on teaching, which is the main reason why they were excluded from this study. The same applies to colleges of art and music although these are predominantly public. The universities included in our analysis are home to about 95% of the scientific staff of all German higher education institutions (Statistisches Bundesamt, 2017a, 2017b).

The German Rectors' Conference, the association of public and government-recognised universities in Germany, counted 398 universities in Germany when the sampling procedure for this study started (Hochschulrektoren-konferenz, 2016). Eighty eight universities and 105 universities of applied sciences are covered in this study. Even though this sample comprises barely half of all listed institutions, 90% of all students in Germany are studying at these 193 universities (Statistisches Bundesamt, 2017b), which shows that these are by far the most important universities in the German higher education landscape.

When it comes to non-university research in the private sector, it is mainly basic research, such as in large industrial companies, where no public engagement with science is expected; hence, these are not part of this study. The research facilities of the federal and state government, such as the ministries, did not seem to be comparable since the government apparatus opens up entirely different possibilities that are not available to the universities or research associations. In addition, the primary purpose of these institutions is 'to support the ministry's activities and provide the necessary scientific basis for the execution of government measures' (Federal Ministry of Education and Research, n.d.) and not to make this research available to the public. For that reason, these research facilities were also excluded.

There remained the four publicly funded research organisations – the Max Planck Society (MPG), Fraunhofer Society, Helmholtz Association of German Research Centres (HGF) and Leibniz Association – with a total of 348 RIs as defined by the MORE-PE project. Most institutes are highly specialised and quite independent, concentrated at one location, with its own buildings and its own press officer. The institutes of the Fraunhofer Society are explicitly dedicated to application-oriented research for the benefit of private companies. The public funding provided by the federal and state governments is comparatively small, with the majority of the research conducted by these institutes being financed in conjunction with industry. The Helmholtz Association comprises 19 research centres. They are largely financed by public budgets of the federal government and the states and conduct research in the fields of natural science and medicine. The Leibniz Association connects 96 independent research institutions that range in focus from natural, engineering and environmental sciences to economics, spatial and social sciences and the humanities. They conduct both basic and applied research. The institutes are mainly financed by the public budgets of the federal government and the states. Finally, the Max Planck Society maintains 86 institutes (known as Max Planck Institutes) and sees itself as the leading institution for basic research in Germany. Most of the institutes keep close relations with individual German universities. The Max Planck institutes are financed almost exclusively by public funds. The main focus of their research is on the natural sciences, but there are also 19 Max Planck institutes for the humanities or social sciences.

Methodology

Sample

Both the university and the non-university sectors make important contributions to research in Germany, but they are too different to merge into one sample and, therefore, have been handled separately throughout the study. Due to the fact that the number of institutes in the non-university sector was known and not too large, a full survey was conducted. The four research organisations keep lists of their facilities. In most cases, these facilities were RIs in the sense of this study. Only in a few cases were they subdivided into further institutes. These were identifiable, for example, by having their own director and administration. The Helmholtz Association unites 'centres', which contain several institutes, for example, the German Aerospace Centre. Seven out of the 348 institutes were 'research museums', which, of course, are more oriented towards the public than other institutes. These research museums were marked in the sample to identify them in the analysis.

The research areas of the institutes were coded by one coder, based on the information on the institutes' websites. Most institutes refer to OECD research areas 1 (natural sciences, n = 155) and 2 (engineering and technology, n = 88), some to research fields 3 (medical and health sciences, n = 44) and 5 (social sciences, n = 38) and only a few to research fields 6 (humanities, n = 15) and 4 (agricultural sciences, n = 8).

Table 13.1 Sampling frame

Research area	Non-university	University	
	Number of RB/RI	Number of RBs	Number of RIs
1. Natural sciences	155	34	234
2. Engineering and technology	88	34	256
3. Medical and health sciences	44	33	542
4. Agricultural sciences	8	24	133
5. Social sciences	38	34	203
6. Humanities	15	34	267
Total	348	193	1,635

For the university sector, there existed only a list of research bodies (RBs) (universities), but not RIs. The number of individual institutes was unknown. To make sure that the desired sample contained institutes from all 193 RBs, a two-step sampling method was applied. In the first step, the list of RBs was randomised, and each RB was assigned an OECD research areas in turn. If a research field was not represented at an RB, the next research field from the list was assigned, and the unallocated research field was assigned to the next RB, to make sure all research fields were equally distributed. In a second step, RB websites were searched to identify all institutes referring to the respective research field. A RI, in the sense of this study, is a unit below a department or faculty[2], with several professors or chairs. Institutes with fewer than three professors or that included more than three research fields have been excluded; affiliated institutes, research clusters, collaborative research centres or similar (temporary) research facilities have been included if they contained three or more professors as well as interdisciplinary units that could be assigned to three or fewer different OECD research areas (e.g. Department of Social Work and Health).

For all OECD research areas except the agricultural sciences, more than the targeted 200 RIs were identified with this procedure (Table 13.1). This can be explained by the fact that agricultural institutes were found at only 24 universities. It can be assumed that the 133 agricultural institutes identified represent the entirety of all agricultural institutes in Germany. As described in the methods section of this book (Chapter 15), a target of N = 200 RIs should be drawn for each OECD research areas in all countries. We elaborated a randomisation procedure to ensure that all 193 RBs were represented as evenly as possible in the final sample. In the first step, all RBs within a research field were put in a random order. Second, all RIs within an RB were put in a random order. This resulted in a twice randomised list of RIs for each research field. To obtain the 200 RIs, for each RB, we selected RIs randomly. This procedure was done automatically by a self-written PHP script.

Initially we contacted 1,133 institutes; yet, after excluding for those institutes who asked to be deleted from the sampling list (reasons were mostly because they did not do any public engagement, or were not reachable, (e.g. emails bounced)), we got to a final sampling frame of n = 1,007 RIs that received the link to the questionnaire.

Questionnaire

The questionnaire was translated into German for both the university and the non-university sample. The translation was done independently by two researchers. Both translations were then merged into a pre-version, which was piloted with two typical addresses from the University of Munster. Their annotations were worked into the final version of the questionnaire.

The data were collected between 2017 and 2018. The achieved response rate of 21.9% in the university sample can be considered satisfactory. In the non-university sample, almost every third RI contacted took part in the survey, resulting in a response rate of 32.8%. In both subsamples, the response rates varied across the OECD research fields, with the lowest response rates in agricultural sciences, at 15.1% (university sample) and 11.1% (non-university sample), and the highest response rates in social sciences, at 26.1% (university sample) and 46.0% (non-university sample). T-tests have shown that these differences between respondents and non-respondents are not significant. It can, therefore, be assumed that no distortions in response behaviour are to be expected.

Public communication activities

Public events

Overall, German RIs show only moderate interest in public communication activities. More than half of the RIs in the university sample 'never' take part in events such as 'science cafés', 'citizen science projects' or international events. Even for less elaborate formats such as public lectures or 'open days', the most frequent response and the median are only in the quarterly range. For the non-university RIs, the trend and order are similar, but with slightly higher frequencies for all items (Figure 13.1).

By means of chi-square tests, it was checked at which events the two samples differ significantly from each other. This is the case for the three most frequently performed events, public lectures ($\chi^2(4) = 9.899$, p <.05), 'open days' ($\chi^2(4) = 27.872$, p <.001) and events by private institutions ($\chi^2(4) = 17.456$, p <.05); the least were international events ($\chi^2(4) = 10.268$, p <.05).

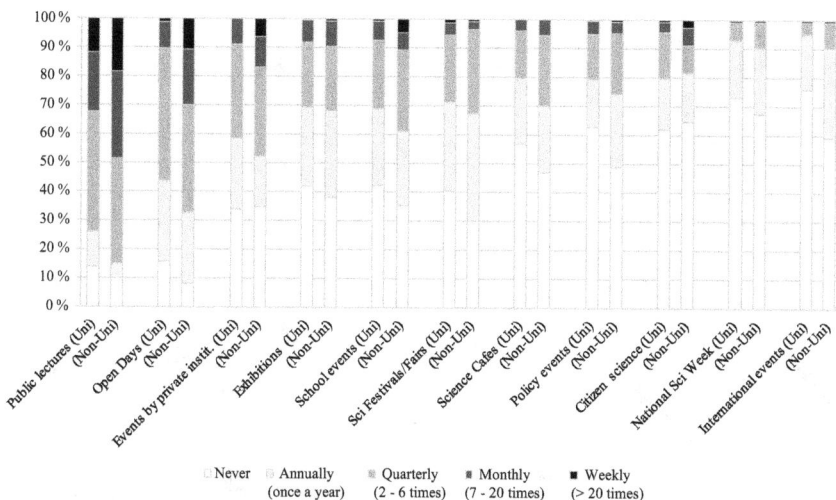

Figure 13.1 Frequency of different events by subsample (uni/non-uni).

Thus, both samples differ significantly from each other in almost half the items, but they show surprisingly similar activity profiles: what university institutes do comparatively often, non-university RIs do even more frequently and events in which non-university RIs rarely participate play an even smaller role for university-owned RIs.

Traditional media channels

A similar picture emerges for the channels used to communicate with the general public. While 7 of the 13 channels queried are predominantly used by non-university institutions at least once a quarter, university institutions use only a single channel, namely, traditional press releases on a quarterly basis. Among the more or less frequently used channels are push channels such as press releases (the most common communication channel) or press conferences (ranked tenth) as well as pull channels where an institute is more likely to respond to requests, such as interviews for newspapers, ranked second or other TV, ranked penultimate in the university sample and fourth-to-last in the non-university sample. It is expected that communication channels that require less effort will be used more frequently than those that require more effort. Press releases are published more often than press conferences; articles for the non-specialist public are published more often than popular books and interviews for newspapers are given more often than interviews for TV (Figure 13.2).

Even with the favoured channels, there are clear differences between the subsamples. This time the chi-square tests are significant for 11 out of the 13 items (p <.05) – except for material for schools and popular books, for which no significant differences exist. It is interesting in this context that 'material

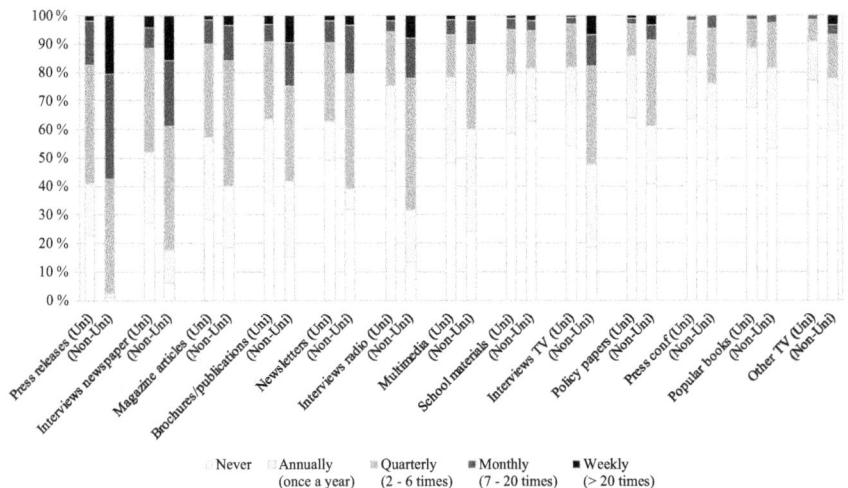

Figure 13.2 Frequency of communication through traditional media channels by subsample (uni/non-uni).

for schools' at non-university RIs is only slightly less frequently communicated than at university RIs although the former do not offer teaching. Overall, however, it is clear that non-university RIs use more channels more frequently than university-based RIs.

New media channels

With regard to the new media channels used, it can be stated that practically only the institutes' own website is used on a regular basis. In the area of non-university RIs, blogs and Twitter are also used by more than half of the institutes, but only about one-third feed those channels at least monthly. In view of the fast pace of social networks and the high frequency with which other communicators post there, it can be assumed that RIs in Germany – with a few exceptions – have practically no presence on the social web. Although this also applies to a large extent to non-university RIs, significant differences (p <.001) between the two subsamples can also be seen in the use of social media channels in that all channels are used more frequently by non-university RIs than by university RIs – with the exception of Google+ and Instagram, which have no significant differences between university and non-university RIs.

Journalists database. Of non-university RIs, 85% maintain a list or database of journalists and media contacts, while such a list exists in only 23% of university RIs ($\chi^2(2)$ = 115.989, p <.001). RIs from the non-university sample had received more than ten media and journalist enquiries in the last 12 months (57%). In the university sample, this was true for only 17% of the RIs ($\chi^2(4)$ = 68.370, p <.001). This finding also fits the picture drawn so far: the large, publicly funded RIs are more active, visible and in greater demand than German universities.

In order to find out the reasons for these differences in the frequency and intensity of public communication, it is worth taking a look at the RIs' self-image with regard to communication with the media. Non-university RIs seem to be much more open in this respect, which is reflected in, for example, the fact that 54% fully agree with the statement: 'Visibility in the media of the research conducted at our unit is important'. In the university sample, only 30% of the respondents agree with these statements. Seven percent answered that they 'do not agree' or 'do not agree at all' with this statement, while in the non-university sample, it is only 4%. These differences are also more than statistical coincidence ($\chi^2(4)$ = 17.608, p <.01). Eighteen percent of the university RIs do not consider it their task to maintain relations with the media. In the non-university sample, the agreement with this statement is significantly lower, at 5%. In contrast, 76% of the respondents from outside the universities completely reject this idea (strongly disagree). This value is more than twice as high as in the university sample (37%; $\chi^2(4)$ = 44.968, p <.001). Therefore, this is not only a question of resources but also of differences in 'institutional mentality' (Figure 13.3).

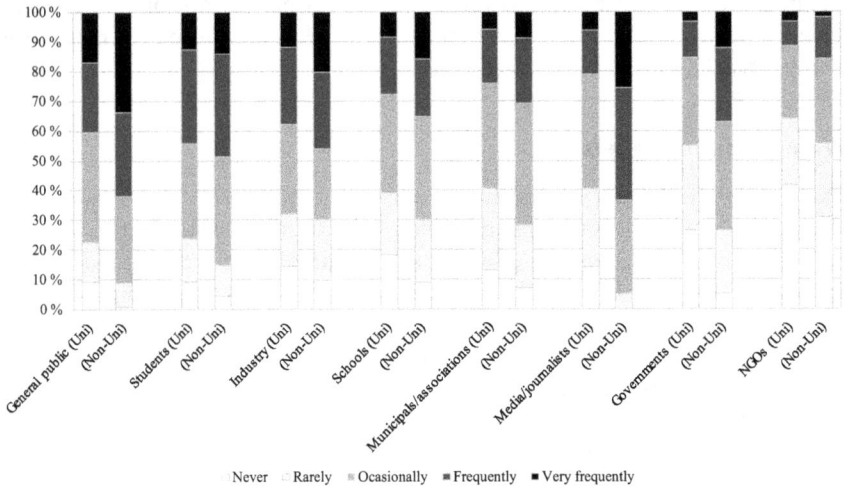

Never ⬚ Rarely ▨ Ocasionally ▪ Frequently ▪ Very frequently

Figure 13.3 Audience types by subsample (uni/non-uni).

Audiences

A greater openness to communication with the public is also underlined by a look at the addressees of RIs' communication. Non-university RIs have had more frequent contact with all the audiences surveyed in the past 12 months than university RIs, even if the differences are not always very large (Figure 13.4). Nevertheless, the differences were significant in contact with governments/politicians/policymakers ($\chi^2(4) = 38.871$, p <.001), media and journalists ($\chi^2(4) = 75.328$, p <.001) and the general public ($\chi^2(4) = 21.000$, p <.001).

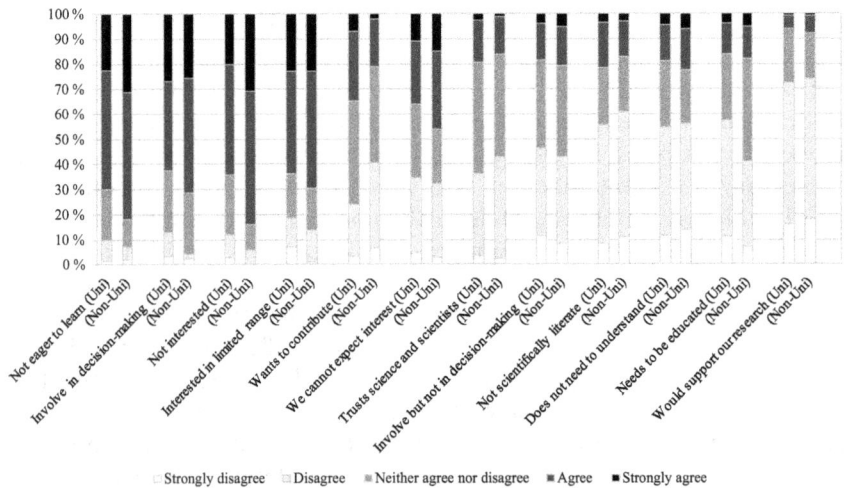

⬚ Strongly disagree ⬚ Disagree ▪ Neither agree nor disagree ▪ Agree ▪ Strongly agree

Figure 13.4 Public perceptions by subsample (uni/non-uni).

Public perceptions

There is little difference between university RIs and non-university RIs in their perception of the public (Figure 13.4). There is a slight tendency in favour of more negative items such as 'the public is not interested in the research conducted at our unit' within the university sample than in the non-university sample ($\chi^2(4) = 15.054$, p <.01), while more emphatic views such as 'the public wants to contribute to science' tend to be more popular in the non-university sample ($\chi^2(4) = 10.959$, p <.05) than in the universities. Beyond these two items, however, the difference in the public's image is not significant. In general, scientific institutions assume a strictly limited interest in research and technology on the part of the non-scientific public and assume that non-scientific target groups have only a limited willingness to get involved. At the same time, they do not consider the level of trust that scientific institutions enjoy in the public sphere to be particularly high, nor do they believe that more knowledge and commitment would lead to more support. At the same time, the basic assumptions of the so-called deficit model, according to which the lay public needs more scientific education, are also well received. All in all, this is a much less optimistic picture of the audience than is assumed in most policy recommendations for ever more science communication.

Rationales for public communication

RIs may communicate with the general public for various reasons. These reasons can be intrinsically motivated, or they can be given from outside. In other words, RIs can develop communication activities of their own accord or in response to social or political demands and expectations. Both university and non-university RIs confirm that RIs communicate mainly on their own initiative. The majority of respondents cited as the most important (university: 37%, non-university: 36%) or second most important (university: 20%, non-university: 24%) reason for communicating with the public that they wanted to 'make their research accessible to the public'. If these statements initially leave, open the actual motive behind them; the second most frequently cited reason, 'we aim to respond to the policy/mission of our host institution/university', explains more about the motivation for communication. Twenty-three percent of university institutes cite this as the most important reason and another 11% as the second most important. In the non-university sample, the scores are significantly higher, between 21% and 33% ($\chi^2(1) = 12.498$, p <.001). Another significant difference is the item 'We want to listen and involve the public in our research'. In the university sample, 10% cited this as the second most important reason and 2% as the most important reason. In the non-university sample, on the other hand, only 3% of the respondents cited this motive as the second most important reason for public communication ($\chi^2(1) = 5.569$, p <.05).

The least important reasons cited were 'We aim to respond to national policies of science communication/scientific culture' in the university sample (26%) and 'We want to raise our research profile' in the non-university sample

(22%). Representatives of university institutes cited the latter significantly less frequently as the least important reason (10%; $\chi^2(1) = 9.339$, p <.01).

Overall, however, it can be stated that the reasons for communicating with the public at university and non-university RIs are largely the same. It is not possible to make a clear distinction between the samples or between internal and external drivers of communication. One possible interpretation of this finding would be that many RIs tend to communicate 'from the gut', without knowing exactly why and for what purpose. Or, to put it another way, they do not follow their own communication strategy but rather routine programmes, and these routines are largely the same at all institutes.

This impression is also supported – especially for university institutes – by RIs' own statements about their motivation. According to this, 62% of the university institutes have 'neither plans nor a policy for public communication/outreach' but nevertheless communicate with the public. At non-university institutes, 28% agree with the statement, while nearly 70% of them report that they have formulated a communication policy. This indicator also reinforces the impression of greater professionalism among the non-university institutes. Apart from this item, all statements of the commitment battery receive significantly more approval within the non-university sample than in the university sample (p <.05). Thus, a large proportion of the German RIs expect their scientists to participate in the communication activities of the institution. How exactly this is to be done, according to which criteria and with which strategies the scientists want to communicate with the public, especially at non-university institutes, is often not specified but remains open to individual initiative.

This could be related to the extent to which communication with the public is already established and institutionalised at the respective institutions. At non-university RIs, communication with the public is usually longer and more firmly established than at university institutes. Nearly, one in three university RIs (31%) has been communicating with non-academic audiences for five years or less. The figure for non-university institutes is only 8% ($\chi^2(3) = 29.461$, p <.001). In three out of four non-university RIs (76%), communication activities have also increased in the past five years, while this is only the case in just over one in two university RIs (54%) ($\chi^2(2) = 14.209$, p <.001).

Public communication resources

Communications staff

In addition to a different self-image and a different degree of openness for communication with the public, material reasons could also be responsible for the different degrees of public engagement of university and non-university institutes. For example, almost all non-university RIs have their own communication staff (95%), while in the university subsample, this was only the case at one in three RIs (36%, $\chi^2(2)$ = 107.018, p <.001). On average, 2.8 part-time and 3.3 full-time employees at non-university RIs with their own communication staff are dedicated to communication. At non-university RIs with their own communication staff, there are only 0.9 employees on average who deal exclusively with communication and two employees who deal with communication among other tasks. The differences are significant both for those who work exclusively in communications ($\chi^2(15)$ = 37.050, p <.01) and those who are responsible for communications ($\chi^2(12)$ = 22.792, p <.05).

At the non-university RIs, there are not only significantly more communication staff but also a difference in the degree of professionalisation. The employees responsible for communication at non-university institutions were previously more often employed in communication professions than those at university RIs. In non-university RIs, an average of 1.0 communications staff member has previously worked in marketing/PR; in university institutes, the average is 0.3. The difference is statistically significant ($\chi^2(7)$ = 20.062, p <.01). Moreover, communication staff at non-university RIs tended to be more likely to have completed workshops or studies in the field of communication than their colleagues at university RIs, but the differences are not significant.

Communications funding

However, there are significant differences in the financial resources available for communication. A large majority, 92% of the RIs in the non-university sample, spend less than 1% or 1–5% of their annual budget on communication activities, according to their own estimates. In the university sample, the figure is only 72%, while 18% of university RIs have no budget at all for communication activities. For non-university RIs, this proportion is significantly lower, at just 4% ($\chi^2(4)$ = 14.816, p <.01). Considering that 96% of non-university RIs have an annual research budget of more than 1 million euros, while the average budget at university institutes is in the range of 250,000–500,000 euros, the financial gap between university and non-university RIs appears all the greater. Nevertheless, representatives of non-university RIs are significantly more likely to believe that their RI should spend more resources on public communication than representatives of university RIs ($\chi^2(2)$ = 7.124, p <.05). At non-university RIs, 54% share this opinion; at university RIs, only 43%. Four percent even think that fewer resources should be spent on communication. In the non-university sample, none of the respondents held this view.

Similar to global findings (Entradas et al., 2020) that more financial resources are associated with more public engagement activities we find in both the university sample and the non-university sample, the sheer amount of communication activities differs significantly between institutes with higher and lower communication budgets, where more money results in more activities (Table 13.2).

Table 13.2 ANOVA for the effects of an RI's communication funding on its communication activities

		df	F	p	η^2
Uni	Index public events	4, 182	5.20	< .001	0.10
	Index traditional media channels	4, 182	5.67	< .001	0.11
	Index new media channels	4, 182	3.25	< .05	0.07
Non-uni	Index public events	3, 81	5.96	< .01	0.18
	Index traditional media channels	3, 81	3.60	< .05	0.12
	Index new media channels	3, 81	5.85	< .01	0.18

For this purpose, individual mean value indices were first formed for the number of events, use of traditional channels and use of new media channels described above.

Similarly, the head count for communication tasks in the university sample and the number of employees exclusively dedicated to communication tasks in the non-university sample make a difference: the more personnel resources are used, and the more activity is observed. The tendency to evaluate one's own communication activities is also strongly correlated with the intensity of the communication activities. From a theoretical point of view, however, it is questionable whether the increased communication is the result of regular evaluation or whether, in cases where there is a lot of communication with the public, routines and structures have already been established that involve more evaluation. This could be interpreted as a professionalisation of communication.

Perceived success of communication

High personnel deployment and the level of financial resources are also reflected in the perceived success of communication efforts. Not a single respondent in the non-university sample believes that their RI is unsuccessful in its communication activities when it comes to supporting the RI's mission. Eighty-six percent are of the opinion that they are 'successful' or 'very successful'. In the university sample, only 44% support this view. Here, as many as 9% of those surveyed believe that they are 'unsuccessful' or 'very unsuccessful' ($\chi^2(4) = 69.149$, p <.001). In both samples, the feeling of being successful with the communication activities is highly significantly correlated with the frequency of the evaluation (university: r =.258, p <.001; non-university: r =.407, p <.001).

Conclusion

The assumption that German RIs as a whole do not communicate with the public very frequently and intensively has been confirmed. It must be noted that the calls for increasingly professional communication between science and the public that have persisted for at least 20 years have largely faded away without consequence at the meso level of the RIs. This is more likely to be the case in universities than in publicly funded federal and state RIs. The core results of the surveys can be summarised as follows: RIs, beyond public lectures, rarely create institutional events for public engagement. This can be explained by the fact that the resources for event management at universities tend to be located at the central level, which also explains the slightly stronger event making of the communication of pure RIs. RIs primarily do classic media work on their own initiative or limit themselves to responding to media inquiries. This finding also applies, by and large, to pure RIs of the federal and state governments, even if they use a more differentiated spectrum of communication channels. As far as online presence is concerned, the classic website dominates. Social media profiles are maintained at many institutes, but only in a few cases are they intensively fed. RIs – unlike many communication professionals, functionaries and science politicians – have a quite realistic picture of the public, which they assume to have a strictly limited and highly selective interest in science. Consequently, the communication activities of German RIs are more motivated by the self-interest of the institutions than by a perceived obligation towards the lay public.

In an international comparison, however, especially as far as neighbouring European countries are concerned, the German institutes are not in such a bad position (see further chapters in this volume). By contrast, there are clear differences between the universities and universities of applied sciences, on the one hand, and the non-university research sector in Germany, on the other hand. For the publicly funded RIs of the Max Planck Society, Fraunhofer Society, Helmholtz Association of German Research Centres and Leibniz Association, the transfer of their work to society is part of the way they see themselves, and communication with the general public is apparently more a part of their self-image than is the case in the universities. The truth is, however, that outside the universities, considerably more resources can be mobilised for communication than are available to a large number of the university institutes. In terms of the ability to mobilise resources for public engagement activities, publicly funded German RIs, both university and non-university, actually live in two different worlds.

Under the current conditions, German RIs can apparently only afford to a very limited extent to 'market' their own work appropriately to the public and enter into a dialogue with society beyond their research tasks. Even if the goodwill of many of the institutes surveyed is evident – a clear majority would like for the public to participate in decisions, and more than a few consider it important to be noticed in the media and count it among their tasks to maintain appropriate relations – this often fails to materialise. In the case of the university institutes, the scientists are already torn between research, more complex teaching duties and self-administration with regard to their time and resource planning. The teaching function does not apply to the non-university RIs, which is another explanation for the fact that considerably more communication activities take place here.

Thus, at the end of a very elaborate study, a simple insight is left: if political decision makers keep assigning more and more new tasks to the universities and, at the same time, the number of their first-year students increases dramatically – as is politically desired – without budgets and staffing increasing at the same rate, then one should not be surprised if the institutions (have to) neglect some of these tasks in order to survive. Thus, the demand for increased scientific communication is obviously the straw that breaks the camel's back.

Notes

1 https://www.bmbf.de/upload_filestore/pub/Grundsatzpapier_zur_Wissenschaftskommunikation.pdf
2 At some research bodies, especially small ones and universities of applied sciences, departments or faculties were not further subdivided. In these cases, the department or faculty as the smallest unit above the individual professor was recorded as an RI.

References

Deutsche Forschungsgemeinschaft. (n.d.). *Zahlen und Fakten: Zweite Programmphase*. Retrieved August 9, 2016, from http://www.dfg.de/dfg_magazin/forschungspolitik/exzellenzinitiative_und_exzellenzstrategie/zahlen_fakten/index.html

Entradas, M., & Bauer, M. W. (2017). Mobilisation for public engagement: Benchmarking the practices of research institutes. *Public Understanding of Science*, 26(7), 771–788.

Entradas, Marta, Bauer, M. W., O'Muircheartaigh, C., Marcinkowski, F., Okamura, A., Pellegrini, G., Besley, Massarani, J. L., Russo, P., Dudo, A., Saracino, B., Silva, C., Kano, K., Amorim, L., Bucchi, M., Suerdem, A., Oyama, T., & Li, Y.-Y. (2020). Public communication by research institutes compared across countries and sciences: Building capacity for engagement or competing for visibility? *PLoS One*, 15(7), e0235191. doi: 10.1371/journal.pone.0235191

Federal Ministry of Education and Research. (n.d.). *Federal institutions*. Retrieved July 2, 2018, from https://www.research-in-germany.org/en/research-landscape/research-organisations/federal-institutions.html

Friedrichsmeier, A., Geils, M., Kohring, M., Laukötter, E., & Marcinkowski, F. (2013). *Organisation und Öffentlichkeit von Hochschulen*. [Forschungsreport 1/2013 des Arbeitsbereichs Kommunikation – Medien – Gesellschaft]. Münster: IfK.

Friedrichsmeier, A., Laukötter, E., & Marcinkowski, F. (2015). Hochschul-PR als Restgröße: Wie Hochschulen in die Medien kommen und was ihre Pressestellen dazu beitragen. In H. Bonfadelli, M. S. Schäfer, & S. Kristiansen (Eds.), *Wissenschaftskommunikation im Wandel* (pp. 128–152). Köln: Herbert von Halem.

Hochschulrektorenkonferenz. (2016). *Hochschulkompass: Übersicht mit allen Hochschulen.* Retrieved August 9, 2016, from https://hs-kompass2.de/kompass/xml/download/hs_liste.txt

Hoffmann, E. (2018). Hochschulkommunikation – Auf die schiefe Bahn geraten? Retrieved from https://www.wissenschaftskommunikation.de/hochschulkommunikation-auf-die-schiefe-bahn-geraten-15939/

Hüther, O. (2010). *Von der Kollegialität zur Hierarchie?: Eine Analyse des New Managerialism in den Landeshochschulgesetzen. Organization & Public Management.* Wiesbaden: VS.

Kohring, M. (2019). Das eigentliche Problem geht nicht von der Bevölkerung aus. Wissenschaftskommunikation.de. Retrieved from https://www.wissenschaftskommunikation.de/das-eigentliche-problem-geht-nicht-von-der-bevoelkerung-aus-32167/

Kohring, M., & Marcinkowski, F. (2015). Währungsrisiken. Die prekären Folgen des Erfolgskriteriums ‚mediale Aufmerksamkeit'. *Forschung und Lehre*, 11/15. Retrieved from https://www.wissenschaftsmanagement-online.de/beitrag/w-hrungsrisiken-die-prek-ren-folgen-des-erfolgskriteriums-mediale-aufmerksamkeit-5913

Kohring, M., Marcinkowski, F., Lindner, C., & Karis, E. (2013): Media orientation of university decision makers and the executive influence of public relations. *Public Relations Review*, 39(3), 171–177.

Laukötter, E. (2014). *Die Sichtbarkeit deutscher Hochschulen in Print- und Online-Medien.* [Forschungsreport 1/2014 des Arbeitsbereichs Kommunikation – Medien – Gesellschaft]. Münster: IfK.

Marcinkowski, F., & Kohring, M. (2014). The changing rationale of science communication: A challenge for scientific autonomy. *Journal of Science Communication* 13(3), C04.

Marcinkowski, F., Kohring, M., Friedrichsmeier, A., & Fürst, S. (2013). Neue Governance und die Öffentlichkeit der Hochschulen. In E. Grande, D. Jansen, O. Jarren, A. Rip, U. Schimank, & P. Weingart (Eds.), *Science Studies. Neue Governance der Wissenschaft: Reorganisation – externe Anforderungen – Medialisierung* (pp. 257–288). Bielefeld: transcript Verlag.

Marcinkowski, F., Kohring, M., Friedrichsmeier, A., & Fürst, S. (2014). Testing the organizational influence on scientists' media contacts. *Science Communication*, 36(1), 56–80.

Peters, H. P. (2013). Gap between science and media revisited: Scientists as public communicators. *Proceedings of the National Academy of Sciences*, 110(Supplement 3), 14102–14109. http://doi.org/10.1073/pnas.1212745110

Peters, H. P., Brossard, D., de Cheveigné, S., Dunwoody, S., Kallfass, M., Miller, S., & Tsuchida, S. (2008a). Science-media interface: It's time to reconsider. *Science Communication*, 30(2), 266–276. http://doi.org/10.1177/1075547008324809

Peters, H. P., Brossard, D., de Cheveigné, S., Dunwoody, S., Kallfass, M., Miller, S., & Tsuchida, S. (2008b). Science Communication: Interactions with the mass media. *Science*, 321(5886), 204–205. http://doi.org/10.1126/science.1157780

Rehländer, J. (2018). *Wer schreibt endlich PUSH Zwei?* Retrieved from https://www.wissenschaftskommunikation.de/wer-schreibt-endlich-push-zwei-14711/

Roessler, I., Duong, S., & Hachmeister, C.-D. (2015). *Welche Missionen haben Hochschulen? Third Mission als Leistung der Fachhochschulen für die und mit der Gesellschaft.* Arbeitspapier/Centrum für Hochschulentwicklung: Nr. 182. Gütersloh: Centrum für Hochschulentwicklung gGmbH.

Schäfer, M. S. (2018). Legitimationsbeschaffung statt Wissensvermittlung. *Forschung und Lehre*, 25(6). Retrieved from https://www.forschung-und-lehre.de/management/es-geht-darum-die-eigene-reputation-zu-staerken-680/

Siggener Kreis (2013). *Siggener Denkanstoß 2013: Zur Zukunft der Wissenschaftskommunikation.* Gut Siggen, Heringsdorf.

Siggener Kreis (2017). *Wissenschaft braucht Courage: Impulse aus der fünften Tagung des Siggener Kreises zur Zukunft der Wissenschaftskommunikation (Siggener Impulse).* Gut Siggen, Heringsdorf.

Siggener Kreis (2018). *WALK THE TALK – Chefsache Wissenschaftskommunikation* (Siggener Impulse). Gut Siggen, Heringsdorf.

Statistisches Bundesamt. (2017a). *Bildung und Kultur: Personal an Hochschulen* (Fachserie 11 Reihe 4.4). Retrieved from https://www.destatis.de/DE/Publikationen/Thematisch/BildungForschungKultur/Hochschulen/PersonalHochschulen2110440167004.pdf

Statistisches Bundesamt. (2017b). *Bildung und Kultur: Private Hochschulen.* Retrieved from https://www.destatis.de/DE/Publikationen/Thematisch/BildungForschungKultur/Hochschulen/PrivateHochschulen5213105167004.pdf

Stifterverband für die Deutsche Wissenschaft et al. (1999). *Memorandum "Dialog Wissenschaft und Gesellschaft".* Retrieved from https://www.hrk.de/positionen/beschluss/detail/memorandum-dialog-wissenschaft-und-gesellschaft/

Chapter 14

'Research Excellence' and Public Communication in Portugal

Fernando Chacón and Marta Entradas

The Portuguese University

The Portuguese university for the 'masses' has a recent history. Despite having one of the oldest universities in the world – the University of Coimbra founded in the 13th century (1290), it was for centuries, an elitist space for the education of a few. Until the 1950s, there were four universities in the country: the University of Coimbra, the University of Lisbon, the University of Porto and the Lisbon Technical University (Teixeira et al., 2007), and only 0.04% of the Portuguese population completed a university degree (Gonçalves, 2002).

University expansion and access to education happened only in the late 20th century. In 1932, Portugal entered a 40-year dictatorship, where science and education had little space in society, political agendas and the public sphere, and scientific culture was low. In 1975, with the end of the dictatorship, new reforms in the education system brought education and science to political attention, reflected mainly in an increase in the number of higher education institutions and education accessibility to broader audiences. For example, the University of Algarve (in the south of the country) was created in 1979, the University of Evora (in Alentejo interior) in 1988 and the University of Beira Interior in 1999 (in the north).

In addition to new public universities, polytechnics also emerged, envisioning a system that supplied more vocationally engineering and technology-oriented programs dedicated to engineering, agriculture, health, management and other fields (e.g. 'Escola Superior de Enfermagem Porto', 'Escola Superior de Hotelaria e Turismo do Estoril') (Teixeira et al., 2007). The reforms in the education system (Veiga et al., 2014), led also to the development of private education with new institutions privately funded emerging such as the Católica University (1970) and the Lusíada University (1980). Nevertheless, while the private sector saw a noticeable growth, public universities continue to be the system's backbone, with both systems contributing to the increasing levels of science education and literacy among the Portuguese population (European Commission, 2005, 2010; OCT, 1996, 2000). Today, there are more

DOI: 10.4324/9781003027133-17

than 60 universities and polytechnic institutes in Portugal, both public and private with 372,000 students enrolled (DGEEC, 2018), and around 18% of the population had a university degree in 2017 (Instituto Nacional de Estatística, 2017). Despite the significant increase over the decades, the levels of 'scientific literacy' of the Portuguese have ranked low compared to European standards (e.g. EC, 2010, 2013).

Research in Portuguese Universities

After joining the European Union in 1986, universities saw pressure to focus on internationalisation and boost productivity and funding for research to be on par with other European countries (Ramos & Sarrico, 2016). In 1995, the Ministry of Science and Technology was created, led by Jose Mariano Gago from 1995 to 2002, and in 1996, the Fundação para a Ciência e a Tecnologia (hereafter, FCT) was created as the national agency for research funding, which led to more public investment in science (Gonçalves & Castro, 2003). At the beginning of the 2000s, the national research funding was around 17,582,744 €, and in 2017, the total investment was around 75,341,580 €. Research expenditure accounted for 0.68% of the GDP in 1999; in 2009, it significantly grew to 1.58% and in 2019, it was reported to be 1.4%. Yet, the investment in research in Portugal has been below the 2.19% average of European countries (Eurostat, 2019).

The push to internationalisation, triggered the government's institutional decentralisation of research activities promoting self-regulation policies and autonomy. The consolidation of this move came in the form of the University Autonomy Law (Law 108/99), specifying decentralisation guidelines and rules to help universities have their own financial, educational, scientific and administrative management of educational programs, and their own research institutes, countering the then standing norm of research being confined to State Laboratories (Gonçalves and Castro, 2002).

With the consolidation of self-governing institutions and increase in science investment, new research institutes emerged and the spectrum of reserach and number of research staff also saw a steady increase. Along with the rise of staff, productivity in the form of publications followed: the annual number of publications almost tripled going from 3,792 in 2000 to 10,081 in 2010, signalling Portugal's commitment to science and a path to achieve 'excellence' in research according to EU standards, which also opened access to more funds (Heitor & Horta, 2011; Ramos & Sarrico, 2016).

At the same time, in the mid-1990s, science communication gained momentum, in a context of full political support, with the government being a significant promoter of initiatives to foster scientific culture in the country (Entradas et al., 2020). The conditions allowed for the creation of a national 'policy for scientific culture' to open science to the Portuguese society and encourage scientists and scientific institutions to engage in societal conversation. Parts of this policy can be seen in government initiatives such as (1) the formulation of legislation governing scientific institutions, teaching and research staff to engage in public communication of science; (2) the creation of fellowships for science communication roles at institutions (called Science and Technology Management Fellowships (BGCT – Bolsa de Gestão de Ciência e Tecnologia) and a specific area in the national PhD calls called PACT (Promotion and Administration of Science and Technology) for training at the postgraduate level; (3) the creation of the *Ciência Viva* program aimed at developing science communication infrastructure and activities in the country (for a detailed description of the Ciência Viva program, see Conceição, 2011).

Competitive funding and evaluation of research

While autonomy was a step forward for the scientific production of research institutions, the government remained the leading funding provider, and the distribution of research budgets triggered national competition. Measuring research 'efficiency' became a pillar concept in the institutional research evaluation system, with multiple parameters used to determine the 'quality' of the research, e.g. resource management, autonomy, funding, policy, research and innovation activities (Ramos & Sarrico, 2016), and particularly the academic productivity of the research institution. Research assessments have become the basis of funding distribution and are embedded in a competitive environment amongst institutions, evaluated by FCT. Since 1996, six R&D evaluations have been conducted, the 2018 evaluation being the latest at the time of writing; as a note, in 1996, Portugal had 270 R&D units certified by FCT, and in 2018, the number grew to 349 units (Table 14.1).

The most significant research funding for research institutes come from the Portuguese government science budget distributed in the form of Research grants to individual research projects (project funding), and Multiple-year funding attributed to recognised R&D units on the basis of evaluation of their excellence in research (basic funding). R&D units are units attached to higher education institutions, and Associated Laboratories, that work independently of an HEI, in any area of research (Decree-Law no. 63/2019). Here, we will refer to these as *research institutes*. Table 14.1 shows the results of the evaluations conducted to R&D units between 1996 (the first evaluation) and 2017/2018 (the most recent). In these evaluations, R&D units are ranked as excellent, very good, good, poor and insufficient (FCT, 2019).

This evaluation system has not come without controversies; for example, the 2013 evaluation had plenty of criticisms because it changed its

Table 14.1 Number of research institutes that received FCT funds over the years. RI split in percentages to show the evaluation and qualification distribution, based on a 5 scale from excellent to weak

Year	Total evaluated R&D units	Evaluated R&D units(%)				
		Excellent (%)	Very good (%)	Good (%)	Poor (%)	Insufficient (%)
1996	270	16	28	31	19	6
1999	354	19	38	27	12	4
2002	437	18	32	28	14	8
2013/2016*	274	22	30	28	10	10
2017/2018	349	30	34	25	10	1

Notes: * In this evaluation, there was a recovery program for 96 units (FCT, 1999, 2005, 2013, 2019).

requirements, number of special panellists, and ultimately reduced RIs approved for funds, triggering a recovery plan in 2016 to RIs that had been poorly evaluated (FCT, 2019). Currently, the latest parameters used in the 2017/2018 evaluation have been simplified in three aspects: (a) Activity developed in the RI has internationalisation relevancy, merit, quality; (b) Scientific merit of the researcher's team and (c) Adequate objectives, strategy, and planning for the following five years.

Requirements for science communication by the funded research

These changes also affected public communication at research institutes. In the past decade, public communication has become a criterion in the evaluation of research institutes and assessments for funding. For example, in 2013, the FCT established that institutions receiving funding from FCT must 'disseminate research to broader audiences and knowledge transfer'. It is then expected that private and public research institutions receiving public funds, promote public engagement and make available updated information from their funded projects. Perhaps influenced by these top-down drivers from government policy to institutions, public communication has over the past years entered the university landscape and is becoming part of research institutes' activities (Entradas, 2015; Entradas et al., 2020).

Based on this backdrop, one could expect a relationship between public communication activity and level of research excellence. Not least as a result of funding requirements. We know from previous work that prestigious institutions are more likely to get research grants funded (Katz and Matter, 2017; Murray et al., 2016). As these often require that the funded research is communicated to the public (European Union, 2013), this could lead to expectations that research funding would associate with public communication activity particularly among 'excellent' institutions.

Yet, this expectation may be challenged by other factors such as organisational views that public communication is less important than research (e.g. Ecklund et al., 2012), with impact on resource allocation and support provided to researchers including time allowed for non-academic activities – lack of time and institutional support have often been reported by scientists as main barriers hampering participation in outreach (e.g. Kassab, 2019; Royal Society, 2006; Wellcome Trust, 2015). That is, outreach could be understood as a competitor to research output (Martin, 2011; Martinez-Conde, 2016) with particular evidence at excellent institutions aspiring to be part of the world rankings. As such, we compare public communication activity of excellent and less-than-excellent institutes for three main aspects of communication: (1) level of activity in various communications means, (2) audiences addressed and (3) professionalisation of staff, funding and policies. We test the following hypotheses:

H1: Public communication is higher among excellent institutes compared to less-than-excellent ones.
H2: Excellent institutes show higher level of interaction with external audiences and also hold more positive views about them.
H3: Excellent institutes show higher professionalisation of science communication than less-than-excellent institutes.

Methods

Sample

This study covered the whole population of research units in Portugal, composed of all non-profit public or private research institutes within and outside universities with active FCT funding, for the period 2007–2014 (N=386). The data were collected online in 2014/2015, and the response rate was 61% (N=234). Institutes cover all areas of research, classified into six research areas according to OECD (Table 14.2). The institutes surveyed were medium to small in size (M=106.2 researchers per institute, SD=140.9; Median=60.0), capturing on average between €250,000 and 500,000 for research annually. The effective working sample size is lower ($N = 224$) because we only considered cases which were fully completed.

Table 14.2 Sample characteristics by research area

	RIs contacted sampling frame		RIs responded sample		Response rate
Area of research	N	%	NI	%	%
Social sciences	98	25.5	59	25.2	60.2
Humanities and the arts	66	17.2	41	17.5	62.1
Medical and health sciences	45	11.7	29	12.4	64.4
Natural sciences	94	24.5	56	23.9	59.6
Engineering and technology	67	17.4	41	17.5	61.2
Agricultural and veterinary sciences	14	3.6	8	3.4	57.1
Total RI	386	100	234	100	61

Questionnaire

One questionnaire was collected for each unit and was completed by a member of staff who could speak for the communications practices of the unit: 47% were directors/coordinators of the units, management/administrative staff (18.8%), researchers (17.4%), communication staff (9.9%) and 6.6% 'other' (e.g. vice-directors, or professors, PhD students and postdoctoral fellows). The questionnaire was composed of 34 questions related to public communication of science including activities, resources, policies, rationales for communication, among others. In this chapter, we focus on a set of those questions that are of interest for our comparison including (1) public communication activities such as public events, traditional media channels and new media channels; (2) audiences and (3) professionalisation questions about communications staff, communications funding and communications policy/guidelines.

Variables

Public communication activities. We asked about institutes' frequency of participation/organisation of communication activities for non-specialists, including public events, traditional media and new media channels. *Events* included 11 items: public lectures, exhibitions, open-days/guided visits, science festivals/fairs, National science week, science cafes/public discussions, UNESCO/researchers night, deliver/participatory events on policymaking, events by private institutions, talks at schools and citizen science. *Traditional media channels* included 13 items: newspapers interviews, radio interviews, TV interviews, other TV interviews, press conferences, press releases, newsletters, brochures/publications, articles in magazines, multimedia, popular books, policy papers and materials for schools; these were measured in an ordinal scale: never (1), annually (2), quarterly (3), monthly (4) and weekly (5). *New media channels* included six items: institutional website, blog, Facebook, Twitter, Podcasts and YouTube; measures were never (1), quarterly (2), monthly (3), weekly (4) and daily (5). We recoded these variables into an estimated number of activities, considering 48 working weeks: never (0), annually (1), quarterly (4), monthly (12), weekly (48) and daily (240).

Audiences addressed. We asked institutes how often they had engaged with external audiences in the year previous to the survey, in a scale 'Never' (1), 'Occasionally' (2), 'Frequently' (=3), 'Don't know' (= 4). The audiences considered were general public, schools, students outside teaching, members of local municipalities/councils/associations, delegates from industry, governments/politicians/policy-makers and NGO's and media journalists.

Perceived images of audiences. Perceptions about these audiences were assessed with 13 statements using a Likert 5-point scale from 'strongly disagree' (1), to 'strongly agree' (5), a 'don't know' (6) option was also included. These variables were recoded into binaries for agreement (1) versus others (disagreement) (=0) ('strongly disagree', 'disagree', 'neither agree nor disagree' and 'don't know' options) (see Table 14.3).

Public communications staff. This variable is coded as (0) for institutes that reported having no communications staff in the institute and (1) for those that had. *Public communication policies/guidelines* include four statements with true/false options: (1) we have a policy, (2) we have action plans, (3) we expect researchers to engage in public communication and (4) we respond to national policies of public communication. And, *public communication funding* refers to the estimated percentage of annual budget spent in the last 12 months in public communication efforts (not considering staff salaries) and is coded none (1), less than 1% (2), between 1% and 5% (3), between 5% and 10% (4), more than 10% (5) and a don't know option (6).

Level of excellence in research. We obtained data on the level of excellence of research institutes from FCT research units' evaluations (2015). Institutes are evaluated as 'excellent', 'very good', 'good' and 'poor' (these do not receive basic funding). We recoded the variable in a binary 'excellent' (1) versus others 'less-than-excellent' (0).

Research funding. RIs were asked to estimate their average research income over the last three years (previous to the time when the study was conducted), in an ordinal variable: (1) less than €100,000; (2) between €100,000 and 250,000; (3) between €250,000 and 500,000; (4) between €500,000 and 1 million€; (5) more than 1 million€.

Analysis. In what follows, we describe how these variables compare among excellent and less excellent institutes and use Chi-Square tests (X^2) and T-tests to determine differences between types of institutes.

Excellent and less-than-excellent institutes

Among the 224 research institutions surveyed, 28% (*N*=58) were 'excellent' and 72% (*N*=150) were 'less-than-excellent' (*N* = 15 RIs were missing values regarding their Excellence level). Given that funding is distributed (by FCT) on the basis of 'excellence', it is then not surprising the disparities in funding levels between excellent and less excellent institutes, with 45% of excellent institutes reporting budgets over 1 million euros, on the contrary, less excellent institutions reported budgets between 100,000€ and 250,000€ and less than 100,000€.

Public communication activities and level of excellence

The average Portuguese institute reported engaging in a variety of public communication activities, including public events (*M*= 31.2, SD=34.8), traditional media (*M*=34.7, SD=46.7) and new media channels to address external audiences (*M*=162.7, SD=235.0). Yet, new media channels were used only sporadically. We find significantly higher activity among excellent institutes for public events (*M*=43.6, SD=38.2), traditional media channels (*M* = 55.1, SD = 62.7) and new media channels (*M* = 255.6, SD = 271.8) compared to less excellent which showed overall lower activity in public events (*M*=26.7, SD=32.5), traditional media channels (*M* = 27.4, SD = 36.9) and new media channels (*M* = 129.5, SD = 211.6) (Figure 14.1). This is shown by the *t* tests for events (*t*(84.44) = −3.233, *p* <.01), traditional media (*t*(72.83) = −3.194, *p* <.01) and new media channels (*t*(89.74)=−3.033, *p* <.01).

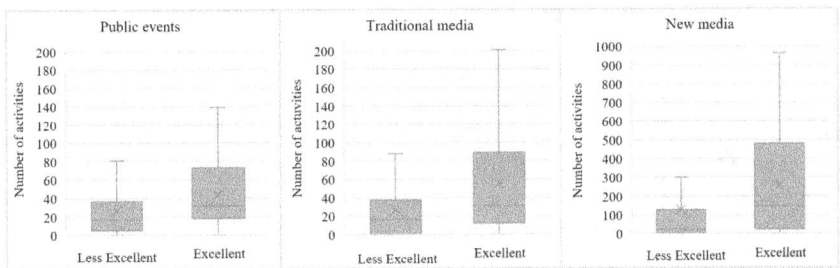

Figure 14.1 Public communication activities reported by Portuguese research institutes by level of excellence. The charts represent the estimated number of activities.

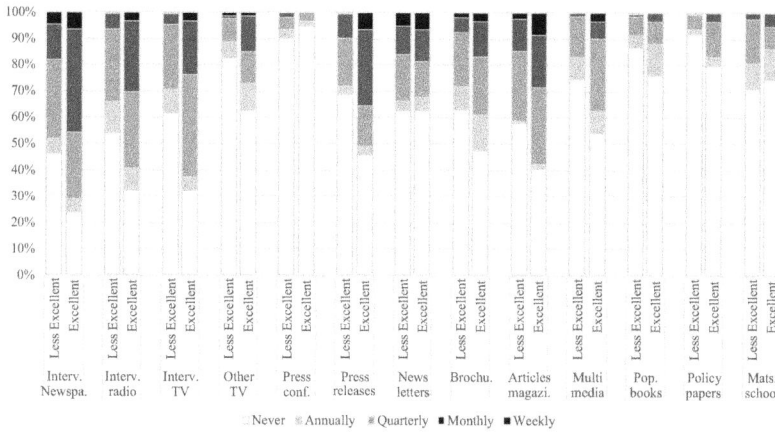

Figure 14.2 Frequency of use of traditional media channels compared between excellent and less-than-excellent institutes (*N* = 208).

When looking at the individual items in each of the three main types of activities, we see that these differences are particularly visible in traditional media channels. For example, excellent institutes reported giving significantly more TV interviews ($t(68.64) = -3.067$, $p <.01$), radio ($t(68.79) = -3.077$, $p <.01$) and newspaper interviews ($t(96.53) = -2.046$, $p <.05$), and writing more press releases ($t(68.95) = -2.963$, $p <.01$), articles in magazines ($t(76.94) = -2.062$, $p <.05$) and policy papers ($t(68.91) = -2.749$, $p <.05$) (see Figure 14.2). Excellent institutes also reported increased use of Facebook ($t(88.21) = -2.799$, $p <.01$) and Twitter ($t(74.94) = -2.037$, $p <.01$) and the institutional website ($t(84.32) = -3.628$, $p <.001$). They also reported organising significantly more public lectures ($t(85.80) = -2.180$, $p <.05$) and talks at schools ($t(69.19) = -3.28$, $p <.01$).

New media channels were overall not much in use. Most institutes reported not using them, except for website updates with the most common answer being weekly ($M = 66.8$ interactions a year, SD = 93.5) and Facebook ($M = 57.5$ interactions a year, SD = 92.24). This pattern is seen also across levels of excellence. Yet, T-tests indicated differences among groups for the frequency of website updates ($t(84.32) = -3.628$, $p <.001$), use of Facebook ($t(88.21) = -2.799$, $p <.01$) and Twitter ($t(74.94) = -2.037$, $p <.05$) that are more frequently used by excellent institutes. These results confirm hypothesis 1 that posed that excellent institutes show higher level of public communication activities. And, also show at a more detailed level that these differences are only for some activities, mostly media channels (Figure 14.3).

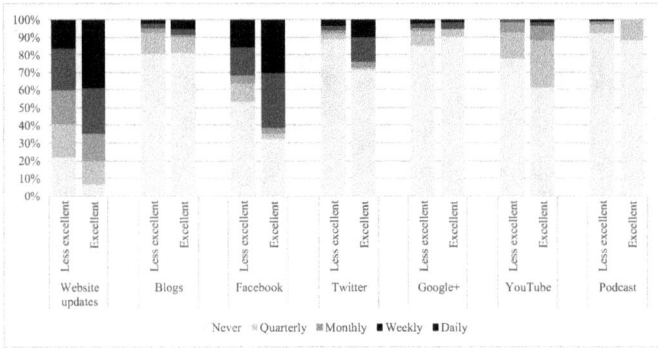

Figure 14.3 New media frequency compared between excellent and less excellent institutes (*N* = 208).

Audiences addressed and levels of excellence

As Figure 14.4 shows, research institutes address a variety of non-academic audiences, most commonly the general public, students, members of local organisations and the media and journalists. Moreover, we find differences with levels of excellence. As indicated by the Chi-square test, excellent institutes are more likely to frequently address some types of audiences such as schools (63% of excellent institutes addressing schools frequently versus 42% of less excellent institutes; Cramer's V = 0.203, *p* <.05), the general public (59% versus 35%; Cramer's V = 0.245, *p* <.01) and media and journalists (38% versus 18%; Cramer's V = 0.201, *p* <.05) when compared to less-than-excellent institutes (Figure 14.4).

The statements on perceived images of the public including interest, knowledge, trust and participation (Table 14.3) provide insights about how

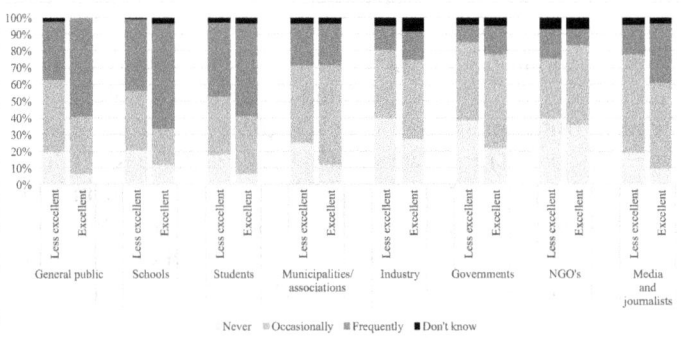

Figure 14.4 Frequency of engagement with various non-academic audiences by level of excellence.

institutes see their audiences. Overall, responses were skewed towards disa-greement with the statements, which were mainly framed with a 'negative' view of publics, suggesting that institutes are unlikely to perceive publics as disinterested, distrustful or disengaged from science. As for differences between levels of excellence, significant differences were found only for two of the statements: 'The public wants to contribute to science' ($\chi^2(1) = 5.386$, p <.05) with around 5% of excellent institutes agreeing compared to 17% of less excellent institutes that agreed, and the statement 'the public is not interested in the research conducted at our unit' ($\chi^2(1) = 4.336$, p <.05), where around 70% of the excellent institutes agreed with the statement compared to 54% of less excellent agreeing. This seems to suggest that less-excellent institutes are more optimistic about the role of the public in research, perhaps a result of their lower engagement with audiences and expcetations to increase it.

Table 14.3 Percentage of agreement with statements on perceived images of the public by level of excellence and significance levels

Perceived image	% Agreement		p-value (X^2)
	Excellent	Less excellent	
The public is not eager to learn about science	58%	46%	0.141
The public is interested in a limited range of research topics such as dinosaurs, dolphins and disasters	63%	50%	0.226
We cannot expect a large public to take interest in the research we do	48%	40%	0.344
The public does not need to understand the full picture; we explain what we think is appropriate	56%	43%	0.091
We would like the public to become more actively involved in decisions about the research conducted at our research unit	32%	28%	0.566
The public does not need to be scientifically literate to discuss the implications of our research	37%	31%	0.374
We would like the public to become more involved in discussing the implications of the research we do, but not necessarily in decisions about our research directions	0%	5%	0.070
If the public knows more about our research, they will be more likely to support it	5%	3%	0.387
The public needs to be educated by those who are knowledgeable	10%	13%	0.606
The public trusts science and scientists	5%	9%	0.306
The public is not interested in the research conducted at our unit	70%	54%	0.037
The public wants to contribute to science	5%	17%	0.020

These results support partly our H2, as excellent institutes did engage with non-specialist audiences more often than less excellent ones particularly the media and journalists despite holding stronger beliefs that the public is not interested and does not want to participate in science.

Professionalisation of public communications activity and levels of excellence

In terms of *communications staff*, about 51% of the institutes reported having specialised communication staff within the unit, while 49% did not. The results of chi-square and T tests showed significant differences between institutes with varying levels of excellence, with excellent institutes being more likely to employ communications staff ($\chi^2(1) = 12.854$, $p <.001$). Indeed, most excellent institutes reported having them (71%) compared to less excellent (44%). And staff is also more likely to be exclusively dedicated to communications in excellent institutes ($t(92.870)=-2.523$, $p <.05$) than in less-than-excellent ones.

As for public communication policies and guidelines in place, overall, the majority of institutes reported having a policy and/or guidelines for public communication in place, having plans, and also expecting their researchers to get involved in such activities, and also respond to national policies of public communication. These patterns are found among both excellent and less excellent institutes, with only little variation, yet not significant. For example, 68% of the excellent institutes say their communication efforts respond to the national policy of public communication and 55% of the less-than-excellent say so.

Table 14.4 also reports important differences in the adoption of policies with the level of excellence for all four statements. Overall, policies are more strongly implemented by excellent institutes. For example, around 80% of the excellent institutes have a policy for public engagement compared to around 60% of the less excellent. Despite these differences, it is important to highlight the fact that most institutes in the Portuguese research landscape follow some type of guidelines for the communications efforts, regardless of excellence level; additionally, none of the differences are statistically significant, so the numerical evidence provides inconclusive evidence of possible differences in terms of policy.

Public communication funding. While we did not find significant differences between the funding allocated to communications by excellent and less excellent institutes, it is important to note that most institutes reported low spending in public communications, between 1% and 5% of their annual research budgets. Perhaps surprising is the fact that more excellent institutes dedicate lower amounts of funding to public communication. For example, around 15% of excellent institutes allocated less than 1% to communications compared to 26% of less excellent ones (Table 14.4).

Table 14.4 Percentage of agreement with statements on communication policy, and percentage of allocated funding per level of excellence and significance levels

	Excellent	Less excellent	p-value (X^2)
Communication policies	%	%	
We have a public communications policy	80%	68%	0.146
We have public communications action plans	86%	74%	0.094
We expect our researchers to be involved in public communication	89%	79%	0.121
Our communication efforts respond to the national policies	68%	55%	0.217
Communications funding	%	%	0.357
None	14%	18%	
<1%	15%	26%	
1–5%	40%	28%	
5–10%	19%	20%	
>10%	12%	10%	

Overall, these results confirm partly H3 that posed higher levels of professionalisation among excellent institutes. As we show, excellent institutes are more likely to have specialised communications staff, respond to policies, yet excellence does not necessarily result in increased funding to public communications.

Concluding remarks

In this chapter, we provided an overview of public communication practices at Portuguese research institutes with a comparative focus on the level of research excellence. We were interested in understanding whether differences in the public communication activity existed between institutes with varying levels of excellence in research.

Portuguese institutes engage in various types of public events, traditional media and new media channels, regardless of the level of excellence. This finding seems to suggest that the level of excellence is unlikely to be an important determinant of the level of public communication. A recent study by authors in this book points in the same direction. Entradas and Santos (2021), using data from five countries, found that the increased activity among more excellent institutes was not a result of a stronger communication infrastructure in place or more researchers engaging, but it is rather related to large amounts of funding that prestigious institutes get and the visibility that comes with it (Entradas and Santos, 2021).

Regardless of its predictive power, we find differences across institutes with varying levels of excellence, supporting partly our three hypotheses that posited that excellent institutions had overall stronger commitment to public communication than less excellent institutes. More excellent institutes reported more interactions with audiences, particularly with the media, more professionalised structures and policies. While this could be a characteristic of Portuguese institutions where strong policies exist in regard to funding that require public engagement plans and national policies for scientific culture (Entradas et al., 2019), Entradas and Santos (2021) study shows that this might be rather a pattern across countries than a characteristic of a single country.

At a first glance, this could indicate that funders requirements are having the desired effect on the outcomes of institutional science communication, or it might be only an 'Mathew effect' (Merton, 1968) that those more excellent institutes attracting more funding are also attracting more media attention; this could also explain the increased media communication and online communication found at excellent institutes. This is not entirely surprising. Research-intensive institutes holding high profiles and reputations are key targets for the media and journalists, which could lead to increased activity among excellent institutes to respond to these demands.

Finally, institutes hold a positive image of publics, a view that is shared both among excellent and less excellent institutes, yet, institutes see audiences with little interest in the research they do, which may compromise their motivations to engage with certain publics. Chapter 4 provides an understanding about how views on the public drive the activities of institutes.

References

Conceição, C. (2011). Promoção de cultura científica: análise teórica e estudo de caso do Programa Ciência Viva.

DGEEC (2018). *Principais resultados do RAIDES 17 – Inscritos 2017/18*. Lisboa: DGEEC.

Ecklund, E. H., James, S. A., & Lincoln, A. E. (2012). How academic biologists and physicists view science outreach. *PLOS ONE, 7*(5), e36240. https://doi.org/10.1371/journal.pone.0036240

Entradas, M. (2015). Envolvimento societal pelos centros de I&D. In M. d. L. Rodrigues & M. Heitor (Eds.) *40 anos de Políticas de Ciência e Ensino Superior em Portugal* (pp 503–551). Almedina.

Entradas, M. (2021). In science we trust: The effects of information sources on COVID-19 risk perceptions. *Health Communication*, 1–9.

Entradas, M., Bauer, M.W., O'Muircheartaigh, C., Marcinkowski, F., Okamura, A., Pellegrini, G., et al. (2020) Public communication by research institutes compared across countries and sciences: Building capacity for engagement or competing for visibility? PLoS ONE 15(7): e0235191. https://doi.org/10.1371/journal.pone.0235191

Entradas, M., Junqueira, L., & Pinto, B. (2020). Portugal: The late bloom of (modern) science communication. In *Communicating science: A global perspective.* Canberra: ANU Press.

European Commission (2005). Eurobarometer 63.1: Europeans, science and technology. Retrieved from www.ec.europa.eu/public_opinion/archives/eb_special_en.htm.

European Commission (2010). Eurobarometer 340/Wave 73.1: Science and technology. Retrieved from www.ec.europa.eu/public_opinion/archives/eb_special_en.htm.

European Commission (2013). Eurobarometer Special 401—Responsible Research and Innovation (RRI), science and technology.

Eurostat (2019). R&D expenditure in the EU at 2.19% of GDP in 2019. EuroStat.

FCT (1999). Avaliação de Unidades de I&D Programa de Financiamento Plurianual 1999.

FCT (2005). Evaluation of research units 2002–2004.

FCT (2013). Avaliação de Unidades de I&D 2013.

FCT. (2017). *FCT — Unidades de I&D — Avaliação 2017.* https://www.fct.pt/apoios/unidades/avaliacoes/2017/index.phtml.pt

FCT (2019). Avaliação de unidades de I&D Programa de Financiamento Plurianual 2018.

Gonçalves, M., & Castro, P. (2003) Science, culture and policy in Portugal: A triangle of changing relationships? *Portuguese Journal of Social Science 1*, 157–173.

Gonçalves, M. E. (2002). *Os portugueses e a ciência.* Lisboa: Dom Quixote.

Heitor, M., & Horta, H. (2011) Science and technology in Portugal: From late awakening to the challenge of knowledge-integrated communities. In: *Higher education in Portugal 1974–2009*, pp. 179–226. Berlin: Springer.

Instituto Nacional de Estatística. (2017). *Instituto Nacional de Estatística.* https://ine.pt/xportal/xmain?xpid=INE&xpgid=ine_main

Kassab, O. (2019). Does public outreach impede research performance? Exploring the 'researcher's dilemma' in a sustainability research center. *Science and Public Policy 46*(5), 710–720. https://doi.org/10.1093/scipol/scz024

Katz, Y., & Matter, U. (2017). On the biomedical elite: Inequality and stasis in scientific knowledge production. SSRN. doi:10.2139/ssrn.3000628.

Martin, B. R. (2011). The research excellence framework and the 'impact agenda': Are we creating a Frankenstein monster? *Research Evaluation 20*(3), 247–254.

Martinez-Conde, S. (2016). Has contemporary academia outgrown the Carl Sagan effect? *Journal of Neuroscience 36*(7), 2077–2082. https://doi.org/10.1523/JNEUROSCI.0086-16.2016

Merton, R. K. (1968). The Matthew effect in science: The reward and communication systems of science are considered. *Science 159*(3810), 56–63.

Murray, D. L. et al. (2016). Bias in research grant evaluation has dire consequences for small universities. *PLoS ONE 11*, e0155876.

OCT (Observatório das Ciências e Tecnologias). (1996). Relatório do Inquérito a Cultura Científica dos Portugueses, Lisboa, OCT.

OCT (Observatório das Ciências e Tecnologias). (2000). Relatório do Inquérito a Cultura Científica dos Portugueses, Lisboa, OCT.

Ramos, A., & Sarrico, C. (2016). Past performance does not guarantee future results: Lessons from the evaluation of research units in Portugal. *Research Evaluation* 25, 94–106.

Royal Society (2006). *Science communication excellence: Survey of factors affecting science communication by scientists and engineers.* London, UK: Royal Society, RCUK & Wellcome Trust.

Teixeira, P., Cardoso, M., Sarrico, C., & Rosa, M. (2007). *The Portuguese Public university system: On the road to improvement?* Cheltenham: Edward Elgar Publishing.

Veiga, A., Magalhães, A. M., Sousa, S., Ribeiro, F. M., & Amaral, A. (2014). A reconfiguração da gestão universitária em Portugal. *Educação, Sociedade & Culturas*, 41, pp.7–23.

Part IV

Methodological
Considerations

Chapter 15

Studying Public Communication of Research Institutes

Sample Design and Data Collection

Marta Entradas, Martin W. Bauer and Colm O'Muircheartaigh

Sample design

Sampling frames

The first step was to define the *target population*, i.e. all RIs in a country, and the *sampling frame* from which to select a sample (e.g lists). Each national partner then listed all RIs in the respective country. When national, official lists were available, these were used; when they were not, lists were built by the national teams. Sampling frames considered all research institutes in Portugal, the Netherlands, Italy, and Japan, mostly countries smaller in size (institutes). In Portugal, Italy, and the Netherlands, official lists from national governments or funding agencies were provided to the teams; in the remaining countries, sampling frames were built from lists of universities. Sampling frames corresponded to complete mapping of all universities in Japan, Germany, and the United Kingdom and a selected sample of research universities in the United States and Brazil (larger research systems in the study). National sampling strategies are described in Table 15.1. For each institute, basic information was collected including area of research following the OECD classification into (1) Natural Sciences, (2) Engineering and Technology, (3) Medical and Health Sciences, (4) Agricultural Sciences, (5) Social Sciences, and (6) Humanities (OECD, 2002) and had a known probability of selection (different from zero) to be part of the survey. In Portugal, Italy, Japan, and the Netherlands, all RIs were included in the study, and with the probability of 1. This was the probability of selection of a unit, considering the sector of research and university institutes belonged to. Brazil, despite not mapping all institutes in the country, selected a representative sample of universities and surveyed all institutes within each university, so probabilities for each research institute were also one.

DOI: 10.4324/9781003027133-19

Table 15.1 Sampling frames and procedures employed in each country

Country	Sampling frame (N)	Target sample (N)	Sampling procedure
Brazil	N=945	N=945	To build the Brazilian target sample, we identified all public Brazilian research institutes linked to Ministries (N= 46) (non-university institutes) and used the census (2015) provided by the INEP (Instituto Nacional de Estudos e Pesquisas Educacionais Anísio Teixeira) for a complete picture of the Brazilian universities (N= 2,368). We selected N=50 best universities using a combination of the lists provided by The Times Higher Education (2017) and the Ranking of the Brazilian newspaper Folha de São Paulo (2017). We listed all research institutes within the selected universities (N=899) and achieved a sampling frame of N=945. We contacted all institutions.
Portugal	N=384	N=384	We identified all research institutes in Portugal from two official lists: the list of all research institutes funded by FCT - Fundação para a Ciência e Tecnologia - (2014), provided by FCT, and a list of the Portuguese Higher Education Institutions provided by the DGEES - Direção-Geral de Estatísticas do Ensino Superior (Estatísticas da Educação2014). The final sampling the frame resulted in N=384 research institutes. This represents at least 80% of the population. We contacted all institutions.
Germany	N=1,900	N=1,358	We identified N=193 state run Research Universities & Universities of Applied Science, from an official list provided by the Association of University Presidents (HRK) (2017) (Liste aller Mitgliederhochschulen der Hochschulrektorenkonferenz); this covers more than 90% of all academic staff in the German Higher education system. We randomly assigned one of the six research fields (OECD classification) to each university so that each research field was covered by approximately 30 organizations. We then listed all research institutes of the respective field for each of the universities. This resulted in at least 10 institutes per University and a total sample of about 1.900 institutes (around 300 per OECD-Field), from which we draw a random sample for each field.
Italy	N=1,120	N=1,120	To build the Italian sample, we collected lists of all Italian Universities and Research Institutes from the Italian Ministry of Education and Research (MIUR) database (2017). We then made a complete list of all institutes within these universities, which corresponded to the Universities Departments and Italian Research Institutes. We classified each unit based on the OECD area of research and contacted the entire population (N=1.120)

Japan	N=1,134	N=1,134	To build the Japanese frame, we used a list from the "National Institute of Science and Technology Policy" (NISTEP) (2016), which covers all R&D conducted at universities and corporations in Japan (N=787); we selected the top 66 universities, based on KAKEN database (this covers around 80% of KAKEN grant in 2016). KAKEN is the grants-in-aid for scientific research, which is operated by Japanese Society for the Promotion of Science (JSPS) and the biggest competitive funding agency in Japan. For each of the 66 universities, we listed all research institutes within each organization (N=1,134). All institutes were contacted.
The Netherlands	N=821	N=821	To build the Netherlands frame, we used a list provided by NARCIS – National Academic Research and Collaborations Information System (2016)* of all universities in the Netherlands (N=14), and Para-university organizations: the Netherlands Organization for Scientific Research (NWO) (N=8) and the Royal Netherlands Academy of Arts and Sciences (KNAW) (N=16). We mapped all institutes within these organizations using information from their websites and classified each of them according to OECD field (N=821). As the total number of Dutch research institutes was smaller than the 200 per group, we contacted the whole population. *NARCIS provides access to datasets of a number of data including research institutes in the Netherlands.
The United Kingdom	N=1,957	N=1,046	For the UK sampling frame, we used the list (census) of UK universities provided by the Research Excellence Framework (2015) (N=150). The REF is the new system for assessing the quality of research in UK higher education institutions. Universities were ranked into HH (high research output; high research impact), HL, LH, and LL, and a N=50 covering all ranks was randomly selected; all universities from the Russel Group were added. All research institutes within the selected universities were mapped using information from universities' websites and classified into OECD field (N=1,957). N=200 from each sector was randomly selected, plus some hand pick of institutes in the agricultural sciences outside universities (N=10), which were underrepresented, resulting in N=1,046.
The United States	N=10,308	N=1,366	To build the American sampling frame, the mapped all research institutes within the "Doctoral Universities: Very High Research Activity" classified by the Carnegie Foundation (2017); this resulted in (N=10,308) research units. Each was classified into OECD field. For each university, a random number generator was used to select two units for each category (i.e., 12 units per university or 230 per OECD group). In cases where a university did not have a relevant center or institute, units were selected from other universities at random. It was not possible to identify enough agricultural science units such that the final contact list included 1,366 potential respondents.

Disproportionately stratified stratum

The sample construction for each country was stratified by areas of research, informed by previous studies that have pointed to differences in public engagement across sciences. This allowed for oversampling those areas which were less represented (such as Agricultural sciences). Weights were used for response rate (RR) calculation. The sample allocation between strata was made disproportionally to the population size. Our goal was to obtain a target of $N=200$ per stratum, for a total sample of 1,200 RIs in each country. Disproportionate stratified sampling was advantageous because the strata being compared differed in size, allowing for RIs in minority to be sufficiently represented (e.g. Agricultural Sciences were oversampled). We then used simple random sampling and systematic sampling to select RIs from each of the six strata. Weights were calculated for each unit to compensate for differential probabilities of selection and RRs among stratum. See Chapter 2 in this book by Colm O'Muircheartaigh for a detailed description on how to sample research institutes.

Data collection

Respondents prenotification. A first contact with institutes was made in February 2017 to announce the study and ask for institutes' participation. All communication with institutes was carried out in maternal languages, which helped establishing a proximity with potential respondents. A cover letter briefly explained the goals, the international scope of the project, and the importance of the study within the national context of public communication in each country. This first contact also served to refine the samples.

Survey distribution and reminders. In June 2017, we launched the first national surveys. The whole period of data collection was carried out online between February 2017, when institutes were first contacted and 2018, when the last survey was closed. Each institute received a unique link to the questionnaire; in Portugal, the preliminary study, the data were collected in late 2015. This difference in the period of data collection does not, however, invalidate our comparison given that institutional changes are unlikely to happen in short periods of time. An average of four reminders was sent per country, varying between countries due to cultural traditions. Overall, the first two reminders turned out to be very efficient in most countries, after the third, response decreased.

Follow up calls and paper surveys. Additional efforts were made to boost RRs. In Portugal, Italy, the United Kingdom, and the Netherlands, nonrespondents were contacted by telephone and encouraged to participate. This strategy leads to increase in RR between 10% and 20%. In the United Kingdom and the United States, a mail survey was conducted with a subsample of nonrespondents ($N=150$) selected using the same selection process as before. A copy of the questionnaire was mailed to institutes. This resulted in an increase of 10% of the RR in both countries.

Response rates and nonresponse bias analysis

We contacted a total of N=8,033 research institutes in the eight countries together. We received a total of N=2,030 valid responses. The overall RR was 25%. This is an acceptable RR for surveys of organizations (see Chapter 2). Response rates varied across countries (see Table 15.2).

Table 15.2 Number of contacted (N), number of that responded (N) by country and areas of research, unweighted (RR), and weighted response rates (WRR). For every country, we present the unweighted RR; for countries where we undertook a nonresponse mitigation approach by subsampling nonrespondents and approaching the subsample again (the United Kingdom, Germany, and the United States), we present the WRR (4).

		Institutions contacted (N)	Institutions responded (N)	Response Rate (RR) (%) (a)	Weighted RR (%) (b)
The United	Nat Sci	185	33	17.8	31.8
Kingdom	Eng & Tech	191	22	11.5	14.8
	Med & Health	187	32	17.1	26.2
	Agric Sci	26	11	42.3	25.4
	Soc Sci	191	50	26.2	53.3
	Hum	187	40	21.4	30.3
	Total	967	188	19.4	31.0
Germany	Nat Sci	357	122	34.2	31.5
	Eng & Tech	258	59	22.9	23.5
	Med & Health	217	42	19.4	16.0
	Agric Sci	120	21	17.5	18.0
	Soc Sci	213	61	28.6	26.2
	Hum	191	53	27.7	26.8
	Total	1,356	358	26.4	23.7
The United	Nat Sci	227	63	27.8	28.1
States	Eng & Tech	221	26	11.8	11.2
	Med & Health	219	42	19.2	21.2
	Agric Sci	221	29	13.1	19.2
	Soc Sci	223	54	24.2	31.5
	Hum	220	48	21.8	30.2
	Total	1,331	262	19.7	23.6
The	Nat Sci	116	30	25.9	
Netherlands	Eng & Tech	105	16	15.2	
	Med & Health	164	25	15.2	
	Agric Sci	12	2	16.7	
	Soc Sci	302	44	14.6	
	Hum	122	25	20.5	
	Total	821	142	17.3	

(Continued)

		Institutions contacted (N)	Institutions responded (N)	Response Rate (RR) (%) (a)	Weighted RR (%) (b)
Japan	Nat Sci	245	66	26.9	
	Eng & Tech	225	70	31.1	
	Med & Health	194	67	34.5	
	Agric Sci	42	16	38.1	
	Soc Sci	318	70	22.0	
	Hum	110	32	29.1	
	Total	1,134	321	28.3	
Italy	Nat Sci	247	98	39.7	
	Eng & Tech	188	70	37.2	
	Med & Health	227	64	28.2	
	Agric Sci	52	16	30.8	
	Soc Sci	252	63	25.0	
	Hum	154	55	35.7	
	Total	1,120	366	32.7	

a The unweighted RR is essentially the ratio of the number of responses to the number of cases approached.
b The weighted response rate (WRR) takes the responses, each weighted by the inverse of its probability of inclusion as a proportion of the sum of the weights across the whole population. This is a better guide to the potential for response bias than the unweighted RR.

There is the possibility of nonresponse bias by overrepresentation of institutes that often engage in public communication (e.g. larger institutions or sciences more likely to engage in outreach). To assess nonresponse bias, we used two methods: (1) compared early and late respondents (responded before and after the third reminder) using T-tests and Chi square for all variables in the questionnaire and (2) compared respondents with nonrespondents for some key variables using data from the sampling frames and the mail surveys. We found no significant differences, suggesting that we do not have overrepresentation of institutes in some disciplines, while supporting the representativeness of our samples. Based on this evidence, we conclude that nonresponse would not greatly affect the results in this study, and weighing was then not considered.

As for size, the overall sample does not seem biased toward larger institutes comprising institutes from a range of sizes including small (31% less than 20 researchers), medium (34% with 20–80 researchers), and large (35% with more than 80 researchers), thus, providing variety for our analyses. Yet, we find differences in sample sizes across countries with Brazil and Japan having somehow larger institutes. This is possibly a characteristic of the meso level, which varies slightly across academic and research systems in the surveyed countries, rather than a sample design bias. Size is used as control variable in all our analyses.

References

Estatísticas da Educação 2013/2014. (2014). Direção-Geral de Estatísticas da Educação e Ciência (DGEEC). https://www.dgeec.mec.pt/np4/96/%7B$clientServletPath%7D/?newsId=145&fileName=EE2014.pdf

OECD. (2002). *Frascati Manual—Proposed standard practice for surveys on research and experimental development.* OECD Publishing. http://dx.doi.org/10.1787/9789264239012-en

Research Excellence Framework. (2015). https://results.ref.ac.uk

Chapter 16

Framework and Indicators of Public Communication of Research Institutes

Marta Entradas

Indicators for public communication of science

The communication activity of research universities and other scientific organizations is a complex activity that occurs in a variety of settings, involving many actors and activities. It is then a complex reality to map empirically? The measurement of this activity is tasked with translating these environments, actors and activities into measurable units, called indicators. One key task of the MORE-PE project was to operationalize such key indicators for measuring public communication of science at the institute level.

There have been some attempts to measure public communication activity at the level of organizations (see Neresini and Bucchi, 2011; Mejlgaard et al. 2012; Entradas and Bauer, 2017; Crettaz von Roten and Entradas, 2018). Our pioneering study in Portuguese institutes (Entradas, 2015; Entradas and Bauer, 2017) was an important contribution to this goal. It tested, in a large-scale study, a series of indicators that were then used in the MORE-PE project, some with minor modifications to accommodate national characteristics.

Framework of indicators for public communication of science

We define a simple, yet useful framework for thinking about and analyzing institutional science communication. We conceptualize public communication as a function of the general organizational context and organizations' dispositions to communicate publicly. That is, the level of public communication activity of an institute (P) can be explained by the combination of the general context of the organization, i.e. contextual factors (C factors), and disposition factors that characterize the specific orientation towards communication activities (D factors) $(P = f(C, D)$. C factors reflect the features of the institution and research environment, such as the country and

DOI: 10.4324/9781003027133-20

CONTEXTUAL INDICATORS	DISPOSITION INDICATORS	OUTPUT INDICATORS
Type of Institution (Public/Private % teaching/research)	Staffing	Audiences
	Funding	
Scientific Area (OECD)	Policies	Activities (events/traditional media, new media)
	Rationales/values/ desirable outcomes	
Size (number of researchers)	Ethos/Perceptions	Media Relations

Figure 16.1 Indicators for measuring public communication of science.

scientific discipline, and the size of their annual research budget – factors known to influence communications (1). D factors reflect the commitment and responsibility at the level of institutes that encourage public communication and supports the development of such activities; it is operationalized by the available funding for communications, guidance by communication policies (policy), recruitment of professional communications staff (staffing), and the rationales, perceptions of publics, and ethos. We consider input indicators the institutional context (C) and the resources (D) that fuel the activity, and outcome indicators (activities, audiences) the communication activities that result from institutional efforts. Figure 16.1 represents the conceptual model and indicators.

Indicators

The MORE-PE project developed a series of indicators of public communication covering five main dimensions. Table 16.1 presents a description of the five dimensions and the indicators used to access each dimension. Dimensions 1 and 2 refer to outcome indicators, dimensions 2–5 refer to input and contextual indicators. Tables 16.2–16.6 present how each indicator was operationalized into questions and measured.

1 Public communication activities
2 Audiences and media relations
3 Resource allocation
4 Rationales, perceptions and ethos
5 General information

Table 16.1 Dimensions and indicators' definitions

Dimension	Indicators	Definition
Public communication activities	Public events	The estimated number of times the institute has engaged in various types of public events as organizer or participant over the past 12 months
	Traditional media channels	The estimated number of times the institute has engaged with various types of traditional media channels over the past 12 months
	New media channels	The estimated number of times the institute has used various types of new media channels over the past 12 months
	Outsourced activities	Indicates whether six different support activities are outsourced or developed in-house
	Time of activity	The estimated amount of time the institute has been conducting activities for nonspecialists
	Evolution over time	Indicates whether the activity has increased, decreased or stayed the same over the last 10 years
	Intention to increase future	Whether the institute intends to increase the communication activity over the next year
	Evaluation efforts	The frequency of evaluation of communication efforts
Audiences & media relations	Audiences	The frequency of engagement with a variety of audiences over the past 12 months
	Journalists contact	The estimated frequency of media enquiries in the last 12 months
	Journalists database	Indicates whether the institution has a list of journalists
	Media contacts *operandus*	Whether institutes initiate contacts themselves
Staffing	Communications staff	Indicates whether the institute has staff dedicated to public communication tasks
	Specialist staff (fully dedicated)	The number of staff that is fully dedicated to communication tasks
	Full time/part time	The estimated number of full-time and part-time employees dedicated to public communication
	Type of contract	The number of staff per type of contract (various types)

	Previous experience	The number of staff by previous job
	Educational background	The number of communications staff by educational background (OECD areas)
	Training in communication-related areas	The number of staff that possess communication training (various types)
	Roles	The roles that science communicators perform
Funding	Communications funding	The estimated percentage of the institute research budget allocated per year to public communication
	Perceived resources	Perceived allocation of the resources
	Perceived 'ideal' funding	The percentage of research funding that should be allocated to communications
Policy	Communications policy, guidelines or plans	Indicates whether the institute has policies for public communication
Researchers involvement	Active researchers	The percentage of researchers taking part in public communication
	Barriers to researchers	Perceived barriers to researcher' engagement with publics
Rationales/goals	Rationales	The stated reasons for undertaking public communication activities (main, second, least)
	Values for communication	The communication goals and choices
	Media perceptions	The perception of media coverage
	Public perceptions	The perception of media publics
	Perceived effectiveness	The perceived effectiveness of public communication activities
General information on institute and respondent	Year of foundation	The year, or estimate, of the institute's foundation
	Research vs teaching	Indicates the relative weight of research and teaching
	Research area (OECD)	The primary research area
	Research discipline (OECD)	The primary discipline
	Size	The estimated number of researchers with a PhD
	Research budget	The estimated amount of budget for research per year
	Job title, time in job	Job title and number of years in the job

Dimension 1 – public communication activities

These indicators relate to the communication activities of institutes. What types of activities institutes are doing and how frequently. The indicators and questions asked in the MORE-PE survey are described in Table 16.2.

Table 16.2 Indicators of communication activities

Indicator	Question
Public events	Q: Roughly, how many times in the past 12 months has your research institute engaged in the following events either as organizer or contributor? (None/Annually/Quarterly (2–6 times a year)/Monthly (7–20 times)/Weekly (>20 times a year)/Don't know): (1) Public lectures, (2) exhibitions, (3) open days/guided visits and similar events, (4) science festivals/fairs, (5) national science week and similar events, (6) science cafés and similar formats of public discussions, (7) international events (e.g., Fame Lab, Researchers' night), (8) deliberative and participatory events (e.g., consultation, policy meetings, citizen juries), (9) events with private organizations (e.g., industry, corporations), (10) talks at schools, (11) citizen science projects, Other (please specify) Note: in the US survey, item 4 was not included because there isn't a national science week, and in JP ad DE, Instagram was excluded.
Traditional media channels	Q: Roughly, how many times in the past 12 months has your research institute engaged, produced or used the following channels to engage with nonspecialist audiences? None/Annually/Quarterly (2–6 times)/Monthly (7–20 times a year)/Weekly (>20 times a year)/Don't know): (1) interviews for newspapers, (2) interviews for the radio, (3) interviews for the TV, (4) Other TV (entertainment shows, programs), (5) press conferences, (6) press releases, (7) newsletters/brochures/nonacademic publications, (8) articles in magazines for the nonspecialist public, (9) multimedia/videos, (10) popular books, (11) policy papers/briefs for industry, (12) materials for schools, Other (please specify)
New media channels	Q: Roughly, how many times in the past 12 months has your research institute used the following social media channels to engage with nonspecialist audiences? None/Quarterly (2–6 times a year)/Monthly (7–20 times)/Weekly (>20 times year)/Daily (Don't know): (1) website updates (events, content…), (2) blogs, (3) Facebook, (4) Twitter, (5) Google+, (6) Instagram, (7) Youtube, (8) Podcasts, Other (please specify)
Outsourced activities	Q: Does your research institute handle the following activities to support public engagement in house or through outsourcing? (1) Formatting/design/layout of materials, (2) Information graphic design (e.g., as graph editing, mapping), (3) Enhanced data visualization (e.g., animation, videos), (4) Event organization/management, (5) Training in communication/media communication, (6) Website construction, (7) Website management, Other (please specify)

Time of activity	Q: For how long has your research institute been carrying out public engagement activities for nonspecialist audiences?(Less than I year/Between I and 5 years/Between 5 and I0 years/More than I0 years/Don't know)
Evolution over time	Over the last 5 years, the total number of public engagement activities for nonspecialist audiences at your research institute has: If the communication structure at your institute was created less than 5 years ago, please consider the period of time it was created. (Decreased/Stayed the same/Increased/Don't know)
Evaluation efforts	Q: How frequently does your research institute evaluate public engagement activities? We evaluate whatever we are doing... (Never/Rarely/Some of the time/Most of the time/Always/Don't Know)

Dimension 2 – audiences and media relations

These indicators relate to the external audiences that institutes address and how frequently, in particular, explore institutional media relations. The indicators and questions asked in the MORE-PE survey are described in Table 16.3.

Table 16.3 Indicators of audiences and media relations

Indicator	Question
Audiences	Q: In the table below, you will find a list of audiences. How often has your research institute/researchers engaged with each of them in the past 12 months? (Never/Rarely/Occasionally/Frequently/ Very frequently/ Don't know): (1) General public (whoever might be interested), (2) Schools Students outside teaching, (3) Members of local municipalities/councils/associations, (4) Delegates from industry, (5) Governments/politicians/policymakers, (6) Nongovernmental organizations (NGOs), (7) Media and journalists, (8) Policymakers, politicians, Other (Please specify)
Media enquiries	Q: How many times has your research institute received media/ journalists enquiries in the last 12 months? (Never /1–2 times/3–5 times/6–10 times/ >10 times/Don't know)
List of journalists	Q: Does your research institute maintain a list of journalists and media contacts? (1) Yes, we have a list/database of journalists and media contacts (2) No, we do not have a list, but we have personal contacts (3) No, we do not have a list/database of journalists and media contacts (4) Don't know
Media contacts operandus	Q: When journalists and the media want to contact researchers at your research institute how do they proceed? (1) They contact our communication/administrative staff first (2) Sometimes, they contact our communication/administrative staff first; other times, they contact the researchers directly (3) They contact the researchers directly (4) They contact the researchers/institutes directly (5) Don't know

Dimension 3 – resource allocation

We explore measures of resource allocation of institutes to public communication activities through a series of questions as described in Table 16.4. The goal of these questions is to characterize the level of professionalization at institutes and resources allocated to the activity including staff and funding as well as the level of commitment to the activity through the adoption of policies.

Table 16.4 Indicators of resource allocation

Indicator	Question
Staffing	
Communications staff	Q: Does your research institute have specialist staff responsible for public engagement activities? Please consider all employees who carry out public engagement tasks as part of their day-to-day responsibilities. This can include staff responsible for maintaining the website, organizing public events, supporting researchers in their public engagement work, producing the newsletter, responding to journalists, etc. For simplicity, we will refer to them as 'communication staff'.
	Yes, our research institute has its own communication taff
	No, our research institute does not have communication staff, but we have access to communication staff within the institution/organization
	No, our research institute does not have its own communication staff and we do not have access to other communication staff within the institution/organization
Specialist staff (fully dedicated)	Q: Please indicate how many of the 'communication staff' at your research institute are either exclusively or/and partly dedicated to public engagement tasks. Please consider only the communication team and not researchers who conduct their own communication activities.
Full time/part time	Q: Please indicate how many of them are full-time/part-time employees. Please count each person only once (part-time employees/full-time employees).
Type of contract	Q: Please indicate what type of contracts the staff at your research institute have: Please count each person only once.
	A temporary contract for a specific research project
	A temporary contract with the research institute or host institution
	A permanent contract with the research institute or host institution
	Other type of contract Please specify

Previous experience	Q: Which of the following most closely matches the previous professional experiences of the 'communication staff' at your research institute? Please count each person only once, based on their most relevant previous job. No previous professional experience Worked as researchers Worked as a teacher Worked as marketing/public relations/communication professionals Worked as design/multimedia professionals Worked as journalists/science writers Worked as administrative professionals Worked as project/finance/human resources officers/managers Other (please specify)
Educational background of staff	Q: Please indicate how many have an education degree in the following research areas: Please count each person only once. (OECD main areas/no university degree)
Training	Q: How many of the 'communication staff' at your research institute have a background and/or training in communication (e.g., media, digital, corporate communications, science communication)? Please count each person only once based on their highest degree of specialization. No formal training in communication Attended workshops/short courses in communication Undergraduate degree related to communication Post-graduate degree related to communication (post-graduation course, masters or PhD)
Roles	Q: Overall how frequently does the 'communication staff' at your research institute conduct the following? Decide on public engagement policies with the leadership of the institute Create/propose public engagement action plans to the leadership of the institute Motivate researchers to get involved in public engagement events Intervene in moments of institutional reputation 'crisis' Compose/edit/print communication materials (press releases, newsletters, leaflets, powerpoint presentations, etc.) Compose/edit/stream audiovisual materials (photographs, videos, etc.) Manage the website and online communication of the institute (social media channels such as Twitter and Facebook) Organize public events (Open days, Science Weeks, talks, workshops, etc.) Organize/offer communication training for researchers (Public speaking, Media training, etc.) Assist researchers on planning/completing research grant applications

(Continued)

Indicator	Question
Funding Communications funding	Q: Please estimate the percentage of the annual budget spent in the last 12 months on the public engagement efforts of your research institute. This can include actions such as maintenance of the website, printing of brochures, organization of public events, etc. Please do not consider salaries of the 'communication staff' (None/< 1%/ 1–5%/5–10%/>10%/Don't know)
Perceived resources	Q: Thinking about the resources your institute devotes to public engagement (funding, staff, etc.), do you think that your research institute: (1) Should devote less resources to public engagement (2) Devotes the right amount of resources to public engagement (3) Should devote more resources to public engagement
Expected future funding	Lastly, what percentage of your institute's annual budget do you think should be allocated to your public engagement efforts? (None/< 1% /1–5%/5–10%/>10%/Don't know)
Policy communications policy	Q: Please tell us whether the following statements about the commitment of your research institute to public engagement are true or false: We have a public engagement policy We have public engagement action plans We expect our researchers to be involved in communication with the public Our communication efforts respond to the national policies on public engagement We have neither plans nor a policy for public engagement, but we nevertheless engage with the public
% researchers participating (active researchers)	Q: Roughly, in the last 12 months, what percentage of researchers at your research institute including PhD students, fellows, visitors and regular research staff, took part in public engagement activities? This can include activities such as public lectures, public debates, activities at schools and National Science Week (None/< 10%/10–20%/20–40%/40–60%/ 60–100% /Don't know)
Barriers to researchers	Q: For those researchers at your institute who do not engage in public engagement activities, what do you think is discouraging them to do it? (5-point scale Definitely not true to Very likely true) They are not enthusiastic about communicating their work to general audiences They do not perceive public engagement as their everyday work/responsibility They do not perceive public engagement as contributing to the progress of their careers They do not have time for it They are not rewarded for their public engagement work They lack institutional support for doing it (e.g., staff, training, funding) They see communication as the responsibility of the communication staff rather than their own They think the public is not interested in the research they do They feel they are not good at it Other reason (please specify)

Dimension 4 – rationales, perceptions, and ethos

Rationales of communication and perceptions of media coverage of the institute's research were measured through the indicators and questions presented in Table 16.5.

Table 16.5 Indicators of rationales, perceived images of media and the public

Indicator	Question
Rationales	Q: What would you say is the most important, second most important and least important reason for your research institute to undertake communication with nonspecialist audiences? We aim to respond to the policy/mission of our host institution/university We aim to respond to the policy of our funding bodies We aim to respond to national policies of public engagement We want to raise our research profile. We want to attract funding We want to get public support for the research we do We want to disseminate our research to the public We want to listen and involve the public in our research We aim to recruit new generations of scientists Other (Please specify)
Criteria for communication	Q: People look to science for different reasons. Please indicate how likely your research institute is to use the following criteria when deciding what research results to communicate with nonspecialist audiences. We communicate what is... (5-point scale, Very unlikely to Very likely) Relevant to daily life Relevant to current debates what people should know Innovative/new developments and findings Entertaining and interesting Other reason (Please specify)
Perceived effectiveness	Q: Overall, how successful do you think your public engagement efforts have been in enhancing the activities of your research institute? (1) Very unsuccessful, (2) Unsuccessful, (3) Neither successful nor unsuccessful, (4) Successful, (5) Very successful)
Media perceptions	Q: To what extent do you agree or disagree with the following statements concerning media coverage of the research conducted at your research institute? (5-point scale, strongly disagree to strongly agree) Visibility in the media of the research conducted at our institute is important Media should give more attention to the research conducted at our institute Journalists have reported badly about our work The research we do is of little interest to journalists To maintain media relations is not our task

(Continued)

Indicator	Question
Public perceptions	Q: The following statements express opinions about the public. To what extent do you agree or disagree with each statement? (5-point scale, strongly disagree to strongly agree). The public:
	is not interested in the research conducted at our institute
	is not eager to learn about science
	is interested in a limited range of research topics such as dinosaurs, dolphins and disasters
	wants to contribute to science
	We cannot expect a large public to take interest in the research we do
	If they know more about our research, they will be more likely to support it
	does not need to understand the full picture, we explain what we think is appropriate
	needs to be educated by those who are knowledgeable
	We communicate with the public very selectively to avoid trouble
	We would like the public to become more actively involved in decisions about the research conducted at our research institute
	We would like the public to become more involved in discussing the implications of the research we do, but not necessarily in decisions about our research directions
	trusts science and scientists
	does not need to be scientifically literate to discuss the implications of our research

Dimension 5 – general information on institutes

This set of questions is aimed at gathering general information on the institute. The indicators and questions used are presented in Table 16.6.

Table 16.6 Indicators of general information

Indicator	Question
Year of foundation	Q: When was your research institute founded? (Please provide the year or an estimate)
Percent research vs teaching	Q: Which of the following best describes the split between research and teaching at your institute? (We only do research/We do more research than teaching/We do more teaching than research/Research and teaching are equally balanced)
Research area and discipline (OECD)	Q: Please select from the list below the main scientific area of your institute and the main discipline.

Size	Q: How many people work at your research institute? Please consider all researchers, post-doctoral fellows and PhD students and technicians; do not count administrative staff.
Research budget	Q: Could you please estimate the average research income of your research institute over the last 3 years [(2014+2015+2016)/3] (< £100,000/£100,000–£250,000/£250,000–£500,000/£500,000–£1M, >1M Euros)
Source of funding	Q: How is your research institute funded? Roughly, what percentage of funds are sourced from recurrent core institutional funding and what percentage comes from external projects' funding? (100% internal funds/80% internal funds – 20% external funds/60% internal funds – 40% external funds/40% internal funds – 60% external funds/20% internal funds – 80% external funds/100% external funds/Don't Know)
Job title, time in job (years)	Q: Which of the following most closely matches your job title? (Researcher/Management/Administrative staff/ Communication staff/Institute's Director/Coordinator/ Head of institute/PhD student/Other (Please specify)) Q: For how long have you been working in this role? (Years)

References

Crettaz von Roten, F., & Entradas, M. (2018). Public engagement measurement. In J. Shin & P. Teixeira (Eds.), *Encyclopedia of international higher education systems and institutions* (pp. 1–4). Berlin, Germany: Springer.

Entradas, M. (2015). Envolvimento societal pelos centros de I&D. In Ridrigues, M. and Heitor, M. (Eds.) (2015). 40 anos de Políticas de Ciência e Ensino Superior em Portugal (pp. 503–516). Almedina.

Entradas, M., & Bauer, M. M. (2017). Mobilisation for public engagement: Benchmarking the practices of research institutes. *Public Understanding of Science, 26*(7), 771–788.

Mejlgaard, N., Bloch, C., Degn, L., Nielsen, M. W., & Ravn, T. (2012). Locating science in society across Europe: Clusters and consequences. *Science and Public Policy, 39*(6), 741–750. doi:10.1093/scipol/scs092)

Neresini, F., & Bucchi, M. (2011). Which indicators for the new public engagement activities? An exploratory study of European research institutions. *Public Understanding of Science, 20*(1), 64–79.

Index

Note: **Bold** page numbers refer to tables, *Italic* page numbers refer to figures and page number followed by "n" refer to end notes.

academic arms race 101
Academic Ranking of World University (ARWU) 214
advanced online media 66
Agostinho, M. 47
Ajzen, I. 81
American Geophysical Union 170
Anderson, M. 168
aristocratic amateur 208
arms race: academic 101; and competition 100; for public communication of science 109–110
Arms Race for Public Communication (ARPC) hypothesis 99; hypotheses derived from 102–104; medialisation as 100–102
arms race model 101–102
artificial competition 60–61
artificial market 60
Association of Universities in the Netherlands (VSNU) 158
attention seeking: competition for 97–100; as element of arms race model 101–102
audiences **143**; addressed and levels of excellence 256–258; communication of research in Italy 143–145; communication with non-specialist *199*; communicative dispositions of British research institutes 219–221; and excellent institutes 256, 258, 260; general and deficit model 70; German research institutes 238; interaction with, by research area *197*; by OECD research areas **144**; perceived images

of 253; public communication in Japanese research institutes 196–197; targets and their image 219–221; types by subsample *238*; US American scholars 176–178
audit system 97, 101
autonomy 208, 214, 248, 249; British universities 63; Dutch universities 63; financial, of RIs 8; institutional 126; regulatory 62; of research university 8, 15–16, 63, 121, 124, 126, 127
Autzen, C. 7

Bacher, J. 121
Backs, S. 117
Baroque: concerns about the misuse of creative communication 113; ornamental art 112–113; overview 111
Barrett, P. 41
Bartlett, W. 60
Bauer, M. W. 4, 5, 7, 17, 37, 81, 97, 98, 111, 128, 133, 168, 186, 195, 212, 213, 217, 222, 231, 272
Bekhradnia, B. 6
benchmark resources 101
Bender Commission 153
Bennett, I. 171
Bennett, N. 93, 168, 184
Berman, E. P. 128
Besley, J. C. 7, 17, 79, 81, 82, 83, 93, 168, 170, 171, 173, 178, 183, 186
Binswanger, M. 61
Blair, T. 59
Bloch, C. 53
blogs 71

Bok, D. 128
Bollen, K. A. 41
Bologna process 208
Borchelt, R. E. 6, 38, 81
Boukaert, G. 59
Boumans, J. 7
Brandenburg, U. 60
Brankovic, J. 214
Brass, K. 7, 8
Braun, D. 59
Brazil 265, 270; sampling frames and procedures employed in **266**
British research institutes 205–225; audiences 219–221; communicative dispositions of 205–225; disposition to communicate more widely 215–216; expanding education sector 210–213; going public 218–219; KEF 213–214; new media channels 216–218; numbers of communicators in 221–222; public events 216–218; REF 213–214; staffing levels 221–223; TEF 213–214; traditional media channels 216–218; university in Britain and elsewhere 205–210; university resilience 213–214; *see also* research institutes (RIs)
British Science Writers Association (BSWA) 212
Bromme, R. 80
Broucker, B. 63
Brown, P. 38, 54
Brown, R. 5, 6
Bucchi, M. 7, 38, 54, 81, 140, 141, 272
Budapest declaration 186
Bühler, H. 7, 39
Burke, E. M. 81
Butterfield, H. 206, 208

Cacciatore, M. A. 82
Carasso, H. 5
Carayannis, E. 124
Cardoso, M. 247
Carnegie Foundation 171
Carroll, L. 100
Carver, R. B. 39
Castro, P. 248
Categorical Principal Components Analysis (CATPCA) 136–137, 143
Cavalier, D. 171
centralised institution 11–12
CERN 209, 211
Chiu, T. 121

Ciencia Viva program 249
Citizen Science 158, 161
Civera, A. 60
Claessens, M. 8, 39
Clark, B. 3
Cochran, W. G. 30
Colquitt, J. A. 80
commercialization of knowledge 117
communication: alignment with competition level 104–109; channels of RIs with nonscientific public 66; disposition variables 43; expanding education sector in need of 210–213; function of universities 97–100; funding 87; intensity of 237, 242; perceived success of 242; rationales for RIs 66–68, **67, 68,** 198; of research institutes 4–5, 37–54; *see also specific types of communication*
communication of research in Italy 133–151; analysis 136–137; audiences 143–145; evaluation of activities 145; public communication activities 137–143; public communication rationales 147; public communication resources 145–147; public perceptions 148–150; sample 136
communications funding 241–242
communication staff 240–241; background 44; characteristics 41–42; characterization of **44–45**; and competition in research units 103–104, *107,* 108–109, *109*; dedicated 163–164; defined 40; level of professionalization 45–50; and policies 197–198; by policy *164*; previous experience 44; professionalization of 41–42; roles 45; training 44; working at research institutes 44
communicative dispositions 205–225; audiences 219–221; education sector, expanding 210–213; new media channels 216–218; public events 216–218; of research institutes 215–216; research institutes going public 218–219; staffing levels 221–223; traditional media channels 216–218; university in Britain and elsewhere 205–210; university resilience in changing context 213–214
Compagnucci, L. 116
comparative design 61–64

competition 101; alignment of communication effort with 104–109; and arms race 100; artificial 60–61; for attention seeking 97–100; resourcing communication lines up with level of 103, 105–107, *106*
competitive funding and evaluation of research 249–250
Conceição, C. 249
conceptual structures 118
Confirmatory Factorial analysis (CFA) model 41, **42**; level of professionalization of staff 45–50
context variables 43
country comparison: RIs and public engagement 72–74, **73**; state 'public duty' 74
Covid-19 pandemic 71, 213
Crettaz von Roten, F. 7, 272
Cunningham, P. 3, 212
Curaj, A. 62

Dalderup, L. 155, 159
Davies, S. R. 57, 150, 184
Dawkins, R. 101
Dearry, A. 169
De Boer, H. F. 6, 59, 60, 62, 63, 157
De Bok, C. 154
decentralisation: defined 9; of public communication function 8–9
decentralisation hypothesis 9; defined 4; overview 13–17; of public communication of science 15; specifying 13–17
decentralised communication: degrees of 9–12, **10**, *11*; and event making 9; indicators at institute level **10**; and new media 9; patterns of *12*, 12–13; and traditional media 9; *see also* communication
de-escalating of arms race 113
Deetman, W. 155
deficit model: defined 58; and general audience 70
De Jonge, J. 154
De La Torre, E. M. 117
democratic rationale: and promotion of science communication 154; and public engagement 154
Diamond, Shari Seidman, 29
diffusionist group 121
Di Franco, G. 117

Dijkstra, A. M. 153, 154, 155, **156,** 159
disciplines: comparative performance of scientists within 28; media activities by research area *196*; OECD 103, 215; PC commitment 105; public event activities by research area *195*; scientific 195, 208; and standardization 32
Doring, S. 205
Dozier, D. M. 79, 80, 173
Drucker, P. F. 6
Dudo, A. 17, 79, 81, 82, 83, 93, 168, 170, 171, 173, 186
Duong, S. 228
Dutch National Research Agenda (NWA) 157
Dutch research agenda 157, 166
Dutch research landscape 156–157

Eatman, T. 170
Ecklund, E. H. 82, 251
economic rationale, and science communication 154–155, 159
education: extra-mural 212; historically unprecedented spending on 211; STEM 189
education sector, UK: expanding science communication 212–213; historically unprecedented spending on education, science and research 211; massive expansion of 210–211; science writing in UK 212–213; university sector 212
Eggebrecht, H. H. 112
Elken, M. 7, 37, 39
Enders, J. 62, 156, 157
engagement *see* public engagement
Engelshoven, I. 154
Engwall, L. 3
Entradas, M. 4, 5, 6, 7, 8, 9, 16, 17, 37, 50, 53, 54, 97, 117, 128, 133, 160, 164, 168, 186, 195, 215, 217, 231, 241, 248, 250, 259–260, 272
entrepreneur 208
'epistemic vigilantism' 213
Esposito, V. 124
Etzkowitz, H. 116
Europe: higher education institutions (HEIs) 57–58; higher education reforms in 59–61; publicly funded universities 57

European higher education:
  communication channels 66;
  comparative design 61–64; institutions
  and public engagement 57–58; and
  NPM 59–60, 66–74; public events 65;
  public perceptions 65; rationales for
  public communication 64–65; reforms
  59–61; research institutes 57–76
European Researchers' Night 134, 138
European Space Agency 211
European Union (EU) 135, 248;
  Framework Programme for Research
  and Innovation 135, 189; Horizon
  2020 189
European universities 75, 208; see also
  European higher education
Evetts, J. 38
Excellence Initiative or the Higher
  Education Pact 2020 60
excellence in research 249, 253, 259
Excellence Theory 79, 80–81
excellent institutes 251, 254; and
  audiences 256, 258, 260; Facebook,
  use of 255; funding 254; and less-
  than-excellent institutes 254; and
  policy for public engagement 258;
  professionalisation among 259; and
  public communication funding 258;
  and traditional media channels 255
extramural activities 207
extra-mural education 212

Facebook 141, 162, 196, 199, 201, 202,
  253, 255
Fähnrich, B. 4, 7, 37
Fahy, D. 39
Federkeil, G. 60
Finkelstein, M. J. 207
Fishbein, M. 81
Fraunhofer Society 232, 243
Friedrichsmeier, A. 228, 231
Frondizi, R. 121, 125
Fukushima accident 204n1
fully decentralised institution 10–11
Fundação para a Ciência e a Tecnologia
  (FCT) 248–251, 253
funded research and science
  communication 250–251
Funk, C. 168
Fürst, S. 228

Gago, J. M. 248
Gascoigne, T. 38

Geils, M. 231
Gelbard, R. 121
General Linear Model regression 88, **92**
general public: communication
  and research institutes (RIs) 70;
  engagement and research institutes
  (RIs) 70
genetically modified organs (GMOs) 186
Gentleman, D. 171
George, M. D. 169
German Aerospace Centre 233
German Rectors' Conference
  228–229, 232
German Research Foundation 228
German research institutes: audiences
  238; methodology 233–235;
  national background 228–231;
  new media channels 237; public
  communication activities 235–238;
  public communication resources
  240–242; public engagement activities
  of 228–244; public events 235–236;
  public perceptions 239; questionnaire
  234–235; rationales for public
  communication 239–240; research
  landscape in Germany 231–233;
  sample 233–234; traditional media
  channels 236–237; see also research
  institutes (RIs)
Germany: number of institutions
  contacted **269**; number of institutions
  that responded **269**; research
  landscape in 231–233; sampling
  frames and procedures employed
  in **266**
Gioia, D. A. 5
Giuri, P. 117
Golan, G. 176
Goldacre, B. 213
Gonçalves, M. E. 247, 248
Google+ 218, 237
governance, of science 229
Grants-in-Aid for Scientific Research
  (KAKEN) 189, 191
Great East Japan Earthquake 186
Green, R. G. 170
Gregory, J. 58, 59, 213
Grunig, J. E. 6, 38, 79, 80, 93, 173
Grunig, L. A. 6, 38, 79, 80, 173
Gutteling, J. M. 159

Habermas, J. 4, 209
Hachmeister, C.-D. 228

Hair, J. F. 41
Hallahan, K. 3
Halsey, A. H. 206–207, 211
Heitor, M. 248
Helmholtz Association of German
  Research Centres (HGF) 232, 233, 243
Hemsley-Brown, J. 5, 97
Hendriks, F. 80
Hicks, D. 117
Higher Education Funding Council for
  England (HEFCE) 214
higher education institutions (HEIs)
  247–248; central funds allocation
  to 60; Europe 57, 59–60; full-time
  researchers employed in 103; R&D
  units 249; and reputation 97; see also
  research institutes (RIs)
Hoffmann, E. 230
Hoffman, S. G. 125
Hon, L. C. 80
Hood, C. 6
Horst, M. 7, 184
Hunt, T. 213
Hüther, O. 228
Hvidtfelt Nielsen, K. 39

Instagram 141, 196, 218, 224, 237
institutional communication:
  and nonspecialist staff 50; and
  professionalisation of science
  communication 50; and specialist staff
  50; see also communication
institutional-context variables 193
institutional institutes 124–125, 126
institutionalisation of Dutch science
  communication 155–156
Integrated Behavioral Model 81
intensity of communication
  237, 242
International Union for the
  Conservation of Science 170
Irwin, A. 58
Italian National Agency for the
  Evaluation of the University and
  Research Systems (ANVUR) 135
Italy: communication of research
  in 133–151; National Research
  Council (CNR) 134, 136; number of
  institutions contacted 271; number of
  institutions responded 271; sampling
  frames and procedures employed
  in 266
ITER 209, 211

James, S. A. 82
Japan: number of institutions contacted
  271; number of institutions that
  responded 271; rationale of the
  MORE-PE survey in 189–190;
  research system 188; sampling frames
  and procedures employed in 267;
  science communities related to social
  issues in 186–187
Japanese research institutes: audiences
  196–197; communications staff and
  policies 197–198; MORE-PE survey
  189–190; policies focused on science
  and society 188–189; policy measures
  189; public communication activities
  194–196; public communication
  in 186–203; research system 188;
  science communities and social issues
  186–187; see also research institutes
  (RIs)
Japan Science and Technology
  Agency 190
Japan Society for the Promotion of
  Science (JSPS) 189, 191
Jensen, E. 186
Jensen, P. 7, 97, 133, 212, 222
Jongbloed, B. 157
Jongbloed, J. 59, 63
journalists database 237

Kalton, G. 28
Kamenetzky, J. R. 169
Kanda, Y. 190
Kantorowicz, E. 205
Kapetaniou, C. 116
Karliczek, A. 230
Kasprzyk, D. 81, 93
Kassab, O. 251
Katz, Y. 251
KEF (knowledge exchange framework)
  213–216, 224–225
Kerkhoven, A. 153
Kerr, C. 3
Kiaer, A. N. 26–27
Kienhues, D. 80
Kim, J. 135
Kirp, D. L. 125
Kish, L. 32–33
Kleimann, B. 5
Kline, R. B. 41
knowledge transfer 208, 213, 216,
  228, 250; one-sided 70; one-way 75;
  traditional 65

Koch, H. A. 3
Kohring, M. 7, 8, 61, 228, 229–231
Koivumäki, K. 7, 39
Kolenikov, S. 41
Kollman, K. 6
Koso, A. 7, 98
Kotosz, B. 117
Krebs, J. R. 101
Krücken, G. 3, 5, 6, 38, 97
Krüger, K. 62
Kucharski, A. 100
Kudo, M. 187

Laredo, P. 3
Laukötter, E. 228, 231
Laursen, K. 117
Lawrence, F. 79, 168
Lee, N. M. 81
Lee, S. H. 116
Le Grand, J. 60
Leibniz Association 232, 243
Leshner, A. I. 128, 168
Leslie, L. 125
less-than-excellent institutes 251, 253,
    254–260, 255, 256, 259
Levey, D. J. 169
Lewis, J. 7, 8, 39
limitation, and arms race model 102
Lincoln, A. E. 82
Linell, P. 5
Lock, S. J. 213
López-Goñi, I. 163
Loroño-Leturiondo, M. 57, 150
Lürsen, M. 154

Maasen, S. 6
Maassen, P. 3
Mcdowell, G. R. 169
Madden, H. 82
Madsen, E. B. 53
Mahoney, J. T. 9
Mandler, P. 97, 208, 211
March for Science 170
Marcinkowski, F. 8, 17, 38, 39, 61, 128,
    228, 229, 231
Marginson, S. 126, 127
marketing functions: professionalisation
    of 6–8; at universities and RIs 38
market-oriented institutes 125–128, 127
Marradi, A. 118
Martin, B. R. 214, 251
Martinez-Conde, S. 251
Martin-Sempere, M. J. 57

mass media channels 66
Matheson, A. 101
'Mathew effect' 260
Matter, U. 251
Max Planck Society (MPG) 232, 243
Meara, K. 170
media activities: by research area 196;
    and R-units 217; social 218
media channels: new 141–143, 162–163,
    216–218, 237; traditional 139–141,
    140, 162, 196, 216–218, 236–237
medialisation: as ARPC hypothesis
    100–102; of science 100
medialisation hypothesis 17, 98, 110
Meier, F. 6, 38, 97
Mejlgaard, N. 8, 10, 37, 53, 272
Merrien, F. 59
Merton, R. K. 260
Metag, J. 7, 71, 98
Metcalfe, J. 38
Michael, R. T. 29
Miller, K. 116
Miller, S. 38, 58
misuse of creative
    communication 113
Modernist Movement 111
Molas-Gallart, J. 135
Montano, D. E. 81, 93
Mora, J. G. 117
MORE-PE (Mobilisation of Resources
    for Public Engagement) 4, 23–24,
    187, 189–190, 201, 232, 272; standard
    protocol for sampling 30; survey
    on scientific community's public
    engagement 80–88; target domains,
    research areas by country 27–28
motivation, and communication of
    research institutes 72
Mulder, H. 154
Muller, J. Z. 214
Murray, D. L. 251

Nadkarni, N. M. 94
National Academy of Sciences 171
National Institute of Science and
    Technology Policy (NISTEP)
    190–191, 204n5
National Institutes of Health (NIH) 169
National Research Council (CNR), Italy
    134, 136
National Science Foundation (NSF)
    169
negative views of the audience index 119

neoliberalism 57; and higher education systems in Europe 59
Neresini, F. 7, 38, 272
Netherlands: Dutch research agenda 157, 166; Dutch research landscape 156–157; number of institutions contacted **269**; number of institutions responded **269**; public communication with science landscape in 153–154; public engagement 153–166; sampling frames and procedures employed in **267**; science communication timeline in **156**
Netherlands Organisation for Scientific Research (NWO) 157
Netherlands Organisation for Technology Assessment 155
Newman, J. 59
Newman, T. 171
new media: activity 162–163; diffusionist group use of 121; frequency of excellent and less excellent institutes *256*; platforms 162; in RIs operations 13
new media channels 42, 50, **51–52,** 141–143, **142,** 162–163, 216–218, 237, 253; activity 217; engagement activities by research institutes/centers **177, 179**; by OECD research areas **142**; by policy *163*
new media channels index 119
New Public Management (NPM) 4, 156; and European higher education 59–60, 66–74; reforms 60, 97; of research universities 5–6; rise of 59
Nielsen, K. H. 6, 38, 81
Nilson, R. 171
Nisbet, M. C. 39
Noam, E. M. 99, 207
nonresponse, in sampling 29–31
nonspecialist staff 46, **46**; and institutional communication 50; and level of research across countries 48; and regression of roles **47**; scores *vs.* specialist scores by country *49*; scores *vs.* specialist scores by research area *49*
non-university institutes 228, 234–244
Norris, P. 101
Nota Wetenschapsbeleid report 154

Observa Science in Society 135
Oever, Selina van den 153
O'Fallon, L. R. 169

O'Hara, K. 81, 173
O'Muircheartaigh, C. 17, 29, 268
online media: advanced 66; standard 66; use by universities 71
Oosthuizen, S. 3
Open Science framework 157–158; as policy-driver for public engagement 158
Oplatka, I. 5, 97
Oppi, C. 124
Organization for Economic Cooperation and Development (OECD) countries 84, 103, 211; classification schema 171; sample characteristics by research area **137**
ornamental art 111
over-adaptation, and attention seeking 98, 100

Pansegrau, P. 8, 98, 213
Paradeise, C. 6, 62
partly decentralised institution 11
Pavel, A. P. 214
Peacock effect 100, 110
Pellegrini, G. 18, 134
PE Rationales 104, 108
perceived successfulness, of public engagement 84–88; analysis using regression modeling 88–89; communication funding 87; communication policy 88; described 84; limitations 92–94; past public engagement 84–87; perceived low public interest 88; perceived public trust 88; public engagement experience 87; research area 84; research results 89–92, **90–91**; specialist staff 87
perception: audience 176–178; of the public 148, **149, 150,** 164, *165, 238, 239*; and public communication 198–201; -related RIs 201; -related variables 193–194
Petersen, S. 170
Peters, H. P. 6, 7, 8, 38, 98, 231
Pfafigan, A. 111
Phillips, D. P. 99
podcasts 71
Poliakoff, E. 79
policies: focused on science and society 188–189; and institutional factors 198–201; perceptions and public communication 198–201
policy-related RIs 199, 201
policy-related variables 193

Pollitt, C. 59
population of interest: defined 23; and research institutes' performance measurement 23–24
Portugal: dictatorship in 247; polytechnic institutes in 248; public communication in 247–260; 'research excellence' in 247–260; sampling frames and procedures employed in **266**
Portuguese universities 247–248; internationalisation of 248; research in 248–249
positive views of the audience index 119
potential bias, in sampling 29–31
Power, M. 6
probability sample 26, 29
professionalisation 102; among excellent institutes 259; of marketing functions 6–8; of public communications activity and excellence 258–259; of public relations (PR) 6–8; research excellence 253; of science communication 258–259
professionalisation level: across countries 48, 49–50; and areas of research 48; and associated roles 47–48; of communication staff 45–50
professionalisation of science communication: and institutional communication 50; methods 40–43; overview 37; at universities 38–40
professionalization thesis 50–54
professional scholar 208
professional science communicators 7, 79, 81, 94
PR/PE ratio index 108
PR rationales 104, 108
public communication 265–270; barriers to researchers' engagement in 164–166, *165*; data collection 268–269; disproportionately stratified stratum 268; follow up calls 268–269; framework and indicators of 272–283; funding for 9; paper surveys 268–269; as public service 57; rationales for 147, 239–240; rationales for RIs 64–65; respondents prenotification 268; response rates and nonresponse bias analysis 269–270; sample design 265–268; sampling frames 265; *vs.* science communication 13; with

science landscape in Netherlands 153–154; survey distribution and reminders 268; at universities 13; and university-affiliated RIs 70–72
public communication activities 194–196, 235–238, 252; and level of excellence 254–256; new media channels 141–143; public events 137–139; traditional media channels 139–141
public communication events *see* public events
public communication function: decentralisation of 8–9; at research universities 3–4
public communication funding 253, 258
public communication in Japanese RIs 186–203; audiences 196–197; communications staff and policies 197–198; institutional context 198, 199; MORE-PE survey 189–190; national background 188–189; perception-related RIs 201; policies focused on science and society 188–189; policy measures 189; policy-related RIs 199, 201; public communication activities 194–196; public events 195; rationales for communication 198; research system 188; sample 190–194; science communities and social issues 186–187; study and research questions 187–188; survey distribution 191–192; traditional media channels 196
public communication of science 4–5; arms race for 109–110; audiences and media relations, indicators of **277**; decentralisation hypothesis 15; dimensions and indicators' definitions **274–275**; framework of indicators for 272–273; general information, indicators of **282–283**; increase in institutes 9; indicators 272–283; indicators for measuring *273*; indicators of rationales **281–282**; input indicators 278, 281, 282; output indicators *273*, 276, 277; perceived images of media and public **281–282**; professionalizing 38–40; resource allocation, indicators of **278–280**
public communication policies/ guidelines **198**, 253, 258

public communication resources 145–147, 240–242; communications funding 241–242; communications staff 240–241; perceived success of communication 242

public communications staff 253

public duty: as motivation of RIs 66, 74; RIs public engagement as 17

public engagement 5; barriers to 164–166; conclusions and future work 166; dedicated communication staff 163–164; democratic rationale 154; Dutch National Research Agenda (NWA) 157; Dutch research landscape 156–157; economic rationale 154–155; and European HEI 57–58; experience 87; institutionalisation of Dutch science communication 155–156; need for resources **183**; new media channels 162–163; open science as policy-driver for 157–158; perceived engagement successfulness **85–86**; perceived reasons for lack of **181–182**; perceived successfulness at research institutes 79–94; public events 161–162; public perceptions 164, *165*; research funding, and policies 158–159; research institutes, sampling of 31–32; at research institutes in Netherlands 153–166; resources for public communication 159; *vs.* science communication 13; traditional media channels 162

public engagement activities: of German research institutes 228–244; perceived reasons for lack of engagement **181–182**; perceived successfulness of **183**; reasons for taking part in **180**

public engagement profiles: and conceptual structures 118; methods 118–120; overview 116–117; and research institutes 116–129; results of developing 120–128

public engagement with science (PE) 58–59, 232

public events 137–139, **138,** 195, 216–218, 235–236; activities by research area *195*; defined 216; by OECD research areas **139**; by policy *161*; public engagement at research institutes 161–162; and RIs 65; and traditional media channels engagement activities **174–175**

public events index 119

public perceptions: communication of research in Italy 148–150; german research institutes 238–239; identified in Dutch institutes *165*; public engagement at research institutes 164, *165*; by subsample *238*; *see also* perception

public perceptions index 119

public relations (PR): officers 38–39; products in research universities 7; professionalisation of 6–8

public relations scholarship 80–81

public service: association 58; organizations 57; public communication as 57

public understanding of science (PUS) 3–5, 75, 212; increase in institutes 9; and science communication function 7

public understanding of science and humanities (PUSH) approach 65

QS World University Ranking 214
questionnaire 234–235, 252

Rainie, L. 168
Ramos, A. 248, 249
Rathenhau Institute 155, 158
rationales for not engaging index 119–120
Red Queen effect 100, 110
REF (research excellence framework) 213–214
Rehländer, J. 229
relentless specialisation 205
reputation 214; competition and RIs 108; and HEIs 97; institutional 230; and marketisation of universities 5; and research-intensive institutes 260; and science communication 230; and scientific community of peers 99; universities and publics public communication 38
reputation management 18, 38, 57, 97, 99, 219, 225
reputation rankings, and universities 97
research: competitive funding and evaluation of 249–250; funding and public engagement policies 158–159; historically unprecedented spending on 211; landscape in Germany 231–233; in Portuguese universities 248–249

research excellence: audiences addressed and levels of 256–258; excellent and less-than-excellent institutes 254; in Portugal 247–260; professionalisation of 258–259; public communication activities and 254–256; questionnaire 252; sample 251–252; variables 252–253

research institutes (RIs): barriers to researchers' engagement 165; communication of 4–5, 37–54; communication staff working at 44–45; communication strategies **122–123**; data of European countries 64–66; in European higher education systems 57–76; excellent and less-than-excellent 254; external communication of 68–69, **69**; and general public engagement 70; going public 218–219; new media channels engagement activities by **177, 179**; organizational system of **24**; overview of **160**; perceived successfulness of public engagement at 79–94; perception-related 201; performance measurement 23; public communication of 265–270; and public engagement profiles 116–129; public engagement sample 31–32; and public events **174–175**; received FCT funds **250**; sampling 23–33; science communication at 4, 6–7; science communication at meso level of 7; target population 24; and third mission 116; and traditional media channels engagement activities **174–175**; university-affiliated, and public communication 70–72; *see also* higher education institutions (HEIs)

research organisations 7, 134, 136, 232–233

research universities: NPM of 5–6; professionalizing public communication of science of 38–40; public communication function at 3–4

resource allocation: to communication, variation in institutes' 104–105, *104–105*; and RIs communication 102–103

response rates, in sampling 29–31

Rhoades, G. 6

Risien, J. 171

Rödder, S. 7, 8

Rodell, J. B. 80

Roessler, I. 228

Rosa, M. 247

Rowe, D. 7, 8

Royal Netherlands Academy of Arts and Sciences (KNAW) 157

Royal Society 212

Rubin, A. 134

Salter, A. 117

Sammut, G. 111

sampling: country samples 27–31; nonresponse 29–31; potential bias 29–31; "representative" sample *26, 25*–27; research institutes 23–33; response rates 29–31; sample design 24–25, *25*; stratification 28–29; target population 24, *25*

Sánchez-Angulo, M. 163

Saracino, B. 18, 140, 141

Sarrico, C. 248, 249

Sataøen, H. L. 5

Schäfer, M. S. 4, 7, 37, 71, 98, 229

Scheu, A. M. 8

Scheufele, D. A. 168, 171

Schimank, U. 62, 156, 157

Schneider, R. 211

Scholl, R. 38

Schröder, G. 59

Schwetje, T. 39

science communication 153; central level of university communication 6–7; and communication professionals 39; and democratic rationale 154; economic rationale 154–155, 159; function analysis 6–7; and funded research 250–251; and medialisation hypothesis 98; at meso level of RIs 7; at micro level 7; professionalization of public 37, 38–40; *vs.* public communication 13; requirements for, by funded research 250–251; at research institutes 4, 6–7; scholars on strengthening 8; and science writing in UK 212–213; specialisation of 205, 209; timeline in the Netherlands **156**

science communicators 7, 94, 155, 229; choices as behaviors 81; and public engagement 79

science communities 186–187

science education 33, 247

Science in Society 135

science news 98
science reportage 98
Science with and for Society (SwafS) 135
scientific area 141–142, 144–146, 148
scientific institutions 37, 58, 98, 111,
    133–134, 144, 147, 229–230, 239, 249
self-interest 243; and governance
    structure of RIs 57; as motivation
    for RIs 17, 67–68, 70; public
    communication and motivation for
    RIs 69; *vs.* public responsibility 64
self-marketing 62
Seydel, E. R. 160
Shipman, M. 8
Shukla, R. 213
Siggen Circle 229
Siggener Kreis 230
Simis, M. J. 82
simple random sampling 26
Slaughter, S. 6, 125
Sloterdijk, P. 111
Smallman, M. 213
Smith, B. 171
Sociale en Geesteswetenschappen
    (SGW) 159
social media 141, 196, 217–218; in
    Anglo-Saxon countries 71; and mass
    media system 98; use by universities
    71; viral messaging 100
social media activities 218
social media channels 7, 199, 237
social media profiles 243
social networks 8; use by universities 71
specialisation of science communication
    205, 209
specialist staff 46, *46*, **47**; and
    institutional communication 50; and
    level of research across countries
    48; and public engagement 87; and
    regression of roles **47**; scores *vs.*
    nonspecialist scores by country *49*;
    scores *vs.* nonspecialist scores by
    research area *49*
Spigarelli, F. 116
'Sputnik Shock' of 1957 210
staffing levels 221–223
standardization: defined 32; as sampling
    technique 32
standard online media 66
Stappers, J. G. 154
Stares, S. 8
STEM education 189

Stensaker, B. 3
Stifterverband für die Deutsche
    Wissenschaft 228
strategic thinking 80
stratification: defined 28; and sampling
    28–29
Straver, G. 154
Super Science High School (SSH)
    programme 187, 204n2
Swiss Neue Zuricher Zeitung (NZZ) 101

target group media 66
Taylor, R. 3
TEF (teaching excellence framework)
    213–214
Teixeira, P. 247
Terra nullius 153–166
Theory of Planned Behavior 81
third mission: and research institutes
    116; and universities 116
Tiffany, L. -A. 171
Toegepaste en Technische
    Wetenschappen (TTW) 159
Tomizawa, H. 190
Toth, E. L. 80
traditional media channels 9, 42, 50,
    51–52, 139–141, **140,** 162, 196,
    216–218, 236–237, 252; activity 217;
    frequency of communication through
    236; by OECD research areas **141**; by
    policy *162*; RIs use of 72
traditional media channels index 119
Trench, B. 38, 54, 58
Trindade, M. 47
Trip, F. 154
Trow, M. A. 206–207, 211
Tukey, J. 117
Twitter 141–142, 162, 196–197, 199, 201,
    202, 237, 253, 255–256
Two-Step Cluster Analysis 118–121

UK education sector: basic figures
    on university sector 212; expansion
    of higher education *211*; massive
    expansion of 210–211; participation
    rates by period of expansion **210**
United Kingdom: number of institutions
    contacted **269**; number of institutions
    responded **269**; sampling frames and
    procedures employed in **267**; science
    communication, expanding 212–213;
    science writing in 212–213

United States: market-oriented institutes 125; modern academic in 207; number of institutions contacted **269**; number of institutions that responded **269**; public engagement in 170–171; sampling frames and procedures employed in **267**; tenure-track faculty in 170

universities: in Britain and elsewhere 205–210; communication function of 97–100; entrepreneurial role of 116–117; online media use by 71; and reputation rankings 97; resilience in changing context 213–214; social networks use by 71; and third mission 116

'university extension movement' 3

University Reform Act of 2010 135

Urdari, C. 116

US American scholars: audience perceptions 176–178; engagement through research institutes 168–184; overall experience 180–183; rationale for engagement 178–180; types of activities American scholars doing 173–176

Vagnoni, E. 124

Vakkuri, J. 116

Van Dam, F. 153, 154, 155

Van der Sanden, M. 153, 154, 155

VanDyke, M. S. 81

variables: institutional-context 193; perception-related 193–194; policy-related 193; research excellence 252–253; by type of activity **194**

Vaubel, R. 112

viral messaging 100

visibility–publicity–popularity 99

Vogler, D. 7, 37, 98

Von Humboldt, W. 124, 125

Wæraas, A. 5

Watermeyer, R. 7, 8, 39, 214

Watts, S. M. 169

Webb, T. L. 79

Weber, M. 205–206, 208

Weetman, W. 154

Weiner, S. 171

Weingart, P. 6, 8, 81, 98

Wiedenhof, N. 159

Wiley, S. L. 169

Wilkinson, C. 7, 39

Willetts, D. 208, 211

Windolf, P. 211

Wissenschaft im Dialog 229

Wölfflin, H. 111

World Conference on Science 186

Wynne, B. 58

Yeo, S. K. 82

Yoo, J. 135

YouTube 71, 141, 163, 176, 196, 199, 201, 202, 218, 253

Yuan, S. 81, 82, 83, 93, 168, 170, 171

Ziman, J. 133

For Product Safety Concerns and Information please contact our EU
representative GPSR@taylorandfrancis.com
Taylor & Francis Verlag GmbH, Kaufingerstraße 24, 80331 München, Germany

www.ingramcontent.com/pod-product-compliance
Lightning Source LLC
Chambersburg PA
CBHW052119230326
41598CB00080B/3881

9 780367 494643